ACCRETION PROCESSES IN ASTROPHYSICS

It has been more than fifty years since the first significant paper on accretion flows was written. In recent years, X-ray satellites capable of identifying accretion disks and radiation jets – indications that accretion has taken place – have significantly advanced our understanding of these phenomena. This volume presents a comprehensive and up-to-date introduction to the major theoretical and observational topics associated with accretion processes in astrophysics. Comprising lectures presented at the twenty-first Winter School of the Canary Islands Institute of Astrophysics, the text emphasizes the physical aspects of accretion, investigating how radiation jets are produced, how accretion power is divided between jets and radiated energy, the geometry of accretion flow, and the accretion processes of active galactic nuclei. Written by an international team of experienced scientists, chapters offer young researchers key analytical tools for supporting and carrying out the next generation of front-line research.

Ignacio González Martínez-País is Lecturer in the Astrophysics Department of La Laguna University and Staff Researcher at the Instituto de Astrofísica de Canarias. His main research interests include the study of compact binaries in which accretion takes place: cataclysmic variables (harboring a white dwarf as the accreting object) and X-ray binaries (harboring a neutron star or black hole). He has extensive experience teaching related subjects such as general physics, fluid mechanics, differential equations, astrophysical instrumentation, observational techniques, and accretion processes.

Tariq Shahbaz is Staff Scientist at the Instituto de Astrofísica de Canarias, where his research focuses on the study of compact objects, such as neutron stars and black holes, and determining their masses by developing methods to model the observed light curves and spectra. His other research interests include high-time-resolution phenomena and X-ray binaries. He is a member of the International Astronomical Union.

Jorge Casares Velázquez is Staff Scientist at the Instituto de Astrofísica de Canarias. His research focuses on the study of galactic black holes, with emphasis on the determination of their dynamical masses. He has promoted novel strategies for deriving fundamental parameters in X-ray binaries by exploiting the reprocessing of X-rays into optical line radiation. His other research interests include the study of gamma-ray binaries, cataclysmic variables, and millisecond pulsars. He is a member of the International Astronomical Union and the Spanish Astronomical Society.

Canary Islands Winter School of Astrophysics

Volume XXI

Accretion Processes in Astrophysics
Series Editor
Francisco Sánchez, *Instituto de Astrofísica de Canarias*

Previous volumes in this series

I. Solar Physics
II. Physical and Observational Cosmology
III. Star Formation in Stellar Systems
IV. Infrared Astronomy
V. The Formation of Galaxies
VI. The Structure of the Sun
VII. Instrumentation for Large Telescopes: A Course for Astronomers
VIII. Stellar Astrophysics for the Local Group: A First Step to the Universe
IX. Astrophysics with Large Databases in the Internet Age
X. Globular Clusters
XI. Galaxies at High Redshift
XII. Astrophysical Spectropolarimetry
XIII. Cosmochemistry: The Melting Pot of Elements
XIV. Dark Matter and Dark Energy in the Universe
XV. Payload and Mission Definition in Space Sciences
XVI. Extrasolar Planets
XVII. 3D Spectroscopy in Astronomy
XVIII. The Emission-Line Universe
XIX. The Cosmic Microwave Background: From Quantum Fluctuations to the Present Universe
XX. Local Group Cosmology

CAMBRIDGE
UNIVERSITY PRESS

University Printing House, Cambridge CB2 8BS, United Kingdom

Cambridge University Press is part of the University of Cambridge.

It furthers the University's mission by disseminating knowledge in the pursuit of education, learning and research at the highest international levels of excellence.

www.cambridge.org
Information on this title: www.cambridge.org/9781107030190

© Cambridge University Press 2014

This publication is in copyright. Subject to statutory exception and to the provisions of relevant collective licensing agreements, no reproduction of any part may take place without the written permission of Cambridge University Press.

First published 2014

A catalogue record for this publication is available from the British Library

Library of Congress Cataloging in Publication data
Canary Islands Winter School of Astrophysics (21st : 2009 : Puerto de la Cruz, Canary Islands)
Accretion processes in astrophysics / [edited by] Ignacio González Martínez-País, Instituto de Astrofísica de Canarias and Universidad de La Laguna, Tariq Shahbaz, Instituto de Astrofísica de Canarias, Jorge Casares Velázquez, Instituto de Astrofísica de Canarias.
    pages cm
Lectures presented at the XXI Canary Islands Winter School of Astrophysics, held in Puerto de la Cruz, Tenerife, Spain, Nov. 2–13, 2009.
Includes bibliographical references.
ISBN 978-1-107-03019-0 (hardback)
1. Accretion (Astrophysics) – Congresses. I. González Martínez-País, Ignacio, 1959– editor of compilation. II. Shahbaz, Tariq, 1970– editor of compilation. III. Casares Velázquez, Jorge, 1964– editor of compilation. IV. Title.
QB466.A25C36 2009
523.01–dc23   2012051629

ISBN 978-1-107-03019-0 Hardback

Cambridge University Press has no responsibility for the persistence or accuracy of URLs for external or third-party internet websites referred to in this publication, and does not guarantee that any content on such websites is, or will remain, accurate or appropriate.

Contents

List of contributors	*page* ix
List of participants	xi
Preface	xiii
Acknowledgments	xv
Abbreviations	xvii
1 Accretion disks *Henk Spruit*	1
2 The evolution of binary systems *Philipp Podsiadlowski*	45
3 Accretion onto white dwarfs *Brian Warner*	89
4 Multiwavelength observations of accretion in low-mass X-ray binary systems *Robert I. Hynes*	117
5 X-ray binary populations in galaxies *Giuseppina Fabbiano*	151
6 Observational characteristics of accretion onto black holes I *Christine Done*	184
7 Observational characteristics of accretion onto black holes II: environment and feedback *Rob Fender*	227
8 Computing black-hole accretion *John F. Hawley*	253
A Piazzi Smyth, the Cape of Good Hope, Tenerife, and the siting of large telescopes *Brian Warner*	291

List of contributors

CHRISTINE DONE University of Durham, UK

GIUSEPPINA FABBIANO Harvard-Smithsonian Center for Astrophysics, USA

ROB FENDER University of Southampton, UK

JOHN F. HAWLEY University of Virginia, USA

ROBERT I. HYNES Louisiana State University, USA

PHILIPP PODSIADLOWSKI University of Oxford, UK

HENK SPRUIT Max-Planck Institut für Astrophysik, Germany

BRIAN WARNER Department of Astronomy, University of Cape Town, South Africa, and School of Physics and Astronomy, University of Southampton, UK

List of participants

Almeida, Leonardo	Instituto Nacional de Pesquisas Espaciais (Brazil)
Armas Padilla, Montserrat	University of Amsterdam (The Netherlands)
Barclay, Thomas	Armagh Observatory (United Kingdom)
Bonfini, Paolo	University of Crete (Greece)
Brodatzki, Katharina Anna	Ruhr-Università Bochum (Germany)
Burmeister, Mari	Tartu Observatory (Estonia)
Candelaresi, Simon	NORDITA, Stockholm (Sweden)
Castelló Mor, Nuria	Instituto de Física de Cantabria (Spain)
Cavecchi, Yuri	University of Amsterdam, University of Leiden (The Netherlands)
Chesnok, Nadiia	Main Astronomical Observatory, National Academy of Science of Ukraine (Ukraine)
Coronado, Yaxk'in Ú Kan	Universidad Nacional Autónoma de México, Instituto de Astronomía (Mexico)
Corral Santana, Jesús M^a	Instituto de Astrofísica de Canarias (Spain)
Dermine, Tyl	Institut d'Astronomie et d'Astrophysique – Université Libre de Bruxelles (Belgium)
Ederoclite, Alessandro	Instituto de Astrofísica de Canarias (Spain)
Falocco, Serena	Instituto de Física de Cantabria (Spain)
Hueyotl-Zahuantitla, Filiberto	Instituto Nacional de Astrofísica Óptica y Electrónica (Mexico)
Kajava, Jari Juha Eemeli	University of Oulu, Department of Physics (Finland)
Kim, Jeong-Sook	Korea Astronomy & Space Science Institute (South Korea)
Kotko, Iwona	Astronomical Observatory of Jagiellonian University (Poland)
Kotze, Marissa	South African Astronomical Observatory, University of Cape Town (South Africa)
Krivosheyev, Yuri M.	Space Research Institute (Russia)
Lasso Cabrera, Néstor Miguel	University of Florida, Department of Astronomy (USA)
Li, Shuang-Liang	Shanghai Astronomical Observatory (China)
Maitra, Chandreyee	Indian Institute of Science, Bangalore (India)
Mederos Gomes da Silva, Karleyne	Instituto Nacional de Pesquisas Espaciais (Brazil)
Nooraee, Nakisa	Dublin Institute for Advanced Studies (Ireland)
Rajoelimanana, Andry	South Africa Astronomical Observatory, University of Cape Town (South Africa)
Ratti, Eva	SRON Netherlands Institute for Space Research (The Netherlands)
Skalicky, Jan	Department of Theoretical Physics & Astrophysics, Masaryk University (Czech Republic)
Somero, Auni	Tuorla Observatory, University of Turku (Finland)
Stalevski, Marko	Astronomical Observatory Belgrade, Sterrenkunding Observatorium (Serbia Republic)

Tremou, Evangelia	University of Cologne, Max Planck Institute für Radioastronomie (Germany)
Tsupko, Oleg Yu	Space Research Institute (Russia)
Valencia-S, Mónica	I. Physics Institute, University of Cologne, IMPRS for Astronomy & Astrophysics (Germany)
van Spaandonk, Lieke	University of Warwick (United Kingdom)
Veledina, Alexandra	University of Oulu, Department of Physics (Finland)
Wu, Qingwen	Korea Astronomy & Space Science Institute (South Korea)

Preface

It was more than 50 years ago when the first significant paper on accretion flows was written. Since then, the subject has grown incredibly, and today many X-ray satellites are engaged in research into observational signatures and tests of theoretical models for accretion processes in astrophysics. Recognizing the continued importance of this field, the Instituto de Astrofísica de Canarias organized the XXIst in its Winter School series around the topic "Accretion Processes in Astrophysics."

The primary aim of the school was to provide a wide-ranging and up-to-date overview of the theoretical, experimental, and analytical tools necessary for carrying out front-line research in the study of accretion processes. The school was particularly designed to offer young researchers guidelines to support their research in these areas.

The 40 lectures presented a fairly comprehensive and up-to-date introduction to the major observational and theoretical topics associated with accretion. With emphasis on the physical processes involved, this includes applications to close binary systems such as cataclysmic variables and X-ray binaries and their evolution, as well as the theory of relativistic accretion flows and the accretion processes in active galactic nuclei. The lectures were given by eight experienced scientists who are actively working on a variety of leading research projects and who have played key roles in the advances made in the field in recent years.

The Editors

Acknowledgments

The editors would like to express their warmest gratitude to all the lecturers for their time in preparing their classes, attending the school, and writing the chapters for this book. We know that this has been a major effort on their part, but we hope that it has been rewarding for them. In particular we would like to thank Prof. John F. Hawley for his entertaining public lecture on "Black Holes" and Prof. Brian Warner for his ad hoc lecture dedicated to Charles Piazzi Smyth, who built the first (temporary) major observatory on Tenerife.

The key to the success of the school has been without any doubt our secretary Lourdes González. Without her help and diligence, the school would not have worked as smoothly as it did. We also thank Nieves Villoslada, who started the preparation and organization of the school; Jesús Burgos of the OTRI, who provided invaluable help with the preparation of applications needed to receive funding; persons at the IAC's Centro de Cálculo for their IT assistance; and Ismael Martínez Delgado, for the technical editing of this book for Cambridge University Press.

We are extremely grateful to the local artist Macu Anelo, who designed the school's poster that depicts an accretion disk around a black hole, superimposed on a background of Guanches sketches, which we hope may entice young scholars to enter this field. The school's poster was prepared by Ramón Castro. We thank Annia Domènech and the Gabinete de Dirección of the IAC for taking the time to conduct interviews and submit press releases for all the lecturers.

We greatly acknowledge the financial assistance from the Spanish Ministerio de Educación y Ciencia and from the Cabildo de Tenerife, who kindly provided the excellent facilities of the Congress Palace of Puerto de la Cruz where the event took place. Last, but not least, we would like to acknowledge all the participants of the school: lecturers, students, and supporting personnel.

The Editors

Abbreviations

AAVSO	American Association of Variable Star Observers
ACF	Auto Correlation Function
ACIS	AXAF CCD Imaging Spectrometer
ADAF	Advection Dominated Accretion Flows
ADC	Accretion Disk Corona
AGB	Asymptotic Giant Branch
AGN	Active Galactic Nucleus
ASCA	Advanced Satellite for Cosmology and Astrophysics
AU	Astronomical Unit
BAL	Broad Absorption Line
BATSE	Burst And Transient Source Experiment
BB	Black Body
BH	Black Hole
BHB	Black Hole Binaries
BHXRT	Black Hole X-Ray Transient
BLR	Broad Line Region
BPS	Binary Population Synthesis
CCD	Charge Coupled Device
CCF	Cross Correlation Function
CE	Common Envelope
CGRO	Compton Gamma-Ray Observatory
CT	Constrained Transport
CV	Cataclysmic Variable
CXB	Cosmological X-ray Background
DD	Double Degenerate
DIM	Disk Instability Model
DN	Dwarf Nova
DNe	Dwarf Novae
DNO	Dwarf Nova Oscillation
DNS	Double Neutron Star
EMF	Electromotive Force
FIR	Far Infra Red
FRED	Fast Rise Exponential Decay
GC	Globular Cluster
GRB	Gamma Ray Burst
GRMHD	General Relativistic Magneto Hydrodynamics
HID	Hardness Intensity Diagram
HMXB	High Mass X-ray Binary
HST	Hubble Space Telescope
IAC	Instituto de Astrofísica de Canarias
IMBH	Intermediate-Mass Black Hole
IMXB	Intermediate-Mass X-ray Binary
IP	Intermediate Polar
IR	Infrared
ISAF	Ion Supported Accretion Flow
ISCO	Innermost Stable Circular Orbit
ISM	Interstellar Medium
IUE	International Ultraviolet Explorer
KG	Kilo Gauss
LARPS	Low Accretion Rate Polars

LGRB	Long-Duration Gamma Ray Burst
LIGO	Laser Interferometer Gravitational-wave Observatory
LINER	Low Ionization Nuclear Emission-line Region
LL	Landau and Lifshitz (1959)
LMC	Large Magellanic Cloud
LMXB	Low-Mass X-ray Binary
LOFAR	Low-Frequency Array
lpDNO	longer-period Dwarf Nova Oscillation
MG	Mega Gauss
MHD	Magneto Hydrodynamics
MRI	Magneto Rotational Instability
MWA	Murchison Widefield Array
NASA	National Aeronautics and Space Administration
NLR	Narrow Line Region
NS	Neutron Star
NSE	Nuclear Statistical Equilibrium
NTT	New Technology Telescope
PPM	Piecewise Parabolic Method
PS	Population Synthesis
PSF	Point Spread Function
QPO	Quasi Periodic Oscillation
RLOF	Roche Lobe Overflow
ROSAT	Röentgen Satellite
RXTE	Rossi X-ray Timing Explorer
SALT	Southern African Large Telescope
SD	Single Degenerate
sdB	subdwarf Binary
SED	Spectral Energy Distribution
SFR	Star Formation Rate
SLE	Shapiro-Lightman-Eardley solutions
SPY	SN Ia Progenitor SurveY
SS	Shakura-Sunyaev
STIS	Space Telescope Imaging Spectrograph
SXT	Soft X-ray Transient
SyS	Symbiotic Star
TZO	Thorne Żytkow Object
UCXB	Ultracompact X-ray Binary
ULX	Ultra Luminous X-ray source
UV	Ultraviolet
VLBI	Very Long Baseline Interferometry
WD	White Dwarf
WRLOF	Wind Roche Lobe Overflow
XLF	X-ray Luminosity Function
XMM	European Space Agency's X-ray Multi-mirror Mission
XRB	X-Ray Binary

1. Accretion disks

HENK SPRUIT

Abstract

In this lecture the basic theory of accretion disks is reviewed, with emphasis on aspects relevant for X-ray binaries and cataclysmic variables. The text gives a general introduction as well as a selective discussion of a number of more recent topics.

1.1 Introduction

Accretion disks are inferred to exist as objects of very different scales: millions of kilometers in low mass X-ray binaries (LMXB) and cataclysmic variables (CV), solar-radius-to-AU–scale disks in protostellar objects, and AU-to-parsec-scale disks in active galactic nuclei (AGN).

An interesting observational connection exists between accretion disks and jets (such as the spectacular jets from AGN and protostars) and outflows (the "CO-outflows" from protostars and the "broad-line regions" in AGN). Lacking direct (i.e., spatially resolved) observations of disks, theory has tried to provide models, with varying degrees of success. Uncertainty still exists with respect to some basic questions. In this situation, progress made by observations or modeling of a particular class of objects has direct impact on the understanding of other objects, including the enigmatic connection with jets.

In this lecture I concentrate on the more basic aspects of accretion disks, but an attempt is made to mention topics of current interest as well. Some emphasis is on those aspects of accretion disk theory that connect to the observations of LMXB and CVs. For other reviews on the basics of accretion disks, see Pringle (1981) and Papaloizou and Lin (1995). For a more extensive introduction, see the textbook by Frank *et al.* (2002). For a comprehensive text on CVs, see Warner (1995).

1.2 Accretion: general

Gas falling into a point mass potential

$$\Phi = -\frac{GM}{r} \qquad (1.1)$$

from a distance r_0 to a distance r converts gravitational into kinetic energy by an amount $\Delta\Phi = GM(1/r - 1/r_0)$. For simplicity, assuming that the starting distance is large, $\Delta\Phi = GM/r$. The speed of arrival, the *free-fall speed* $v_{\rm ff}$, is given by

$$\frac{1}{2}v_{\rm ff}^2 = \frac{GM}{r}. \qquad (1.2)$$

If the gas is then brought to rest, for example at the surface of a star, the amount of energy e dissipated per unit mass is

$$e = \frac{1}{2}v_{\rm ff}^2 = \frac{GM}{r} \qquad ({\rm rest}). \qquad (1.3)$$

If, instead, it goes into a circular Kepler orbit at distance r:

$$e = \frac{1}{2}\frac{GM}{r} \qquad ({\rm orbit}). \qquad (1.4)$$

The dissipated energy may go into internal energy of the gas, and into radiation, which escapes to infinity (usually in the form of photons, but neutrino losses can also play a role in some cases).

1.2.1 Adiabatic accretion

Consider first the case when radiation losses are neglected. Any mechanical energy dissipated stays locally in the flow. This is called an *adiabatic* flow (not to be confused with *isentropic* flow). For an ideal gas with constant ratio of specific heats γ, the internal energy per unit mass is

$$e = \frac{P}{(\gamma-1)\rho}. \tag{1.5}$$

With the equation of state

$$P = \frac{\mathcal{R}\rho T}{\mu}, \tag{1.6}$$

where $\mathcal{R}$ is the gas constant and μ the mean atomic weight per particle, we find the temperature of the gas after the dissipation has taken place (assuming that the gas goes into a circular orbit):

$$T = \frac{1}{2}(\gamma-1)T_{\rm vir}, \tag{1.7}$$

where $T_{\rm vir}$, the *virial temperature*, is defined as

$$T_{\rm vir} = \frac{GM\mu}{\mathcal{R}r} = \frac{g\,r\,\mu}{\mathcal{R}}, \tag{1.8}$$

where g is the acceleration of gravity at distance r. In an atmosphere with temperature near $T_{\rm vir}$, the sound speed $c_{\rm s} = (\gamma \mathcal{R} T/\mu)^{1/2}$ is close to the escape speed from the system, and the hydrostatic pressure scale height, $H \equiv \mathcal{R}T/(\mu g)$, is of the order of r. Such an atmosphere may evaporate on a relatively short time scale in the form of a stellar wind. This is as expected from energy conservation: if no energy is lost through radiation, the energy gained by the fluid while falling into a gravitational potential is also sufficient to move it back out again.

A simple example is spherically symmetric adiabatic accretion (Bondi, 1952). An important result is that such accretion is possible only if $\gamma \leq 5/3$. The lower γ, the lower the temperature in the accreted gas (eq. 1.7), and the easier it is for the gas to stay bound in the potential. A classical situation where adiabatic and roughly spherical accretion takes place is a supernova implosion: when the central temperature becomes high enough for the radiation field to start disintegrating nuclei, γ drops and the envelope collapses onto the forming neutron star via an accretion shock. Another case is Thorne-Zytkow objects (e.g., Cannon *et al.*, 1992), where γ can drop to low values due to pair creation, initiating an adiabatic accretion onto the black hole.

Adiabatic spherical accretion is fast, taking place on the dynamical time scale: something on the order of the free fall time scale, or Kepler orbital time scale,

$$\tau_{\rm d} = \frac{r}{v_{\rm K}} = \Omega_{\rm K}^{-1} = \left(\frac{r^3}{GM}\right)^{1/2}, \tag{1.9}$$

where $v_{\rm K}$ and $\Omega_{\rm K}$ are the Kepler orbital velocity and angular frequency, respectively.

When radiative loss becomes important, the accreting gas can stay cool irrespective of the value of γ, and Bondi's critical value $\gamma = 5/3$ plays no role. With such losses, the temperatures of accretion disks are usually much lower than the virial temperature.

1.2.2 Temperature near compact objects

For accretion onto a neutron star surface, $R = 10$ km, $M = 1.4$ M$_\odot$, we have a free fall speed $v_{\rm ff}/c \approx 0.4\,c$ (this in Newtonian approximation; the correct value in general relativity is quantitatively somewhat different). The corresponding virial temperature would be $T_v \sim 2 \ 10^{12}$ K, equivalent to an average energy of 150 MeV per particle.

This is not the actual temperature we should expect, since other things happen before such temperatures are reached. If the accretion is adiabatic, one of these is the creation of a very dense radiation field. The energy liberated per infalling particle is still the same, but it is shared among a large number of photons. At temperatures above the electron rest mass (≈ 0.5 MeV), electron-positron pairs $e^{\pm}$ are produced in addition to photons. These can take up most of the accretion energy, limiting the temperature typically to a few MeV.

In most observed disks, however, temperatures do not get even close to 1 MeV, because accretion is rarely adiabatic. Energy loss takes place by escaping photons (or, under more extreme conditions, neutrinos). Exceptions are the radiatively inefficient accretion flows discussed in Section 1.13.

Radiative loss

Next to the adiabatic temperature estimates, a useful characteristic number is the *blackbody effective temperature*. Here, the approximation made is that the accretion energy is radiated away from an optically thick surface of some geometry, under the assumption of a balance between the heating rate by release of accretion energy and cooling by radiation. For a specific example, consider the surface of a star of radius R and mass M accreting via a disk. Most of the gravitational energy is released close to the star, in a region with surface area of the order (let us call it) $4\pi R^2$. If the surface radiates approximately as a blackbody of temperature T (make a note of the fact that this is a bad approximation if the opacity is dominated by electron scattering), the balance would be

$$\dot{M}\frac{GM}{R} \approx 4\pi R^2 \sigma_r T^4, \tag{1.10}$$

where σ_r is the Stefan-Boltzmann constant, $\sigma_r = a_r c/4$, if a_r is Planck's radiation constant. For a neutron star with $M = 1.4 \ M_\odot = 3 \times 10^{33}$ g, $R = 10$ km, accreting at a typical observed rate (near Eddington rate; see later), $\dot{M} = 10^{18}$ g s$^{-1} \approx 10^{-8} \ M_\odot$ yr^{-1}, this temperature would be $T \approx 10^7$ K, or ≈ 1 keV per particle. Radiation with this characteristic temperature is observed in accreting neutron stars and black holes in their so-called soft X-ray states (as opposed to their hard states, in which the spectrum is very far from a blackbody).

The kinds of processes involved in radiation from accretion flows form a large subject in itself and are not covered here. For introductions, see Rybicki and Lightman (1979) and Frank *et al.* (2002). At the moderate temperatures encountered in protostars and white dwarf accretors, the dominant processes are the ones known from stellar physics: molecular and atomic transitions and Thomson scattering. Up to photon energies around 10 keV (the Lyman edge of an iron nucleus with one electron left), these processes also dominate the spectra of neutron star and black hole accretors. Above this energy, observed spectra become dominated by Compton scattering. Cyclotron/synchrotron radiation plays a role when a strong magnetic field is present, which can happen in most classes of accreting objects.

1.2.3 Critical luminosity

Objects of high luminosity have a tendency to blow their atmospheres away because of the radiative force exerted when the outward-traveling photons are scattered or absorbed. Consider a volume of gas on which a flux of photons is incident from one side. Per gram of matter, the gas presents a scattering (or absorbing) surface area of κ cm^2 to the escaping radiation. The force exerted by the radiative flux F on 1 g is $F\kappa/c$. The force of gravity pulling back on this 1 g of mass is GM/r^2. The critical flux at which the two forces balance (energy per unit area and time) is

$$F_{\rm E} = \frac{c}{\kappa}\frac{GM}{r^2}. \tag{1.11}$$

Assuming that this flux is *spherically symmetric*, it can be converted into a luminosity,

$$L_{\rm E} = \frac{4\pi GMc}{\kappa}, \tag{1.12}$$

the Eddington critical luminosity, popularly called the Eddington limit (e.g., Rybicki and Lightman, 1979). If the gas is fully ionized, its opacity is dominated by electron scattering, and for solar composition, κ is then of the order 0.3 cm^2 g^{-1} (about a factor of 2 lower than for fully ionized helium). With these assumptions,

$$L_{\rm E} \approx 1.7\,10^{38}\frac{M}{M_\odot}\ {\rm erg\ s^{-1}} \approx 4\,10^4 \frac{M}{M_\odot}\ {\rm L}_\odot. \tag{1.13}$$

This number is different if the opacity is not dominated by electron scattering. In partially ionized gases of solar composition, bound-bound and bound-free transitions can increase the opacity by a factor up to 10^3; the Eddington flux is then correspondingly lower.

If the luminosity results from accretion, one can define a corresponding Eddington characteristic accretion rate $\dot M_{\rm E}$:

$$\frac{GM}{r}\dot M_{\rm E} = L_{\rm E} \quad \rightarrow \quad \dot M_{\rm E} = \frac{4\pi rc}{\kappa}. \tag{1.14}$$

With $\kappa = 0.3$:

$$\dot M_{\rm E} \approx 1.3\,10^{18} r_6\ {\rm g\ s^{-1}} \approx 2\,10^{-8} r_6\ {\rm M}_\odot\ {\rm yr}^{-1}, \tag{1.15}$$

where r_6 is the radius of the accreting object in units of 10^6 cm. The characteristic accretion rate thus scales with the *size* of the accreting object, while the critical luminosity scales with *mass*.

Whereas $L_{\rm E}$ is a critical value that in several circumstances plays the role of a limit, the Eddington characteristic accretion rate is less of a limit. For more on exceptions to $L_{\rm E}$ and $\dot M_{\rm E}$, see Section 1.2.3.

Eddington luminosity at high optical depth

The Eddington characteristic luminosity was derived earlier under the assumption of a radiation flux passing through an optically thin medium surrounding the radiation source. What changes if the radiation passes through an optically thick medium, such as a stellar interior? At high optical depth, the radiation field can be assumed to be nearly isotropic, and the *diffusion approximation* applies (cf. Rybicki and Lightman, 1979). The radiative heat flux can then be written in terms of the radiation pressure $P_{\rm r}$ as

$$F_{\rm r} = c\frac{{\rm d}P_{\rm r}}{{\rm d}\tau}, \tag{1.16}$$

where τ is the optical depth

$$ {\rm d}\tau = \kappa\rho\,{\rm d}s \tag{1.17}$$

along a path s, and κ an appropriate frequency-averaged opacity (such as the Rosseland mean). Balancing the gradient of the radiation pressure against the force of gravity gives the maximum radiation pressure that can be supported:

$$\nabla P_{\rm r,max} = {\bf g}\rho, \tag{1.18}$$

where $\bf g$ is the acceleration of gravity. With eq. 1.16, this yields the maximum radiation flux at a given point in a static gravitating object:

$${\bf F}_{\rm r,max} = \frac{c\,{\bf g}}{\kappa}, \tag{1.19}$$

that is, the same as the critical flux in the optically thin case.

Limitations of the Eddington limit

The derivation of F_E assumed that the force relevant in the argument is gravity. Other forces can be larger. An example would be a neutron star with a strong magnetic field. The curvature force $B^2/(4\pi r_c)$ in a loop of magnetic field (where r_c is the radius of curvature of the field lines) can balance a pressure gradient $\sim P/r_c$, so the maximum pressure that can be contained in a magnetic field[1] is of order $P \sim B^2/8\pi$. If the pressure is due to radiation, assuming an optical depth $\tau \geq 1$ so the diffusion approximation can be used, the maximum radiative energy flux is then of the order

$$F_{\text{r,max}} \approx \frac{cP_r}{\tau} \approx c\frac{B^2}{8\pi\tau}. \qquad (1.20)$$

In the range of validity of the assumptions made this has its maximum for an optical depth of order unity: $F_{\text{r,max}} \approx cB^2/8\pi$. For a neutron star of radius $R = 10^6$ cm and a field strength of 10^{12} G, this gives $L_{\text{r,max}} \approx 10^{46}$ erg s^{-1}, many orders of magnitude higher than the Eddington value L_E. (This explains the enormous luminosities that can be reached in so-called magnetar outbursts; Hurley et al., 2005.)

The Eddington argument considers only the radiative flux. Larger energy fluxes are possible if energy is transported by other means, for example by convection.

Since L_E depends on opacity, it can happen that L_E is lower in the atmosphere of a star than in its interior. A luminous star radiating near its (internal) Eddington rate will then blow off its atmosphere in a *radiatively driven stellar wind*; this happens, for example, in Wolf-Rayet stars. In the context of protostellar accretion, the opacity in the star-forming cloud from which the protostar accretes is high due to atomic and molecular transitions. As a result, the radiation pressure from a massive (proto-)star, with a luminosity approaching equation 1.13, is able to clear away the molecular cloud from which it formed. This is believed to set a limit on the mass that can be reached by a star formed in a molecular cloud.

Neutron stars versus black hole accretors

In deriving the critical accretion rate, it was assumed that the gravitational energy liberated is emitted in the form of radiation. In the case of a black hole accretor, this is not necessary, since mass can flow through the hole's horizon, taking with it all energy contained in it. Instead of being emitted as radiation, the energy adds to the mass of the hole. This becomes especially important at high accretion rates, $\dot{M} > \dot{M}_E$ (and in the ion-supported accretion flows discussed in Section 1.13). The parts of the flow close to the hole then become optically thick; the radiation stays trapped in the flow and, instead of producing luminosity, is swallowed by the hole. The accretion rate on a black hole can thus be arbitrarily large in principle (see also Section 1.11, and chapter 10 in Frank et al., 2002).

A neutron star cannot absorb this much energy (only a negligible amount is taken up by conduction of heat into its interior), so $\dot{M}_E$ is more relevant for neutron stars than for black holes. It is not clear to what extent it actually limits the possible accretion rate, however, since the limit was derived under the assumption of spherical symmetry. It is possible that accretion takes place in the form of an optically thick disk, while the energy released at the surface produces an outflow along the axis, increasing the maximum possible accretion rate (cf. discussion in Section 1.11). This has been proposed (e.g., King, 2004) as a possible conservative interpretation of the so-called ultraluminous X-ray sources (ULX), rare objects with luminosities of 10^{39} to 10^{41} erg s^{-1}. These are alternatively and more excitingly suggested to harbor intermediate mass black holes (above ≈ 30 M$_\odot$).

---

[1] Depending on circumstances the actual maximum is less than this because a magnetically contained plasma tends to "leak across" field lines through MHD instabilities.

1.3 Accretion with angular momentum

When the accreting gas has a nonzero angular momentum with respect to the accreting object, it cannot accrete directly. A new time scale then plays a role: the time scale for outward transport of angular momentum. Because this is in general much longer than the dynamical time scale, much of what was said about spherical accretion needs modification for accretion with angular momentum.

Consider the accretion in a close binary consisting of a compact (white dwarf, neutron star, or black hole) primary of mass M_1 and a main sequence companion of mass M_2. The mass ratio is defined as $q = M_2/M_1$ (note: in the literature, q is just as often defined the other way around).

If M_1 and M_2 orbit each other in a circular orbit and their separation is a, the orbital frequency Ω is

$$\Omega^2 = \frac{G(M_1 + M_2)}{a^3}. \tag{1.21}$$

The accretion process is most easily described in a coordinate frame that corotates with this orbit, and with its origin in the center of mass. Matter that is stationary in this frame experiences an effective potential, the *Roche potential* (chapter 4 in Frank *et al.*, 2002), given by

$$\phi_R(\mathbf{r}) = -\frac{GM}{r_1} - \frac{GM}{r_2} - \frac{1}{2}\Omega^2 \varpi^2, \tag{1.22}$$

where $r_{1,2}$ are the distances of point $\mathbf{r}$ to stars 1, 2, and ϖ is the distance from the rotation axis (the axis through the center of mass, perpendicular to the orbit). Matter that does *not* corotate experiences a very different force (due to the Coriolis force). The Roche potential is therefore useful only in a rather limited sense. For non-corotating gas, intuition based on the Roche geometry can be misleading. Keeping in mind this limitation, consider the equipotential surfaces of equation 1.22. The surfaces of stars $M_{1,2}$, assumed to corotate with the orbit, are equipotential surfaces of equation 1.22. Near the centers of mass (at low values of ϕ_R), they are unaffected by the other star; at higher Φ, they are distorted; and at a critical value Φ_1, the two parts of the surface touch. This is the critical Roche surface S_1 whose two parts are called the *Roche lobes*.

Binaries lose angular momentum through gravitational radiation and a magnetic wind from the secondary (if it has a convective envelope). Through this loss, the separation between the components decreases and both Roche lobes decrease in size. Mass transfer starts when M_2 fills its Roche lobe and continues as long as the angular momentum loss from the system lasts. Mass transfer can also be due to expansion of the secondary in the course of its evolution, and mass transfer can be a runaway process, depending on mass and internal structure of the secondary. This is a classical subject in the theory of binary stars; for an introduction, see Warner (1995).

A stream of gas then flows through the point of contact of the two parts of S_1, the inner Lagrange point L_1. If the force acting on it were derivable entirely from equation 1.22, the gas would just fall in radially onto M_1. As soon as it moves, however, it no longer corotates, and its orbit under the influence of the Coriolis force is different (Fig. 1.1).

Since the gas at L_1 is usually very cold compared with the virial temperature, the velocity it acquires already exceeds the sound speed after moving a small distance from L_1. The flow into the Roche lobe of M_1 is therefore highly *supersonic*. Such hypersonic flow is essentially ballistic, that is, the stream flows approximately along the path taken by freely falling particles.

Though the gas stream on the whole follows a path close to that of a free particle, a strong shock develops at the point where the path intersects itself.[2] After this, the gas

---

[2]In practice, shocks develop shortly after passing the pericenter at M_1, when the gas is decelerated again. Supersonic flows that are decelerated, by whatever means, in general develop

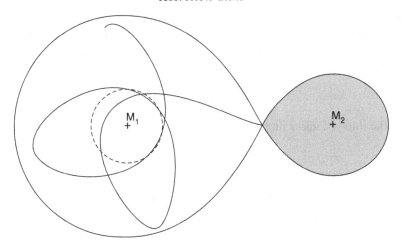

FIGURE 1.1. Roche geometry for $q = 0.2$, with free particle orbit from L_1 (as seen in a frame corotating with the orbit). Dashed: circularization radius.

settles into a ring, into which the stream continues to feed mass. If the mass ratio q is not too small, this ring forms fairly close to M_1. An approximate value for its radius is found by noting that near M_1, the tidal force due to the secondary is small, so that the angular momentum of the gas with respect to M_1 is approximately conserved. If the gas continues to conserve angular momentum while dissipating energy, it settles into the minimum-energy orbit with the specific angular momentum j of the incoming stream. The value of j is found by a simple integration of the orbit starting at L_1 and measuring j at some point near the pericenter. The radius of the orbit, the nominal *circularization radius*, r_c, is then defined through $(GM_1 r_c)^{1/2} = j$. In units of the orbital separation a, r_c and the distance r_{L1} from M_1 to L_1 are functions of the mass ratio only. As an example, for $q = 0.2$, $r_{L1} \approx 0.66a$ and the circularization radius $r_c \approx 0.16a$. In practice, the ring forms somewhat outside r_c, because there is some angular momentum redistribution in the shocks that form at the impact of the stream on the ring. The evolution of the ring depends critically on nature and strength of the angular momentum transport processes. If sufficient "viscosity" is present, it spreads inward and outward to form a disk.

At the point of impact of the stream on the disk, the energy dissipated is a significant fraction of the orbital kinetic energy; hence, the gas heats up to a significant fraction of the virial temperature. For a typical system with $M_1 = 1$ M$_\odot$, $M_2 = 0.2$ M$_\odot$, and an orbital period of 2 hours, the observed size of the disk (e.g., Wood *et al.*, 1989b; Rutten *et al.*, 1992) $r_d/a \approx 0.3$, the orbital velocity at r_d is about 900 km s^{-1}, and the virial temperature at r_d is $\approx 10^8$ K. The actual temperatures at the impact point are much lower, due to rapid cooling of the shocked gas. Nevertheless, the impact gives rise to a prominent "hot spot" in many systems, and an overall heating of the outermost part of the disk.

1.4 Thin disks: properties

1.4.1 *Flow in a cool disk is supersonic*

Ignoring viscosity, the equation of motion in the potential of a point mass is

$$\frac{\partial \mathbf{v}}{\partial t} + \mathbf{v} \cdot \nabla \mathbf{v} = -\frac{1}{\rho} \nabla P - \frac{GM}{r^2} \hat{\mathbf{r}}, \qquad (1.23)$$

shocks (e.g., Courant and Friedrichs, 1948; Massey, 1968). The effect can be seen in action in the video published in Różyczka and Spruit (1993).

where $\hat{\mathbf{r}}$ is a unit vector in the spherical radial direction r. To compare the order of magnitude of the terms, choose a position r_0 in the disk and choose as typical time and velocity scales the orbital time scale $\Omega_0^{-1} = (r_0^3/GM)^{1/2}$ and velocity $\Omega_0 r_0$. For simplicity of the argument, assume a fixed temperature T. The pressure gradient term is then

$$\frac{1}{\rho}\nabla P = \frac{\mathcal{R}}{\mu} T \nabla \ln P. \tag{1.24}$$

In terms of the dimensionless quantities

$$\tilde{r} = \frac{r}{r_0}, \qquad \tilde{v} = \frac{v}{(\Omega_0 r_0)}, \tag{1.25}$$

$$\tilde{t} = \Omega_0 t, \qquad \tilde{\nabla} = r_0 \nabla, \tag{1.26}$$

the equation of motion becomes

$$\frac{\partial \tilde{\mathbf{v}}}{\partial \tilde{t}} + \tilde{\mathbf{v}} \cdot \tilde{\nabla} \tilde{\mathbf{v}} = -\frac{T}{T_{\rm vir}} \tilde{\nabla} \ln P - \frac{1}{\tilde{r}^2} \hat{\mathbf{r}}. \tag{1.27}$$

All terms and quantities in this equation are of order unity by the assumptions made, except the pressure gradient term, which has the coefficient $T/T_{\rm vir}$ in front. If cooling is important, so that $T/T_{\rm vir} \ll 1$, the pressure term is negligible to first approximation. Equivalent statements are also that the gas moves hypersonically on nearly Keplerian orbits, and that the disk is thin, as is shown next.

1.4.2 Disk thickness

The thickness of the disk is found by considering its equilibrium in the direction perpendicular to the disk plane. In an axisymmetric disk, using cylindrical coordinates (ϖ, ϕ, z), consider the forces at a point $\mathbf{r}_0$ $(\varpi, \phi, 0)$ in the midplane, in a frame rotating with the Kepler rate Ω_0 at that point. The gravitational acceleration $-GM/r^2 \hat{\mathbf{r}}$ balances the centrifugal acceleration $\Omega_0^2 \hat{\boldsymbol{\varpi}}$ at this point, but not at some distance z above it because gravity and centrifugal acceleration work in different directions. Expanding both accelerations near $\mathbf{r}_0$, one finds a residual acceleration toward the midplane of magnitude

$$g_z = -\Omega_0^2 z. \tag{1.28}$$

Assuming an isothermal gas at temperature T, the condition for equilibrium in the z-direction under this acceleration yields a hydrostatic density distribution

$$\rho = \rho_0(\varpi) \exp\left(-\frac{z^2}{2H^2}\right). \tag{1.29}$$

$H(\varpi)$, called the *scale height* of the disk or simply the disk thickness, is given in terms of the isothermal sound speed $c_{\rm i} = (\mathcal{R}T/\mu)^{1/2}$ by

$$H = \frac{c_{\rm i}}{\Omega_0}. \tag{1.30}$$

We define $\delta \equiv H/r$, the *aspect ratio* of the disk; it can be expressed in several equivalent ways:

$$\delta = \frac{H}{r} = \frac{c_{\rm i}}{\Omega r} = \frac{1}{\rm Ma} = \left(\frac{T}{T_{\rm vir}}\right)^{1/2}, \tag{1.31}$$

where Ma is the Mach number of the orbital motion.

1.4.3 Viscous spreading

The shear flow between neighboring Kepler orbits in the disk causes friction if some form of viscosity is present. The frictional torque is equivalent to exchange of angular momentum between these orbits. But because the orbits are close to Keplerian, a change in angular momentum of a ring of gas also means it must change its distance from the central mass. If the angular momentum is increased, the ring moves to a larger radius. In a thin disk, angular momentum transport (more precisely, a nonzero divergence of the angular momentum flux) therefore automatically implies redistribution of mass in the disk.

A simple example (Lüst, 1952; Lynden-Bell and Pringle, 1974) is a narrow ring of gas at some distance r_0. If at $t = 0$ this ring is released to evolve under the viscous torques, one finds that it first spreads into an asymmetric hump. The hump quickly spreads inward onto the central object while a long tail spreads slowly outward to large distances. As $t \to \infty$, almost all the *mass* of the ring accretes onto the center, while a vanishingly small fraction of the gas carries almost all the *angular momentum* to infinity.

As a result of this asymmetric behavior, essentially all the mass of a disk can accrete even if the total angular momentum of the disk is conserved. In practice, however, there is often an external torque removing angular momentum from the disk: when the disk results from mass transfer in a binary system. The tidal forces exerted by the companion star take up angular momentum from the outer parts of the disk, limiting its outward spread.

1.4.4 Observational evidence of disk viscosity

Evidence for the strength of the angular momentum transport processes in disks comes from observations of variability time scales. This evidence is not good enough to determine whether the processes really behave in the same way as viscosity, but if this is assumed, estimates can be made of its magnitude.

Observations of cataclysmic variables (CVs) give the most detailed information. These are binaries with white dwarf (WD) primaries and (usually) main sequence companions (for reviews, see Meyer-Hofmeister and Ritter, 1993; Warner, 1995). A subclass of these systems, the dwarf novae, show semiregular outbursts. In the currently most developed theory, these outbursts are due to an instability in the disk (Smak, 1971; Meyer and Meyer-Hofmeister, 1981; King, 1995; Hameury et al., 1998). The outbursts are episodes of enhanced accretion of mass from the disk onto the primary, involving a significant part of the whole disk. The decay time of the burst is thus a measure of the viscous time scale of the disk (the quantitative details depend on the model; see Cannizzo et al., 1988; Hameury et al., 1998):

$$t_{\text{visc}} = \frac{r_d^2}{\nu}, \qquad (1.32)$$

where r_d is the size of the disk, $\sim 10^{10}$ cm for a CV. With decay times on the order of days, this yields viscosities of the order 10^{15} cm^2 s^{-1}, some 14 orders of magnitude above the microscopic "molecular" viscosity of the gas.

Other evidence comes from the inferred time scale on which disks around protostars disappear, which is of the order of 10^7 years (e.g., Strom et al., 1993).

1.4.5 α-Parameterization

Several processes have been proposed to account for these short time scales and the large apparent viscosities inferred from them. One of these is that accretion does not in fact take place through a viscous-like process as described earlier at all, but results from angular momentum loss through a magnetic wind driven from the disk surface (Bisnovatyi-Kogan and Ruzmaikin, 1976; Blandford, 1976), much in the way sunlike stars spin down by angular momentum loss through their stellar winds. The extent to which this plays a role in accretion disks is still uncertain. It would be a "quiet" form of accretion, since it can do without energy-dissipating processes such as viscosity. It has been proposed as the explanation for the low ratio of X-ray luminosity to jet power in many radio sources (see Migliari and Fender, 2006, and references therein). This low ratio, however, is also plausibly attributed to the low efficiency with which X-rays are produced in the "ion-supported accretion flows" (see Section 1.13) that are expected to be the source of the jet outflows observed from X-ray binaries and AGN.

The most quantitatively developed mechanism for angular momentum transport is a form of magnetic viscosity (anticipated already in Shakura and Sunyaev, 1973). This is discussed below in Section 1.8.1. It requires the accreting plasma to be sufficiently electrically conducting. This is often the case, but not always: it is questionable for example in the cool outer parts of protostellar disks. Other mechanisms thus still play a role in the discussion, for example spiral shocks (Spruit et al., 1987) and self-gravitating instabilities (Paczyński, 1978; Gammie, 1997).

In order to compare the viscosities of disks in objects with (widely) different sizes and physical conditions, introduce a dimensionless viscosity α:

$$\nu = \alpha \frac{c_i^2}{\Omega}, \qquad (1.33)$$

where c_i is the isothermal sound speed as before. The quantity α was introduced by Shakura and Sunyaev (1973) as a way of parametrizing our ignorance of the angular momentum transport process in a way that allows comparison between systems of very different size and physical origin. (Their definition of α differs a bit, by a constant factor of order unity.)

1.4.6 Causality limit on turbulent viscosity

How large can the value of α be, on theoretical grounds? As a simple model, let's assume that the shear flow between Kepler orbits is unstable to the same kind of shear instabilities found for flows in tubes, in channels, near walls, and in jets. These instabilities occur so ubiquitously that the fluid mechanics community considers them an automatic consequence of a high Reynolds number:

$$\mathrm{Re} = \frac{LV}{\nu} \qquad (1.34)$$

where L and V are characteristic length and velocity scales of the flow. If this number exceeds about 1,000 (for some forms of instability much less), instability and turbulence are generally observed. It has been argued (e.g., Zel'dovich, 1981) that for this reason hydrodynamic turbulence is the cause of disk viscosity. Let's look at the consequences of this assumption. If an eddy of radial length scale l develops because of shear instability, it will rotate at a rate given by the rate of shear in the flow, σ, in a Keplerian disk:

$$\sigma \approx r \frac{\partial \Omega}{\partial r} = -\frac{3}{2}\Omega. \qquad (1.35)$$

The velocity amplitude of the eddy is $V = \sigma l$, and a field of such eddies would produce a turbulent viscosity of the order (leaving out numerical factors of order unity)

$$\nu_{\mathrm{turb}} \approx l^2 \Omega. \qquad (1.36)$$

In compressible flows, there is a maximum to the size of the eddy set by causality considerations. The force that allows an instability to form a coherently overturning eddy is the pressure, which transports information about the flow at the speed of sound. The eddies formed by a shear instability can therefore not rotate faster than c_i; hence, their size does not exceed $c_i/\sigma \approx H$ (eq. 1.31). At the same time, the largest eddies formed also have the largest contribution to the exchange of angular momentum. Thus we should expect that the turbulent viscosity is given by eddies with size of the order H:

$$\nu < H^2 \Omega, \qquad (1.37)$$

or
$$\alpha < 1. \tag{1.38}$$

The dimensionless number α can thus be interpreted as the effective viscosity in units of the maximum value expected if a disk were hydrodynamically turbulent.

Large-scale vortices?

The small size of hydrodynamic eddies expected, $L < H = c_\mathrm{i}/\Omega$, is due to the high Mach number of the flow in an accretion disk; this makes a disk behave like a very *compressible* fluid. Attempts to construct large-scale vortices $L \gg H$ in disks using *incompressible* fluid analogies continue to be made, both analytically and experimentally. Expansions in disk thickness as a small parameter then suggest themselves, in analogy with large-scale flows such as weather systems and tropical storms in Earth's atmosphere. The size of these systems is large compared with the height of Earth's atmosphere, and expansions making use of this can be effective. This tempting analogy is misleading for accretion disks, however, because in contrast with the atmosphere, all flows with a horizontal scale exceeding the vertical thickness are supersonic, and the use of incompressible fluid models is meaningless.

1.4.7 Hydrodynamic turbulence?

Does hydrodynamic turbulence along these lines actually exist in disks? In the astrophysical community, a consensus has developed that it does *not*, certainly not along the simple lines suggested by a Reynolds number argument. An intuitive argument is that the flow in a cool disk is close to Kepler orbits, and these are quite stable. Numerical simulations of disklike flows with different rotation profiles do not show the expected shear flow instabilities in cases where the angular momentum increases outward, whereas turbulence put in by hand as initial condition decays in such simulations (Hawley et al., 1999).

Analytical work on the problem has not been able to demonstrate instability in this case either (for recent work and references, see Lesur and Longaretti, 2005; Rincon et al., 2007). Existing proposals for disk turbulence involve ad-hoc assumptions about the existence of hydrodynamic instabilities (e.g. Dubrulle, 1992). However, a subtle form of hydrodynamic angular momentum transport has been identified more recently (Lesur and Papaloizou, 2010); it depends on vertical stratification of the disk as well as the presence of a convectively unstable *radial* gradient.

A laboratory analogy is the rotating Couette flow, an experiment with water between differentially rotating cylinders. Recent such experiments have demonstrated that turbulence is absent in cases where angular momentum increases outward (as in Keplerian rotation) at Reynolds numbers as high as 10^6 (the Princeton Couette experiment; Ji et al., 2006).

In view of these negative results, the popular mechanism for angular momentum transport in accretion disk has become magnetic: "MRI turbulence" (Section 1.8.1). But beware: in the fluid mechanics community it is considered crackpot to suggest that at Reynolds numbers like our 10^{14} a flow (no matter which or where) can be anything but "fiercely turbulent." If your career depends on being friends with this community, it would be wise to avoid discussions about accretion disks or Couette experiments.

1.5 Thin disks: equations

Consider a thin (= cool, nearly Keplerian, cf. Section 1.4.2) disk, axisymmetric but not stationary. Using cylindrical coordinates (r, ϕ, z) (note that we have changed notation from ϖ to r compared with 1.4.2), we define the *surface density* Σ of the disk as

$$\Sigma = \int_{-\infty}^{\infty} \rho \, \mathrm{d}z \approx 2H_0\rho_0, \tag{1.39}$$

where ρ_0 and H_0 are the density and scale height at the midplane, respectively. The approximate sign is used to indicate that the coefficient in front of H in the last expression actually depends on details of the vertical structure of the disk. Conservation of mass, in terms of Σ, is described by

$$\frac{\partial}{\partial t}(r\Sigma) + \frac{\partial}{\partial r}(r\Sigma v_r) = 0 \tag{1.40}$$

(derived by integrating the continuity equation over z). Because the disk is axisymmetric and nearly Keplerian, the radial equation of motion reduces to

$$v_\phi^2 = \frac{GM}{r}. \tag{1.41}$$

The ϕ-equation of motion is

$$\frac{\partial v_\phi}{\partial t} + v_r \frac{\partial v_\phi}{\partial r} + \frac{v_r v_\phi}{r} = F_\phi, \tag{1.42}$$

where F_ϕ is the azimuthal component of the viscous force. By integrating this over height z and using (1.40), one gets an equation for the angular momentum balance:

$$\frac{\partial}{\partial t}(r\Sigma\Omega r^2) + \frac{\partial}{\partial r}\left(r\Sigma v_r \Omega r^2\right) = \frac{\partial}{\partial r}\left(Sr^3 \frac{\partial \Omega}{\partial r}\right), \tag{1.43}$$

where $\Omega = v_\phi/r$, and

$$S = \int_{-\infty}^{\infty} \rho\nu \, \mathrm{d}z \approx \Sigma\nu. \tag{1.44}$$

The second approximate equality in (1.44) holds if ν can be considered independent of z. The right-hand side of (1.43) is the divergence of the viscous angular momentum flux and is derived most easily with a physical argument, as described in, for example, Pringle (1981) or Frank *et al.* (2002).[3]

Assume now that ν can be taken constant with height. For an isothermal disk (T independent of z), this is equivalent to taking the viscosity parameter α independent of z. As long as we are not sure what causes the viscosity, this is a reasonable simplification. Note, however, that recent numerical simulations of magnetic turbulence suggest that the effective α and the rate of viscous dissipation per unit mass are higher near the disk surface than near the midplane (c.f. Section 1.8). Although equation 1.43 is still valid for rotation rates Ω deviating from Keplerian (only the integration over disk thickness must

---

[3] If you prefer a more formal derivation, the fastest way is to consult Landau and Lifshitz (1959), chapter 15 (hereafter LL). Noting that the divergence of the flow vanishes for a thin axisymmetric disk, the viscous stress σ becomes (LL eq. 15.3)

$$\sigma_{ik} = \eta\left(\frac{\partial v_i}{\partial x_k} + \frac{\partial v_k}{\partial x_i}\right), \tag{1.45}$$

where $\eta = \rho\nu$. This can be written in cylindrical or spherical coordinates using LL equations 15.15 through 15.18. The viscous force is

$$F_i = \frac{\partial \sigma_{ik}}{\partial x_k} = \frac{1}{\eta}\frac{\partial \eta}{\partial x_k}\sigma_{ik} + \eta\nabla^2 v_i. \tag{1.46}$$

Writing the Laplacian in cylindrical coordinates, the viscous torque is then computed from the ϕ-component of the viscous force by multiplying by r and is then integrated over z.

be justifiable), we now use the fact that $\Omega \sim r^{-3/2}$. Then equations 1.40 through 1.43 can be combined into a single equation for Σ:

$$r\frac{\partial \Sigma}{\partial t} = 3\frac{\partial}{\partial r}\left[r^{1/2}\frac{\partial}{\partial r}(\nu\Sigma r^{1/2})\right]. \tag{1.47}$$

Under the same assumptions, equation 1.42 yields the mass flux $\dot{M}$ at any point in the disk:

$$\dot{M} = -2\pi r \Sigma v_r = 6\pi r^{1/2}\frac{\partial}{\partial r}(\nu\Sigma r^{1/2}). \tag{1.48}$$

Equation 1.47 is the standard form of the *thin disk diffusion equation*. An important conclusion from this equation is: in the thin disk limit, all the physics that influences the time-dependent behavior of the disk enters through one quantity only: the viscosity ν. This is the main attraction of the thin disk approximation.

1.5.1 Steady thin disks

In a steady disk ($\partial/\partial t = 0$), the mass flux $\dot{M}$ is constant through the disk and equal to the accretion rate onto the central object. From equation 1.48 we get the surface density distribution:

$$\nu\Sigma = \frac{1}{3\pi}\dot{M}\left[1 - \beta\left(\frac{r_i}{r}\right)^{1/2}\right], \tag{1.49}$$

where r_i is the inner radius of the disk and β is a parameter appearing through the integration constant. It is related to the flux of angular momentum F_J through the disk:

$$F_J = -\dot{M}\beta\Omega_i r_i^2, \tag{1.50}$$

where Ω_i is the Kepler rotation rate at the inner edge of the disk. If the disk accretes onto an object with a rotation rate Ω_* *less* than Ω_i, the rotation rate $\Omega(r)$ as a function of distance r jumps from Ω_* close to the star to the Kepler rate $\Omega_K(r)$ in the disk. It then has a maximum at some distance close to the star (distance of the order of the disk thickness H). At this point the viscous stress (proportional to $\partial_r \Omega$) vanishes and the angular momentum flux is just the amount carried by the accretion flow. In (1.50) this corresponds to $\beta = 1$, *independent* of Ω_* (Shakura and Sunyaev, 1973; Lynden-Bell and Pringle, 1974). This is referred to as the standard or "accreting case." The angular momentum flux (equal to the torque on the accreting star), is then inward, causing the rotation rate of the star to increase (spinup).

For stars rotating near their maximum rate ($\Omega_* \approx \Omega_i$) and for accretion onto magnetospheres, which can rotate faster than the disk, the situation is different (Siuniaev and Shakura, 1977, translation in Syunyaev and Shakura, 1977; Popham and Narayan, 1991; Paczyński, 1991; Bisnovatyi-Kogan, 1993; Rappaport et al., 2004). If the inner edge of the disk is at the *corotation radius* r_{co}, defined by

$$\Omega_K(r_{co}) = \Omega_*, \tag{1.51}$$

the viscous stress cannot be assumed to vanish there, and the thin disk approximation does not give a unique answer for the angular momentum flux parameter β. Its value is then determined by details of the hydrodynamics at r_{co}, which need to be investigated separately. Depending on the outcome of this investigation, values varying from 1 (spinup, standard accreting case) to negative values (spindown) are possible. As equation 1.50 shows, the surface density at the inner edge of the disk depends sensitively on the value of β (see also Fig. 1.2). This plays a role in the cyclic accretion process discussed in the next subsection.

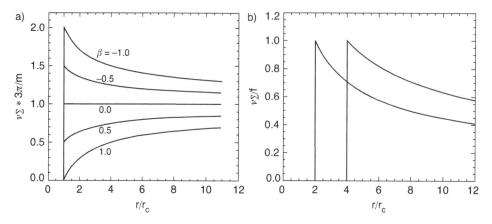

FIGURE 1.2. Surface density $\nu\Sigma$ of a thin disk as a function of distance from the corotation radius r_c, for a steady, thin viscous disk. (a) Steady accretion at a fixed accretion rate $\dot m$, for inner edge of the disk at corotation. β measures the angular momentum flux, $\beta=-1$, corresponding to the standard case of accretion onto a slowly rotating object. For $\beta<0$, the angular momentum flux is outward (spindown of the star). (b) "Quiescent disk" solutions with $\dot m=0$ and a steady outward angular momentum flux due to a torque f applied at the inner edge.

1.5.2 Magnetospheres, "propellering," and "dead disks"

Stars with magnetospheres, instead of spinning up by accretion of angular momentum from the disk, can actually *spin down* by interaction with the disk, even while accretion is still going on. The thin disk approximation just discussed covers this case as well (cf. Fig. 1.2). The surface density distribution is then of the form given in equation 1.49, but with $\beta<0$ (see also Spruit and Taam, 1993; Rappaport et al., 2004). The angular momentum flux through the disk is outward, and the accreting star spins down. This is possible even when the interaction between the disk and the magnetosphere takes place *only* at the inner edge of the disk. Magnetic torques due to interaction between the disk and the magnetosphere may exist at larger distances in the disk as well, but are not necessary for creating an outward angular momentum flux. Numerical simulations of disk-magnetosphere interaction (Miller and Stone, 1997; for recent work, see Long et al., 2008, and references therein) give an interesting view of how such interaction may take place, presenting a picture that is very different from what is assumed in the previous "standard" models. Among other things, they show the interaction region to be quite narrow.

Rapidly rotating magnetospheres play a role in some CVs and X-ray binaries, and probably also in protostellar accretion disks. When the rotation velocity of the magnetosphere is larger than the Kepler velocity at the inner edge of the disk (i.e., when $r_i>r_{co}$), the literature often assumes that the mass transferred from the secondary must be "flung out" of the binary system. This idea is called *propellering* (Illarionov and Sunyaev, 1975). The term is then used as synonymous with the condition $r_i>r_{co}$. Although a process like this is likely to happen when the rotation rate of the star is sufficiently large (the CV AE Aqr being an example), it is not necessary. It is not possible, either, when the difference in rotation velocity is too small to put the accreting mass on an escape orbit. This was realized early on (Syunyaev and Shakura, 1977). Instead of being flung out, accretion is halted and mass accumulates in the disk. The buildup of mass in the disk then leads to a following episode of accretion. In this phase the magnetosphere is compressed such that $r_i<r_{co}$. Instead of propellering, the result is *cyclic* accretion. For quantitative models, see D'Angelo and Spruit (2010); an example is shown in Figure 1.3. The phase with $r_i>r_{co}$, during which accretion is halted, is called the "dead disk" phase in Syunyaev and Shakura (1977).

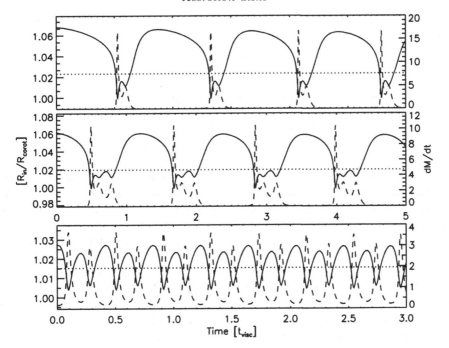

FIGURE 1.3. Cyclic accretion due to interaction between a magnetosphere and a disk with inner edge near corotation. Solid: inner edge radius of the disk in units of the corotation radius; dashed: accretion rate (from D'Angelo and Spruit, 2010).

1.5.3 Disk temperature

In this section, I assume accretion onto not-too-rapidly rotating objects, so that $\beta = 1$ (eq. 1.49). The surface temperature of the disk, which determines how much energy it loses by radiation, is governed by the energy dissipation rate in the disk, which in turn is given by the accretion rate. From the first law of thermodynamics, we have

$$\rho T \frac{dS}{dt} = -\text{div}\mathbf{F} + Q_\text{v}, \tag{1.52}$$

where S is the entropy per unit mass, $\mathbf{F}$ the heat flux (including radiation and convection), and Q_v the viscous dissipation rate. For thin disks, where the advection of internal energy (or entropy) can be neglected, and for changes that happen on time scales longer than the dynamical time Ω^{-1}, the left-hand side is small compared with the terms on the right-hand side. Integrating over z, the divergence turns into a surface term, and we get

$$2\sigma_\text{r} T_\text{s}^4 = \int_{-\infty}^{\infty} Q_\text{v} dz, \tag{1.53}$$

where T_s is the surface temperature of the disk, σ_r is Stefan-Boltzmann's radiation constant $\sigma_\text{r} = a_\text{r} c/4$, and the factor 2 comes about because the disk has two radiating surfaces (assumed to radiate like blackbodies). Thus, the energy balance is *local* (for such slow changes): what is generated by viscous dissipation inside the disk at any radius r is also radiated away from the surface at that position. The viscous dissipation rate is equal to $Q_\text{v} = \sigma_{ij} \partial v_i / \partial x_j$, where σ_{ij} is the viscous stress tensor (see footnote in Section 1.5), and this works out[4] to be

$$Q_\text{v} = \frac{9}{4} \Omega^2 \nu \rho. \tag{1.54}$$

---

[4]Using, for example, LL equation 16.3.

Equation 1.53, using equation 1.49, then gives the surface temperature in terms of the accretion rate:

$$\sigma_r T_s^4 = \frac{9}{8}\Omega^2 \nu \Sigma = \frac{GM}{r^3}\frac{3\dot{M}}{8\pi}\left[1-\left(\frac{r_i}{r}\right)^{1/2}\right]. \tag{1.55}$$

This shows that the surface temperature of the disk, at a given distance r from a steady accretor, depends *only* on the product $M\dot{M}$, and not on the highly uncertain value of the viscosity. For $r \gg r_i$, we have

$$T_s \sim r^{-3/4}. \tag{1.56}$$

These considerations tell us only about the surface temperature. The internal temperature in the disk is quite different and depends on the mechanism transporting energy to the surface. Because it is the internal temperature that determines the disk thickness H (and probably also the viscosity), this transport needs to be considered in some detail for realistic disk models. This involves a calculation of the vertical structure of the disk. Because of the local (in r) nature of the balance between dissipation and energy loss, such calculations can be done as a grid of models in r, without having to deal with exchange of energy between neighboring models. Schemes borrowed from stellar structure computations are used (e.g., Meyer and Meyer-Hofmeister, 1982; Cannizzo et al., 1988).

An approximation to the temperature in the disk can be found when a number of additional assumptions is made. As in stellar interiors, the energy transport is radiative rather than convective at high temperatures. Assuming local thermodynamic equilibrium (LTE, e.g., Rybicki and Lightman, 1979), the temperature structure of a radiative atmosphere is given, in the so-called Eddington approximation of radiative transfer (not to be confused with the Eddington limit), by

$$\frac{d}{d\tau}\sigma_r T^4 = \frac{3}{4}F. \tag{1.57}$$

The boundary condition that there is no incident flux from outside the atmosphere yields the approximate condition

$$\sigma_r T^4\left(\tau = \frac{2}{3}\right) = F, \tag{1.58}$$

where $\tau = \int_z^\infty \kappa \rho dz$ is the optical depth at geometrical depth z, and F the energy flux through the atmosphere. Assuming that most of the heat is generated near the midplane (which is the case if ν is constant with height), F is approximately constant with height and equal to $\sigma_r T_s^4$, given by (1.55). Equation 1.57 then yields

$$\sigma_r T^4 = \frac{3}{4}\left(\tau + \frac{2}{3}\right)F. \tag{1.59}$$

Approximating the opacity κ as constant with z, the optical depth at the midplane is $\tau = \kappa\Sigma/2$. If $\tau \gg 1$, the temperature at the midplane is then

$$T^4 = \frac{27}{64}\sigma_r^{-1}\Omega^2\nu\Sigma^2\kappa. \tag{1.60}$$

With the equation of state 1.6, valid when radiation pressure is small, we find for the disk thickness, using equation 1.49:

$$\frac{H}{r} = \left(\frac{\mathcal{R}}{\mu}\right)^{2/5}\left(\sigma_r \frac{64}{3}\pi^2\right)^{-1/10}\left(\frac{\kappa}{\alpha}\right)^{1/10}(GM)^{-7/20}r^{1/20}(f\dot{M})^{1/5}$$
$$= 5\,10^{-3}\alpha^{-1/10}r_6^{1/20}\left(\frac{M}{M_\odot}\right)^{-7/20}(f\dot{M}_{16})^{1/5}, \qquad (P_r \ll P) \tag{1.61}$$

where $r_6 = r/(10^6 \text{ cm})$, $\dot{M}_{16} = \dot{M}/(10^{16}\text{g s}^{-1})$, and

$$f = 1 - \left(\frac{r_i}{r}\right)^{1/2}. \tag{1.62}$$

From this we conclude (i) that the disk is thin in X-ray binaries, $H/r < 0.01$, and (ii) the disk thickness is relatively insensitive to the parameters, especially α, κ, and r. It must be stressed, however, that this depends fairly strongly on the assumption that the energy is dissipated in the disk interior. If the dissipation takes place close to the surface, such as in some magnetic reconnection models, the internal disk temperature will be much closer to the surface temperature (Haardt et al., 1994; Di Matteo et al., 1999, and references therein). The midplane temperature and H are even smaller in such disks than calculated from equation 1.61.

1.5.4 A subtlety: viscous vs. gravitational energy release

The viscous dissipation rate per unit area of the disk, $W_v = (9/4)\Omega^2\nu\Sigma$ (cf. eq. 1.55), can be compared with the local rate W_G at which gravitational energy is liberated in the accretion flow. Because half the gravitational energy stays in the flow as orbital motion, we have

$$W_G = \frac{1}{2\pi r} \frac{GM\dot{M}}{2r^2}, \qquad (1.63)$$

so that, in a steady thin disk accreting on a slowly rotating object (using eq. 1.49),

$$W_v/W_G = 3f = 3\left[1 - \left(\frac{r_i}{r}\right)^{1/2}\right]. \qquad (1.64)$$

At large distances from the inner edge, the dissipation rate is *3 times larger than the rate of gravitational energy release*. This may seem odd, but it becomes understandable when it is realized that there is a significant flux of energy through the disk associated with the viscous stress.[5] Integrating the viscous energy dissipation over the whole disk, one finds

$$\int_{r_i}^{\infty} 2\pi r W_v \, dr = \frac{GM\dot{M}}{2r_i}, \qquad (1.65)$$

as expected. That is, globally but not locally, half of the gravitational energy is radiated from the disk while the other half remains in the orbital kinetic energy of the material accreting on the star.

What happens to this remaining orbital energy depends on the nature of the accreting object. If the object is a slowly rotating black hole, the orbital energy is just swallowed by the hole. If it has a solid surface, the orbiting gas slows down until it corotates with the surface, dissipating the orbital energy into heat in a boundary layer. Unless the surface rotates close to the orbital rate ("breakup"), the energy released in this way is of the same order as the total energy released in the accretion disk. The properties of this boundary layer are therefore critical for accretion onto neutron stars and white dwarfs. See also Section 1.9.1 and Inogamov and Sunyaev (1999).

1.5.5 Radiation pressure–dominated disks

In the inner regions of disks in XRB, the radiation pressure can dominate over the gas pressure, which results in a different expression for the disk thickness. The total pressure P is

$$P = P_r + P_g = \frac{1}{3}aT^4 + P_g. \qquad (1.66)$$

Defining a "total sound speed" by $c_t^2 = P/\rho$, the relation between temperature and disk thickness (1.30) still holds, so $c_t = \Omega H$. For $P_r \gg P_g$, we get from equation 1.60, with

---

[5] See Landau and Lifshitz, section 16.

1.55 and $\tau \gg 1$,

$$cH = \frac{3}{8\pi}\kappa f \dot{M}, \tag{1.67}$$

(where the rather approximate relation $\Sigma = 2H\rho_0$ has been used). Thus,

$$\frac{H}{R} \approx \frac{3}{8\pi}\frac{\kappa}{cR}f\dot{M} = \frac{3}{2}f\frac{\dot{M}}{\dot{M}_\mathrm{E}}, \tag{1.68}$$

where R is the stellar radius and $\dot{M}_\mathrm{E}$ the Eddington rate for this radius. It follows that the disk becomes thick near the star, if the accretion rate is near Eddington (though this is mitigated somewhat by the factor f). Accretion near the Eddington limit is evidently not geometrically thin any more. In addition, other processes such as angular momentum loss from the disk by "photon drag" in the dense radiation field of the accreting star have to be taken into account.

1.6 Time scales in thin disks

A number of time scales play a role in the behavior of disks. In a thin disk, they differ by powers of the large factor r/H. The shortest of these is the dynamical time scale t_d:

$$t_\mathrm{d} = \Omega_\mathrm{K}^{-1} = \left(\frac{GM}{r^3}\right)^{-1/2}. \tag{1.69}$$

The time scale for radial drift through the disk over a distance of order r is the viscous time scale:

$$t_\mathrm{v} = \frac{r}{(-v_r)} = \frac{2}{3}\frac{rf}{\nu} = \frac{2f}{3\alpha\Omega}\left(\frac{r}{H}\right)^2 \tag{1.70}$$

(using eqs. 1.48 and 1.49, valid for steady accretion). This is of the order $(r/H)^2$ longer than the dynamical time scale, except near the inner edge of the disk, where $f \downarrow 0$, and this time scale formally drops to zero as a consequence of the thin disk approximations. If more physics was included, the viscous time scale would stay above the dynamical time scale.

Intermediate are *thermal* time scales. If E_t is the thermal energy content (enthalpy) of the disk per unit of surface area, and $W_\mathrm{v} = (9/4)\Omega^2\nu\Sigma$ the heating rate by viscous dissipation, we can define a heating time scale:

$$t_\mathrm{h} = \frac{E_\mathrm{t}}{W_\mathrm{v}}. \tag{1.71}$$

In the same way, a cooling time scale is defined by the energy content and the radiative loss rate:

$$t_\mathrm{c} = \frac{E_\mathrm{t}}{(2\sigma_\mathrm{r}T_\mathrm{s}^4)}. \tag{1.72}$$

For a thin disk, the two are equal because the viscous energy dissipation is locally balanced by radiation from the two disk surfaces. In thick disks (ADAFs), this balance does not hold, because the advection of heat with the accretion flow is not negligible. In this case $t_\mathrm{c} > t_\mathrm{h}$ (see Section 1.12). Thus, we can replace both time scales by a single thermal time scale t_t and find, with equation 1.54,

$$t_\mathrm{t} = \frac{1}{W_\mathrm{v}}\int_{-\infty}^{\infty}\frac{\gamma P}{\gamma - 1}\mathrm{d}z, \tag{1.73}$$

where the enthalpy of an ideal gas of constant ratio of specific heats γ has been used. Leaving out numerical factors of order unity, this yields

$$t_\mathrm{t} \approx \frac{1}{\alpha\Omega}. \tag{1.74}$$

That is, the thermal time scale of the disk is independent of most of the disk properties and of the order $1/\alpha$ times longer than the dynamical time scale. This independence is a consequence of the α-parameterization used. If α is not a constant, but dependent on disk temperature, for example, the dependence of the thermal time scale on disk properties will become apparent again.

If, as seems likely from observations, α is generally < 1, we have in thin disks the ordering of time scales:

$$t_v \gg t_t > t_d. \tag{1.75}$$

1.7 Comparison with CV observations

The number of meaningful quantitative tests between the theory of disks and observations is somewhat limited because, in the absence of a theory for ν, it is a bit meager on predictive power. The most detailed information perhaps comes from modeling of CV outbursts.

Two simple tests are possible (nearly) independently of ν. These are the prediction that the disk is geometrically quite thin (eq. 1.61) and the prediction that the surface temperature $T_s \sim r^{-3/4}$ in a steady disk. The latter can be tested in a subclass of the CVs that do not show outbursts, the nova-like systems, which are believed to be approximately steady accretors. If such a system is also eclipsing, eclipse mapping techniques can be used to derive the brightness distribution with r in the disk (Horne, 1985, 1993). If this is done in a number of colors so that bolometric corrections can be made, the results (e.g., Rutten *et al.*, 1992) show in general a *fair* agreement with the $r^{-3/4}$ prediction. Two deviations occur: (i) a few systems show significantly flatter distributions than predicted, and (ii) most systems show a "hump" near the outer edge of the disk. The latter deviation is easily explained, because we have not taken into account that the impact of the stream additionally heats the outer edge of the disk. Though not important for the total light from the disk, it is an important local contribution near the edge.

Eclipse mapping of dwarf novae in quiescence gives a quite different picture. Here, the inferred surface temperature profile is often nearly flat (e.g., Wood *et al.*, 1989a, 1992). This is understandable, however, since in quiescence the mass flux depends strongly on r. In the inner parts of the disk, it is small; near the outer edge, it is close to its average value. With equation 1.55, this yields a flatter $T_s(r)$. The lack of light from the inner disk is compensated during the outburst, when the accretion rate in the inner disk is higher than average (see Mineshige and Wood, 1989, for a more detailed comparison). The effect is also seen in the two-dimensional (2-D) hydrodynamic simulations of accretion in a binary by Różyczka and Spruit (1993). These simulations show an outburst during which the accretion in the inner disk is enhanced, between two episodes in which mass accumulates in the outer disk.

1.7.1 *Comparison with LMXB observations: irradiated disks*

In low-mass X-ray binaries, a complication arises because of the much higher luminosity of the accreting object. Because a neutron star is roughly 1,000 times smaller than a white dwarf, it produces 1,000 times more luminosity for a given accretion rate.

Irradiation of the disk by the central source (the accreting star plus inner disk) leads to a different surface temperature than predicted by equation 1.55. The central source radiates nearly the total accretion luminosity $GM\dot{M}/R$ (assuming sub-Eddington accretion; see Section 1.2). If the disk is *concave*, it will intercept some of this luminosity. If the central source is approximated as a point source, the irradiating flux on the disk surface is

$$F_{\text{irr}} = \frac{\epsilon}{2} \frac{GM\dot{M}}{4\pi R r^2},\tag{1.76}$$

where ϵ is the angle between the disk surface and the direction from a point on the disk surface to the central source:

$$\epsilon = \frac{dH}{dr} - \frac{H}{r}.\tag{1.77}$$

The disk is concave if ϵ is positive. We have

$$\frac{F_{\text{irr}}}{F} = \frac{1}{3}\frac{\epsilon}{f}\frac{r}{R},\tag{1.78}$$

where F is the flux generated internally in the disk, given by equation 1.55. On average, the angle ϵ is of the order of the aspect ratio $\delta = H/r$. With $f \approx 1$, and our fiducial value $\delta \approx 5\,10^{-3}$, we find that irradiation in LMXB dominates for $r \gtrsim 10^9$ cm. This is compatible with observations (for reviews, see van Paradijs and McClintock, 1995), which show that the optical and UV are dominated by reprocessed radiation from the innermost regions.

When irradiation by an external source is included in the thin disk model, the surface boundary condition of the radiative transfer problem, equation 1.58 becomes

$$\sigma_r T_s^4 = F + (1-a)F_{\text{irr}},\tag{1.79}$$

where a is the X-ray albedo of the surface, that is, $1-a$ is the fraction of the incident flux that is absorbed in the *optically thick* layers of the disk (photons absorbed higher up only serve to heat up the corona of the disk).[6] The surface temperature T_s increases in order to compensate for the additional incident heat flux. The magnitude of the incident flux is sensitive to the assumed disk shape $H(r)$, as well as to the assumed shape (plane or spherical, for example) of the central X-ray emitting region.

The disk thickness depends on temperature, and thereby also on the irradiation. It turns out, however, that this dependence on the irradiating flux is small, if the disk is optically thick and the energy transport is by radiation (Lyutyi and Syunyaev 1976a, translation in Lyutyi and Syunyaev 1976b). To see this, integrate (1.57) with the modified boundary condition (1.79). This yields

$$\sigma_r T^4 = \frac{3}{4}F\left(\tau + \frac{2}{3}\right) + (1-a)F_{\text{irr}}.\tag{1.80}$$

Thus the irradiation adds an additive constant to $T^4(z)$. At the midplane, this constant has much less effect than at the surface. For the midplane temperature and the disk thickness to be affected significantly, it is necessary that

$$\frac{F_{\text{irr}}}{F} \gtrsim \tau.\tag{1.81}$$

The reason for this weak dependence of the midplane conditions on irradiation is the same as in radiative envelopes of stars, which are also insensitive to the surface boundary condition. The situation is different for convective disks. As in fully convective stars, the adiabatic stratification then causes the conditions at the midplane to depend much more directly on the surface temperature. The outer parts of the disks in LMXB with wide orbits are in fact convective; hence, their thickness is more directly affected by irradiation. This leads to the possibility of *shadowing*, with irradiated, vertically extended regions blocking irradiation of the disk outside. This plays an observable role in protostellar disks (Dullemond et al., 2001).

---

[6] Incorrect derivations exist in the literature, in which the effect of irradiation is treated like an energy flux added to the internal viscous dissipation inside the disk rather than incident on the surface.

1.7.2 Observational evidence of disk thickness

From the paucity of sources in which eclipses of the central source by the companion are observed, one deduces that the companion is barely or not at all visible as seen from the inner disk. Apparently some parts of the disk are much thicker than expected from the previous arguments. This is consistent with the observation that a characteristic modulation of the optical light curve indicative of irradiation of the secondary's surface by the X-rays is not very strong in LMXB (an exception being Her X-1, which has an atypically large companion). The place of the eclipsing systems is taken by the so-called accretion disk corona (ADC) systems, where shallow eclipses of a rather extended X-ray source are seen instead of the expected sharp eclipses of the inner disk (for reviews of the observations, see Lewin et al., 1995). The conclusion is that there is an extended X-ray-scattering "corona" above the disk. It scatters a few per cent of the X-ray luminosity.

What causes this corona and the large inferred thickness of the disk? The thickness expected from disk theory is a rather stable small number. To "suspend" matter at the inferred height of the disk, forces are needed that are much larger than the pressure forces available in an optically thick disk. A thermally driven wind, produced by X-ray heating of the disk surface, has been invoked (Begelman et al., 1983; Schandl and Meyer, 1994; Meyer-Hofmeister et al., 1997). For other explanations, see van Paradijs and McClintock (1995). Perhaps a magnetically driven wind from the disk (Bisnovatyi-Kogan and Ruzmaikin, 1976; Blandford and Payne, 1982), such as that inferred for protostellar objects (cf. Lee et al., 2000), can explain both the shielding of the companion and the scattering. Such a model would resemble magnetically driven wind models for the broad-line region in AGN (e.g., Emmering et al., 1992; Königl and Kartje, 1994). A promising possibility is that the reprocessing region consists of matter "kicked up" at the disk edge by the impact of the mass transferring stream (Meyer-Hofmeister et al., 1997; Armitage and Livio, 1998; Spruit and Rutten, 1998). This produces qualitatively the right dependence of X-ray absorption on orbital phase in ADC sources, and the light curves of the so-called supersoft sources.

1.7.3 Transients

Soft X-ray transients (also called X-ray novae) are believed to be binaries similar to the other LMXB, but somehow the accretion is episodic, with very large outbursts recurring on time scales of decades (sometimes years). Most of these transients turn out to be black hole candidates (see Lewin et al., 1995, for a review). As with the dwarf novae, the time dependence of the accretion can in principle be exploited in transients to derive information on the disk viscosity, assuming that the outburst is caused by an instability in the disk. The closest relatives of soft transients among the white dwarf plus main sequence star systems are probably the WZ Sge stars (van Paradijs and Verbunt, 1984; Kuulkers et al., 1996), which show (in the optical) similar outbursts with similar recurrence times (cf. Warner, 1987; O'Donoghue et al., 1991). Like the soft transients, they have low mass ratios ($q < 0.1$). For a given angular momentum loss, systems with low mass ratios have low mass transfer rates, so the speculation is that the peculiar behavior of these systems is somehow connected in both cases with a low mean accretion rate.

Transients in quiescence

X-ray transients in quiescence (i.e., after an outburst) usually show a very low X-ray luminosity. The mass transfer rate from the secondary in quiescence can be inferred from the optical emission. This shows the characteristic "hot spot," known from other systems to be the location where the mass transferring stream impacts on the edge of an accretion disk (e.g., van Paradijs and McClintock, 1995). These observations thus show that a disk is present in quiescence, while the mass transfer rate can be measured from the brightness of the hot spot. If this disk were to extend to the neutron star with constant mass flux, the predicted X-ray luminosity would be much higher than observed. This has traditionally been interpreted as a consequence of the fact that in transient systems, the accretion is not steady. Mass is stored in the outer parts and released by a disk instability (e.g., King, 1995; Meyer-Hofmeister and Meyer, 1999), producing the X-ray outburst. During quiescence, the accretion rate onto the compact object is much smaller than the mass transfer from the secondary to the disk.

1.7.4 Disk instability

The most developed model for outbursts is the disk instability model of Osaki (1974), Hōshi (1979), Smak (1971, 1984), and Meyer and Meyer-Hofmeister (1981); see also King (1995) and Osaki (1994). In this model, the instability that gives rise to cyclic accretion is due to a temperature dependence of the viscous stress. In any local process that causes an effective viscosity, the resulting α-parameter will be a function of the main dimensionless parameter of the disk, the aspect ratio H/r. If this is a sufficiently rapidly increasing function, such that α is large in hot disks and low in cool disks, an instability results by the following mechanism. Suppose we start the disk in a stationary state at the mean accretion rate. If this state is perturbed by a small temperature increase, α goes up, and by the increased viscous stress the mass flux $\dot{M}$ increases. By equation 1.55, this increases the disk temperature further, resulting in a runaway to a hot state. Since $\dot{M}$ is larger now than the average, the disk partly empties, reducing the surface density and the central temperature (eq. 1.60). A cooling front then transforms the disk to a cool state with an accretion rate below the mean. The disk in this model switches back and forth between hot and cool states. By adjusting the value of α in the hot and cool states, or by adjusting the functional dependence of α on H/r, outbursts are obtained that agree reasonably with the observations of soft transients (Lin and Taam, 1984; Mineshige and Wheeler, 1989). A rather strong increase of α with H/r is needed to get the observed long recurrence times.

Another possible mechanism for instability has been found in 2-D numerical simulations of accretion disks (Blaes and Hawley, 1988; Różyczka and Spruit, 1993). The outer edge of a disk is found, in these simulations, to become dynamically unstable to an oscillation that grows into a strong eccentric perturbation (a crescent-shaped density enhancement that rotates at the local orbital period). Shock waves generated by this perturbation spread mass over most of the Roche lobe; at the same time, the accretion rate onto the central object is strongly enhanced. This process is different from the Smak-Osaki-Hōshi mechanism, since it requires two dimensions and does not depend on the viscosity (instead, the internal dynamics in this instability *generates* the effective viscosity that causes a burst of accretion).

1.7.5 Other instabilities

Instability to heating/cooling of the disk can be due to several effects. The cooling rate of the disk, if it depends on temperature in an appropriate way, can cause a thermal instability like that in the interstellar medium. Other instabilities may result from the dependence of viscosity on conditions in the disk. For a general treatment, see Piran (1978); for a shorter discussion, see Treves et al. (1988).

1.8 Sources of viscosity

The high Reynolds number of the flow in accretion disks (of the order 10^{14} in the outer parts of a CV disk) would, to most fluid dynamicists, seem an amply sufficient condition for the occurrence of hydrodynamic turbulence (see also the discussion in Section 1.4.7). A theoretical argument against such turbulence often used in astrophysics (Kippenhahn and Thomas, 1981; Pringle, 1981) is that in cool disks the gas moves almost on Kepler orbits, which are quite stable (except for the orbits that get close to the companion or near a black hole). This stability is related to the known stabilizing effect that rotation has on hydrodynamical turbulence (Bradshaw, 1969; Lesur and Longaretti, 2005). A (not very strong) observational argument is that hydrodynamical turbulence as described earlier would produce an α that does not depend on the nature of the disk, so that all objects should have the same value, which is not what observations show. From the modeling of CV outbursts one knows, for example, that α probably increases with temperature (more accurately, with H/r; see previous section). Also, there are indications from the inferred lifetimes and sizes of protostellar disks (Strom et al., 1993) that α may be rather small there, $\sim 10^{-3}$, whereas in outbursts of CVs one infers values of the order 0.1 to 1.

Among the processes that have been proposed repeatedly as sources of viscosity is convection due to a vertical entropy gradient (e.g., Kley et al., 1993), which may have some limited effect in convective parts of disks. Rotation rate varies somewhat with height in a disk, due to the radial temperature gradient ("baroclinicity"). Modest amounts of turbulence have been reported from instability of this form of differential rotation (Klahr and Bodenheimer, 2003; Lesur and Papaloizou, 2010).

Another class are *waves* of various kinds. Their effect can be global, that is, not reducible to a local viscous term, because by traveling across the disk they can communicate torques over large distances. For example, waves set up at the outer edge of the disk by tidal forces can travel inward and, by dissipating there, can effectively transport angular momentum *outward* (e.g., Narayan et al., 1987; Spruit et al., 1987). A nonlinear version of this idea are self-similar spiral shocks, observed in numerical simulations (Sawada et al., 1987) and studied analytically (Spruit, 1987). Such shocks can produce accretion at an effective α of 0.01 in hot disks, but are probably not very effective in disks as cool as those in CVs and XRB.

A second nonlocal mechanism is provided by a magnetically accelerated *wind* originating from the disk surface (Blandford, 1976; Bisnovatyi-Kogan and Ruzmaikin, 1976; Lovelace, 1976; for an introduction, see Spruit, 1996). In principle, such winds can take care of *all* the angular momentum loss needed to make accretion possible in the absence of a viscosity (Blandford, 1976; Königl, 1989). The attraction of this idea is that magnetic winds are a strong contender for explaining the strong outflows and jets seen in some protostellar objects and AGN. It is not yet clear, however, if, even in these objects, the wind is actually the main source of angular momentum loss.

In sufficiently cool or massive disks, self-gravitating instabilities of the disk matter can produce internal friction. Paczyński (1978) proposed that the resulting heating would limit the instability and keep the disk in a well-defined moderately unstable state. The angular momentum transport in such a disk has been studied numerically (e.g., Gammie, 1997; Ostriker et al., 1999). Disks in CVs and XRB are too hot for self-gravity to play a role, but it can be important in protostellar disks (cf. Rafikov, 2009).

1.8.1 *Magnetic viscosity*

Magnetic forces can be very effective at transporting angular momentum. If it can be shown that the shear flow in the disk produces some kind of small scale fast dynamo process, that is, some form of magnetic turbulence, an effective $\alpha \sim O(1)$ would be expected (Shakura and Sunyaev, 1973). Numerical simulations of initially weak magnetic fields in accretion disks show that this does indeed happen in sufficiently ionized disks (Hawley et al., 1995; Brandenburg et al., 1995; Balbus, 2003). These show a small-scale magnetic field with azimuthal component dominating (due to stretching by differential rotation). The angular momentum transport is due to magnetic stresses; the fluid motions induced by the magnetic forces contribute only a little to the angular momentum transport. In a perfectly conducting plasma, this turbulence can develop from an arbitrarily small initial field through magnetic shear instability (also called magnetorotational instability; Velinikhov, 1959; Chandrasekhar, 1961; Balbus and Hawley, 1991). The significance of this instability is that it shows that at large conductivity, accretion disks must be magnetic.

The actual form of the highly time dependent small scale magnetic field which develops can only be found from numerical simulations. The simplest case thought to be representative of the process considers a vertically unstratified disk (component of gravity perpendicular to the disk ignored), of which only a radial extent of a few times H is included (cf. the causality argument above). Different simulations have yielded somewhat different values for the effective viscosity, contrary to the expectation that at least in this simplest form the process should yield a unique value.

The nature of these differences has been appreciated only recently. Fromang *et al.* (2007) show that the results do not converge with increasing numerical resolution (the effective α found appears to decrease indefinitely). The result also depends on the magnetic Prandtl number $P_{\rm m} = \nu/\eta_{\rm m}$, the ratio of viscosity to magnetic diffusivity $\eta_{\rm m}$. No turbulence is found for $P_{\rm m} < 1$ (Fromang *et al.*, 2007). The significance of these findings is not quite obvious yet, but they clearly contradict the commonly assumed "cascade" picture of MHD turbulence (taken over from what is assumed in hydrodynamic turbulence, where the large-scale behavior of turbulence appears to converge with numerical resolution).

Perhaps the behavior found in this case is related to the highly symmetric nature of the idealized problem. This is suggested by the finding (Davis *et al.*, 2010; Shi *et al.*, 2010) that effective viscosity appears to converge again with increasing numerical resolution when vertical stratification is included. By inference, the process defining the state of magnetic turbulence is different in this case from that in the unstratified case. It can plausibly be attributed to magnetic buoyancy instabilities (Shi *et al.*, 2010). This would make the process somewhat analogous to the mechanism operating the solar cycle (Spruit, 2010a). Magnetic turbulence also behaves differently if the net vertical magnetic flux crossing the simulated box does not vanish. (This flux is a conserved quantity set by the initial conditions.) Magnetic turbulence then develops more easily (at lower numerical resolution) and appears to converge as the numerical resolution is increased.

The consequences of these new findings for angular momentum transport in disks are still to be settled (c.f. the insightful discussion in Lesur and Ogilvie, 2008).

1.8.2 Viscosity in radiatively supported disks

A disk in which the radiation pressure $P_{\rm r}$ dominates must be optically thick (otherwise the radiation would escape). The radiation pressure then adds to the total pressure. The pressure is larger than it would be, for a given temperature, if only the gas pressure were effective. If the viscosity is then parametrized by equation 1.33, it turns out (Lightman and Eardley, 1974) that the disk is locally unstable. An increase in temperature increases the radiation pressure, which increases the viscous dissipation and the temperature, leading to a runaway. This has raised the question whether the radiation pressure should be included in the sound speed that enters equation 1.33. If it is left out, a lower viscosity results, and there is no thermal-viscous runaway. Without knowledge of the process causing the effective viscous stress, this question can not be answered. Sakimoto and Coroniti (1989) have shown, however, that if the stress is due to some form of magnetic turbulence, it most likely scales with the gas pressure alone, rather than the total pressure. Now that it seems likely, from the numerical simulations, that the stress is indeed magnetic, there is reason to believe that in the radiation pressure–dominated case the effective viscosity will scale as $\nu \sim \alpha P_{\rm g}/(\rho\Omega)$, making such disks thermally stable, as indicated by the results of Hirose *et al.* (2009).

1.9 Beyond thin disks

Ultimately, much of the progress in developing useful models of accretion disks will depend on detailed numerical simulations in two or three dimensions. In the disks one is interested in, there is usually a large range in length scales (in LMXB disks, from less than the 10-km neutron star radius to the more than 10^5-km orbital scale). Correspondingly, there is a large range in time scales that have to be followed. This is not technically possible at present or in the foreseeable future. In numerical simulations, one is therefore limited to studying in an approximate way aspects that are either local or of limited dynamic range in r, t (for examples, see Hawley, 1991; Armitage, 1998; De Villiers and Hawley, 2003; Hirose et al., 2004). For this reason, approaches have been used that relax the strict thin disk framework somewhat without resorting to full simulations. Some of the physics of thick disks can be included in a fairly consistent way in the "slim disk" approximation (Abramowicz et al., 1988). The so-called advection-dominated accretion flows (ADAFs) are related to this approach (for a review, see Yi, 1999). They are discussed in Sections 1.11, 1.12, and 1.13.

1.9.1 Boundary layers

In order to accrete onto a star rotating at the rate Ω_*, the disk matter must dissipate an amount of energy given by

$$\frac{GM\dot{M}}{2R}\left[\frac{1-\Omega_*}{\Omega_k(R)}\right]^2. \tag{1.82}$$

The factor in brackets measures the kinetic energy of the matter at the inner edge of the disk ($r = R$), in the frame of the stellar surface. Because of this dissipation, the disk inflates into a "belt" at the equator of the star, of thickness D and radial extent of the same order. Equating the radiation emitted from the surface of this belt to equation 1.82, one gets for the surface temperature $T_{\rm sb}$ of the belt, assuming optically thick conditions and a slowly rotating star ($\Omega_*/\Omega_k \ll 1$),

$$\frac{GM\dot{M}}{8\pi R^2 D} = \sigma_{\rm r} T_{\rm sb}^4 \tag{1.83}$$

To find the temperature inside the belt and its thickness, use equation 1.59. The value of the surface temperature is higher, by a factor of the order $(R/D)^{1/4}$, than the simplest thin disk estimate (1.55, ignoring the factor f). In practice, this works out to a factor of a few. The surface of the belt is therefore not very hot. The situation is quite different if the boundary layer is not optically thick (Pringle and Savonije, 1979). It then heats up to much higher temperatures.

Analytical methods to obtain the boundary layer structure have been used by Regev and Hougerat (1988), numerical solutions of the slim disk type by Narayan and Popham (1994) and Popham (1997), and 2-D numerical simulations by Kley (1991). These considerations are primarily relevant for CV disks; in accreting neutron stars, the dominant effects of radiation pressure have to be included. More analytic progress on the structure of the boundary layer between a disk and a neutron star and the way in which it spreads over the surface of the star has been reported by Inogamov and Sunyaev (1999).

1.10 Radiative efficiency of accretion disks

In a thin accretion disk, the time available for the accreting gas to radiate away the energy released by the viscous stress is the accretion time,

$$t_{\rm acc} \approx \frac{1}{\alpha\Omega_{\rm K}}\left(\frac{r}{H}\right)^2, \tag{1.84}$$

where α is the dimensionless viscosity parameter, Ω_K the local Keplerian rotation rate, r the distance from the central mass, and H the disk thickness (see Frank et al., 2002, or Section 1.6). For a thin disk, $H/r \ll 1$, this time is much longer than the thermal time scale $t_t \approx 1/(\alpha\Omega)$ (Section 1.6). There is then enough time for a local balance to hold between viscous dissipation and radiative cooling. For the accretion rates implied in observed systems, the disk is then rather cool, which then justifies the starting assumption $H/r \ll 1$.

This argument is somewhat circular, of course, because the accretion time is long enough for effective cooling only if the disk is assumed to be thin to begin with. Other forms of accretion disks may exist, even at the same accretion rates, in which the cooling is ineffective compared with that of standard (geometrically thin, optically thick) disks. In the following sections, we consider such forms of accretion and the conditions under which they are to be expected.

Since radiatively inefficient disks tend to be thick, $H/r \sim O(1)$, they are sometimes called "quasispherical." However, this does *not* mean that a spherically symmetric accretion model would be a reasonable approximation. The crucial difference is that the flow has angular momentum. The inward flow speed is governed by the rate at which angular momentum can be transferred outward, rather than by gravity and pressure gradient as in the Bondi accretion problem mentioned earlier. With $H/r \sim O(1)$, the accretion time scale, $t_{\rm acc} \sim 1/(\alpha\Omega)$, is still longer than the accretion time scale in the spherical case, $1/\Omega$ (unless the viscosity parameter α is as large as $O(1)$). The dominant velocity component is azimuthal rather than radial, and the density and optical depth are much larger than in the spherical case.

It turns out that there are two kinds of radiatively inefficient disks, the optically thin and optically thick varieties. A second distinction occurs because accretion flows are different for central objects with a solid surface (neutron stars, white dwarfs, main sequence stars, planets), and those without (i.e., black holes). We start with optically thick flows.

1.11 Radiation-supported radiatively inefficient accretion

If the energy loss by radiation is small, the gravitational energy release $W_{\rm grav} \approx GM/(2r)$ is converted into enthalpy of the gas and radiation field[7]

$$\frac{1}{2}\frac{GM}{r} = \frac{1}{\rho}\left[\frac{\gamma}{\gamma-1}P_{\rm g} + 4P_{\rm r}\right], \tag{1.85}$$

where an ideal gas of constant ratio of specific heats γ has been assumed, and $P_{\rm r} = \frac{1}{3}aT^4$ is the radiation pressure. In terms of the virial temperature $T_{\rm vir} = GM/(\mathcal{R}r)$, and assuming $\gamma = 5/3$, appropriate for a fully ionized gas (see Section 1.2.1), this can be written as

$$\frac{T}{T_{\rm vir}} = \left[5 + 8\frac{P_{\rm r}}{P_{\rm g}}\right]^{-1}. \tag{1.86}$$

Thus, for radiation pressure dominated accretion, $P_{\rm r} \gg P_{\rm g}$, the temperature is much less than the virial temperature: much of the accretion energy goes into photon production instead of heating. By hydrostatic equilibrium, the disk thickness is given by (cf. Section 1.4.2)

$$H \approx \frac{[(P_{\rm g} + P_{\rm r})/\rho]^{1/2}}{\Omega}, \tag{1.87}$$

With equation 1.86, this yields

$$\frac{H}{r} \sim O(1). \tag{1.88}$$

---

[7]This assumes that a fraction ~ 0.5 of the gravitational potential energy stays in the flow as orbital kinetic energy. This is only an approximation; see also Section 1.12.

In the limit $P_r \gg P_g$, the flow is therefore geometrically thick. Radiation pressure then supplies a nonnegligible fraction of the support of the gas in the radial direction against gravity (the remainder being provided by rotation).

For $P_r \gg P_g$, equation 1.85 yields

$$\frac{GM}{2r} = \frac{4}{3}\frac{aT^4}{\rho}. \tag{1.89}$$

The radiative energy flux, in the diffusion approximation, is (eq. 1.57)

$$F = \frac{4}{3}\frac{d}{d\tau}\sigma_r T^4 \approx \frac{4}{3}\frac{\sigma_r T^4}{\tau}. \tag{1.90}$$

Hence,

$$F = \frac{1}{8}\frac{GM}{rH}\frac{c}{\kappa} = F_E \frac{r}{8H}, \tag{1.91}$$

where $F_E = L_E/(4\pi r^2)$ is the local Eddington flux. Since $H/r \approx 1$, a radiatively inefficient, radiation pressure–dominated accretion flow has a luminosity of the order of the Eddington luminosity.

The temperature depends on the accretion rate and the viscosity ν assumed. The accretion rate is of the order $\dot{M} \sim 3\pi\nu\Sigma$ (cf. eq. 1.49), where $\Sigma = \int \rho dz$ is the surface mass density. In units of the Eddington rate $\dot{M}_E$, equation 1.14, we get

$$\dot{m} \equiv \frac{\dot{M}}{\dot{M}_E} \approx \frac{\nu \rho \kappa}{c}, \tag{1.92}$$

where $H/r \approx 1$ has been used.[8] Assume that the viscosity scales with the gas pressure:

$$\nu = \alpha \frac{P_g}{\rho \Omega_K}, \tag{1.93}$$

instead of the total pressure $P_r + P_g$. This is the form that is likely to hold if the angular momentum transport is due to a small-scale magnetic field (Sakimoto and Coroniti, 1989; Turner, 2004). Then, with equation 1.89 and 1.92, we have (up to a numerical factor of $O(1)$)

$$T^5 \approx \frac{(GM)^{3/2}}{r^{5/2}} \frac{\dot{m}c}{\alpha \kappa a \mathcal{R}}, \tag{1.94}$$

or

$$T \approx 10^8 r_6^{-1/5} \left(\frac{r}{r_g}\right)^{3/10} \dot{m}^{1/5}, \tag{1.95}$$

where $r = 10^6 r_6$ and $r_g = 2GM/c^2$ is the gravitational radius of the accreting object, and the electron scattering opacity of 0.3 cm^2 g^{-1} has been assumed. The temperatures expected in radiation-supported advection dominated flows are therefore quite low compared with the virial temperature (if the viscosity is assumed to scale with the total pressure instead of P_g, the temperature is even lower). The effect of electron-positron pairs can be neglected (Schultz and Price, 1985), since they are present only at temperatures approaching the electron rest mass energy, $T \gtrsim 10^9$ K.

In order for the flow to be dominated by radiation pressure and advection, the optical depth has to be sufficiently large that the radiation does not leak out. The energy density in the flow, vertically integrated, is of the order

$$E \approx aT^4 H, \tag{1.96}$$

---

[8] The definition of $\dot{M}_E$ differs by factors of order unity between different authors. It depends on the assumed efficiency η of conversion of gravitational energy GM/R into radiation. For accretion onto black holes, a more realistic value is of order $\eta = 0.1$; for accretion onto neutron stars, $\eta \approx 0.4$, depending on the radius of the star.

and the energy loss rate per cm² of disk surface is given by equation 1.91. The cooling time is therefore

$$t_c = \frac{E}{F} = 3\tau \frac{H}{c}. \tag{1.97}$$

This is to be compared with the accretion time, which can be written in terms of the mass in the disk at radius r, of the order $2\pi r^2 \Sigma$, and the accretion time:

$$t_{\rm acc} \approx 2\pi r^2 \frac{\Sigma}{\dot M}. \tag{1.98}$$

This yields

$$\frac{t_c}{t_{\rm acc}} \approx \frac{\kappa}{\pi r c}\dot M = \frac{4}{\eta}\dot m \frac{R}{r} \tag{1.99}$$

(where a factor $3/2\, H/r \sim O(1)$ has been neglected). Since $r > R$, this shows that accretion has to be around the Eddington rate or larger in order to be both radiation and advection dominated.

This condition can also be expressed in terms of the so-called trapping radius r_t (e.g., Rees, 1978). Equating $t_{\rm acc}$ and t_c yields

$$\frac{r_t}{R} \approx 4\dot m. \tag{1.100}$$

Inside r_t, the flow is advection dominated: the radiation field produced by viscous dissipation stays trapped inside the flow, instead of being radiated from the disk, as happens in a standard thin disk. Outside the trapping radius, the radiation field cannot be sufficiently strong to maintain a disk with $H/r \sim 1$; it must be a thin form of disk instead. Such a thin disk can still be radiation supported (i.e., $P_r \gg P_g$), but it cannot be advection dominated.

Flows of this kind are called "radiation-supported accretion tori" (or radiation tori, for short) by Rees et al. (1982). They must accrete at a rate above the Eddington value to exist. The converse is not quite true: a flow accreting above Eddington is an advection-dominated flow, but it need not necessarily be radiation dominated. Advection-dominated optically thick accretion flows exist in which radiation does not play a major role (see Section 1.12.2).

That an accretion flow above $\dot M_E$ is advection dominated, not a thin disk, also follows from the fact that in a thin disk, the energy dissipated must be radiated away locally. Since the local radiative flux cannot exceed the Eddington energy flux F_E, the mass accretion rate in a thin disk cannot significantly exceed the Eddington value (eq. 1.14).

The gravitational energy, dissipated by viscous stress in differential rotation and advected with the flow, ends up on the central object. If this is a black hole, the photons, particles, and their thermal energy are conveniently swallowed at the horizon and do not react back on the flow. Radiation tori are therefore mostly relevant for accretion onto black holes. They are convectively unstable (Bisnovatyi-Kogan and Blinnikov, 1977): the way in which energy is dissipated, in the standard α-prescription, is such that the entropy (entropy of radiation, $\sim T^3/\rho$) decreases with height in the disk. Numerical simulations (see Section 1.14) show the effects of this convection.

1.11.1 Super-Eddington accretion onto black holes

As the accretion rate onto a black hole is increased above $\dot M_E$, the trapping radius moves out. The total luminosity increases only slowly and remains of the order of the Eddington luminosity. Such supercritical accretion has been considered by Wang and Zhou (1999); they show that the flow has a radially self-similar structure.

Abramowicz et al. (1988, 1989) studied accretion onto black holes at rates near $\dot M_E$. They used a vertically integrated approximation for the disk, but included the advection terms. The resulting models were called "slim disks." They show how with increasing accretion rate, a standard thin Shakura-Sunyaev disk turns into a radiation-supported advection flow. The nature of the transition depends on the viscosity prescription used and can show a nonmonotonic dependence of $\dot{M}$ on surface density Σ (Honma et al., 1991). This suggests the possibility of instability and cyclic behavior of the inner disk near a black hole, at accretion rates near and above $\dot{M}_{\rm E}$ (for an application to GRS 1915+105, see Nayakshin et al., 1999).

1.11.2 Super-Eddington accretion onto neutron stars

In the case of accretion onto a neutron star, the energy trapped in the flow, plus the remaining orbital energy, settles onto its surface. If the accretion rate is below $\dot{M}_{\rm E}$, the energy can be radiated away by the surface, and steady accretion is possible. A secondary star providing the mass may, under some circumstances, transfer more than $\dot{M}_{\rm E}$, since it does not know about the neutron star's Eddington value. The outcome of this case is still somewhat uncertain; it is generally believed on intuitive grounds that the "surplus" (the amount above $\dot{M}_{\rm E}$) somehow gets expelled from the system.

One possibility is that, as the transfer rate is increased, the accreting hot gas forms an extended atmosphere around the neutron star like the envelope of a giant. If it is large enough, the outer parts of this envelope are partially ionized. The opacity in these layers, due to atomic transitions of the CNO and heavier elements, is then much higher than the electron scattering opacity. The Eddington luminosity based on the local value of the opacity is then smaller than it is near the neutron star surface. Once an extended atmosphere forms, the accretion luminosity is thus large enough to drive a wind from the envelope (see Kato, 1997, where the importance of this effect is demonstrated in the context of novae).

This scenario is somewhat dubious, however, because it assumes that the mass transferred from the secondary continues to reach the neutron star and generate a high luminosity there. This is not at all obvious, since the mass-transferring stream may instead dissipate inside the growing envelope of the neutron star. The result of this could be a giant (more precisely, a Thorne-Zytkow star), with a steadily increasing envelope mass. Such an envelope is likely to be large enough to engulf the entire binary system, which then develops into a common-envelope (CE) system. The envelope mass is then expected to be ejected by CE hydrodynamics (for reviews, see Taam, 1996; Taam and Sandquist, 2000).

A more speculative proposal, suggested by the properties of SS 433, is that the "surplus mass" is ejected in the form of jets. The binary parameters of Cyg X-2 are observational evidence for mass ejection in super-Eddington mass transfer phases (King and Ritter, 1999; King and Begelman, 1999; Podsiadlowski and Rappaport, 2000).

1.12 ADAF hydrodynamics

The hydrodynamics of radiatively inefficient flows (or "advection-dominated accretion flows") can be studied by starting, at a very simple level, with a generalization of the thin disk equations. Making the assumption that quantities integrated over the height z of the disk give a fair representation (though this is justifiable only for thin disks), and assuming axisymmetry, the problem reduces to a one-dimensional time-dependent one. Further simplifying this by restriction to a steady flow yields the equations

$$2\pi r \Sigma v_r = \dot{M} = {\rm cst}, \qquad (1.101)$$

$$r \Sigma v_r \partial_r (\Omega r^2) = \partial_r (\nu \Sigma r^3 \partial_r \Omega), \qquad (1.102)$$

$$v_r \partial_r v_r - (\Omega^2 - \Omega_{\rm K}^2) r = -\frac{1}{\rho} \partial_r P, \qquad (1.103)$$

$$\Sigma v_r T \partial_r S = q^+ - q^-, \qquad (1.104)$$

where S is the specific entropy of the gas, Ω the local rotation rate, now different from the Keplerian rate Ω_K, and

$$q^+ = \int Q_v dz, \qquad q^- = \int \text{div} F_r dz \qquad (1.105)$$

are the height-integrated viscous dissipation rate and radiative loss rate, respectively. In the case of thin disks, equations 1.101 and 1.102 are unchanged, but 1.103 simplifies to $\Omega^2 = \Omega_K^2$ – that is, the rotation is Keplerian – while 1.104 simplifies to $q^+ = q^-$, expressing local balance between viscous dissipation and cooling. The left-hand side of equation 1.104 describes the radial advection of heat and is perhaps the most important deviation from the thin disk equations at this level of approximation (hence the name "advection-dominated flows"). The characteristic properties are seen most clearly when radiative loss is neglected altogether, $q^- = 0$. The equations are supplemented with expressions for ν and q^+:

$$\nu = \frac{\alpha c_s^2}{\Omega_K}; \qquad q^+ = (r\partial_r \Omega)^2 \nu \Sigma. \qquad (1.106)$$

If α is taken constant, $q^- = 0$, and an ideal gas is assumed with constant ratio of specific heats, so that the entropy is given by

$$S = c_v \ln\left(\frac{p}{\rho^\gamma}\right), \qquad (1.107)$$

then equations 1.101 through 1.104 have no explicit length scale in them. This means that a special so-called self-similar solution exists, in which all quantities are powers of r. Such self-similar solutions were apparently described first by Gilham (1981), but reinvented several times (Spruit *et al.*, 1987; Narayan and Yi, 1994). The dependences on r are

$$\Omega \sim r^{-3/2}; \qquad \rho \sim r^{-3/2}, \qquad (1.108)$$

$$H \sim r; \qquad T \sim r^{-1}. \qquad (1.109)$$

In the limit $\alpha \ll 1$, one finds

$$v_r = -\alpha \Omega_K r \left(9\frac{\gamma - 1}{5 - \gamma}\right), \qquad (1.110)$$

$$\Omega = \Omega_K \left(2\frac{5 - 3\gamma}{5 - \gamma}\right)^{1/2}, \qquad (1.111)$$

$$c_s^2 = \Omega_K^2 r^2 \frac{\gamma - 1}{5 - \gamma}, \qquad (1.112)$$

$$\frac{H}{r} = \left(\frac{\gamma - 1}{5 - \gamma}\right)^{1/2}. \qquad (1.113)$$

The precise from of these expressions depends somewhat on the way in which vertical integrations such as in equation 1.105 are done (which are only approximate).

The self-similar solution can be compared with numerical solutions of equations 1.101 through 1.104 with appropriate conditions applied at inner (r_i) and outer (r_o) boundaries (Nakamura *et al.*, 1996; Narayan *et al.*, 1997). The results show that the self-similar solution is valid in an intermediate regime $r_i \ll r \ll r_o$. That is, the solutions of equations 1.101 through 1.104 approach the self-similar solution far from the boundaries, as is characteristic of self-similar solutions.

The solution exists only if $1 < \gamma \leq 5/3$, a condition satisfied by all ideal gases. As $\gamma \downarrow 1$, the disk temperature and thickness vanish. This is understandable, since a γ close to 1 means that the particles making up the gas have a large number of internal degrees of freedom. In thermal equilibrium, the accretion energy is shared between all degrees of freedom, so that for a low γ, less is available for the kinetic energy (temperature) of the particles.

Second, the *rotation rate vanishes* for $\gamma \to 5/3$. As in the case of spherical accretion, no accreting solutions exist for $\gamma > 5/3$ (cf. Section 1.2). Since a fully ionized gas has $\gamma = 5/3$, it is a relevant value for the hot, ion-supported accretion flows discussed later. Apparently, steady advection-dominated accretion cannot have angular momentum in this case. To see how this comes about, consider the entropy of the flow. For accretion to take place in a rotating flow, there has to be friction and increase of entropy. Accretion is then necessarily accompanied by an inward increase of entropy. In a radially self-similar flow, the scalings 1.108 and 1.109 yield $P \sim r^{-5/2}$, so entropy scales as $S \sim \ln P/\rho^\gamma \sim (5 - 3\gamma) \ln r$. This increases with decreasing distance r only for $\gamma < 5/3$, and an accretion flow with $\gamma = 5/3$ cannot be both rotating and adiabatic. (With energy losses by radiation, the constraint on γ disappears again.)

The question then arises how an adiabatic flow with $\gamma = 5/3$ will behave if one starts it as a rotating torus around a black hole. In the literature, this problem has been circumvented by arguing that real flows would have magnetic fields in them, which would change the effective compressibility of the gas. Even if a magnetic field of sufficient strength is present, however, (energy density comparable to the gas pressure), the effective γ is not automatically lowered. If the field is compressed mainly perpendicular to the field lines, for example, the effective γ is closer to 2. Also, this does not solve the conceptual problem of what would happen to a rotating accretion flow consisting of a more weakly magnetized ionized gas.

1.12.1 *The case of the vanishing rotation rate*

This conceptual problem has been solved by Ogilvie (1999), who showed how a gas cloud initially rotating around a point mass settles to the slowly rotating self-similar solutions of the steady problem discussed previously. He constructed similarity solutions to the time-dependent version of equations 1.101 through 1.104, in which distance and time occur in the combination $rt^{-2/3}$. This solution describes the asymptotic behavior (in time) of a viscously spreading disk, analogous to the viscous spreading of thin disks (see Section 1.4.3). As in the thin disk case, all the mass accretes asymptotically onto the central mass, while all the angular momentum travels to infinity together with a vanishing amount of mass. For all $\gamma < 5/3$, the rotation rate at a fixed r tends to a finite value as $t \to \infty$, but for $\gamma = 5/3$ it tends to zero. The size of the slowly rotating region expands as $r \sim t^{2/3}$. The typical slow rotation of ADAFs at γ near $5/3$ is thus a real physical property. In such a flow, the angular momentum gets expelled from the inner regions almost completely, and the accretion flow becomes purely radial, as in Bondi accretion.

1.12.2 *Other optically thick accretion flows*

The radiation-dominated flows discussed in Section 1.11 are not the only possible optically thick advection-dominated flows. From the discussion of the hydrodynamics, it is clear that disklike (i.e., rotating) accretion is possible whenever the ratio of specific heats is less than $5/3$. A radiation-supported flow satisfies this requirement, since radiation pressure scales with volume like a gas with $\gamma = 4/3$, but it can also happen in the absence of radiation if energy is taken up in the gas by internal degrees of freedom of the particles. Examples are the rotational and vibrational degrees of freedom in molecules, and the energy associated with dissociation and ionization. If the accreting object has a gravitational potential not too far from the $2.3 + 13.6$ eV per proton for dissociation plus ionization, a gas initially consisting of molecular hydrogen can stay bound at arbitrary accretion rates. This translates into a limit $M/M_\odot \, R_\odot/R < 0.01$. This is satisfied approximately by the giant planets, which are believed to have gone through a phase of rapid adiabatic gas accretion (e.g., Podolak *et al.*, 1993).

A more remotely related example is the core-collapse supernova. The accretion energy of the core mass falling onto the growing proto–neutron star at its center is lost mostly to internal degrees of freedom represented by photodisintegration of the nuclei. If the precollapse core rotates sufficiently rapidly, the collapse will form an accretion torus (inside the supernova envelope), with properties similar to advection-dominated accretion flows (but at extreme densities and accretion rates, by X-ray binary standards). Such objects have been invoked as sources of gamma-ray bursts (Woosley, 1993; Paczyński, 1998; Popham *et al.*, 1999).

A final possibility for optically thick accretion is through *neutrino losses*. If the temperature and density near an accreting neutron star become large enough, additional cooling takes place through neutrinos (as in the cores of giants). This is relevant for the physics of Thorne-Zytkow stars (neutron stars or black holes in massive supergiant envelopes; compare Bisnovatyi-Kogan and Lamzin, 1984; Cannon *et al.*, 1992), and perhaps for the spiral-in of neutron stars into giants (Chevalier, 1993; Taam and Sandquist, 2000).

1.13 Optically thin radiatively inefficient flows (ISAFs)

The optically thin case has received most attention, because of the promise it holds for explaining the (radio to X-ray) spectra of X-ray binaries and the central black holes in galaxies, including our own. For a review, see Yi (1999). This kind of flow occurs if the gas is optically thin and radiation processes are sufficiently weak. The gas then heats up to near the virial temperature. Near the last stable orbit of a black hole, this is of the order 100 MeV, or 10^{12} K. At such temperatures, a gas in thermal equilibrium would radiate at a fantastic rate, even if it were optically thin, because the interaction between electrons and photons has already become very strong near the electron rest mass of 0.5 MeV. In a remarkable early paper, Shapiro *et al.* (1976) noted that this, however, is not what will happen in an optically thin plasma accreting on a hole. They showed that instead thermal equilibrium between ions and electrons breaks down and a *two-temperature plasma* forms. We call such a flow an ion-supported accretion flow (ISAF), following the nomenclature suggested by Rees *et al.* (1982). The argument is as follows.

Suppose that the energy released by viscous dissipation is distributed equally among the carriers of mass, that is, mostly to the ions and $\sim 1/2,000$ to the electrons. Most of the energy then resides in the ions, which radiate very inefficiently (their high mass prevents the rapid accelerations that are needed to produce electromagnetic radiation). Their energy is transferred to the electrons by Coulomb interactions. These interactions are slow, however, under the conditions mentioned. They are slow because of the low density (on account of the assumed low optical thickness) and because they decrease with increasing temperature. The electric forces that transfer energy from an ion to an electron act only as long as the ion is within the electron's Debye sphere (e.g., Spitzer, 1965). The interaction time between proton and electron, and thus the momentum transferred, therefore decrease as $1/v_{\rm p} \sim T_{\rm p}^{-1/2}$ where $T_{\rm p}$ is the proton temperature.

In this way, an optically thin plasma near a compact object can be in a two-temperature state, with the ions being near the virial temperature and the electrons, which are doing the radiating, at a much lower temperature around 50 to 200 keV. The energy transfer from the gravitational field to the ions is fast (by some form of viscous or magnetic dissipation, say), from the ions to the electrons slow, and finally the energy losses of the electrons fast (by synchrotron radiation in a magnetic field or by inverse Compton scattering off soft photons). Such a flow would be radiatively inefficient because the

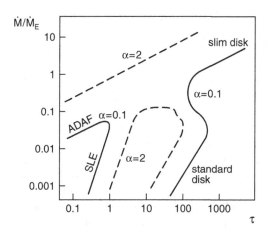

FIGURE 1.4. Branches of advection-dominated and thin disks for two values of the viscosity parameter α, as functions of accretion rate and (vertical) optical depth of the flow (schematic, after Chen et al., 1995; Zdziarski, 1998). Optically thin branches are the ISAF and SLE (Shapiro-Lightman-Eardley) solutions, optically thick ones the radiation dominated ("slim disk" or "radiation torus") and SS (Shakura-Sunyaev or standard thin disk). Advection-dominated are the ISAF and the radiation torus, geometrically thin are the SLE and SS. The SLE solution is a thermally unstable branch. Reproduced by permission of the AAS.

receivers of the accretion energy, the ions, get swallowed by the hole before having a chance to transfer their energy to the electrons. Most of the accretion energy thus gets lost into the hole, and the radiative efficiency is much less than for a cool disk. The first disk models that take into account this physics of advection and a two-temperature plasma were developed by Ichimaru (1977).

It is clear from this description that both the physics of such flows and the radiation spectrum to be expected depend critically on the details of the ion-electron interaction and radiation processes assumed. This is unlike the case of the optically thick advection-dominated flows, where gas and radiation are in approximate thermodynamic equilibrium. This is a source of uncertainty in the application of ISAFs to observed systems, since their radiative properties depend on poorly known quantities such as the strength of the magnetic field in the flow.

The various branches of optically thin and thick accretion flows are summarized in Figure 1.4. Each defines a relation between surface density Σ (or optical depth $\tau = \kappa \Sigma$) and accretion rate. ISAFs require low densities, which can result because of either low accretion rates or large values of the viscosity parameter. The condition that the cooling time of the ions by energy transfer to the electrons is longer than the accretion time yields a maximum accretion rate (Rees et al., 1982),

$$\dot{m} \lesssim \alpha^2, \qquad (1.114)$$

where $\dot{m}$ is the accretion rate in units of the Eddington value. If $\alpha \approx 0.05$, as suggested by current simulations of magnetic turbulence, the maximum accretion rate would be a few 10^{-3}. If ISAFs are to be applicable to systems with higher accretion rates, such as Cyg X-1, the viscosity parameter must be larger, on the order of 0.3.

1.13.1 Application: hard spectra in X-ray binaries

In the hard state, the X-ray spectrum of black hole and neutron star accretors is characterized by a peak in the energy distribution (νF_ν or $E\,F(E)$) at photon energies around 100 keV. This is to be compared with the typical photon energy of ~ 1 keV expected from a standard optically thick thin disk accreting near the Eddington limit. The standard, and by far most likely, explanation is that the observed hard photons are softer photons (around 1 keV) that have been up-scattered through inverse Compton scattering on hot electrons. Fits of such Comptonized spectra (e.g., Sunyaev and Titarchuk, 1980; Zdziarski, 1998, and references therein) yield an electron scattering optical depth around unity and an electron temperature of 50 to 100 keV. The scatter in these parameters is rather small between different sources. The reason may lie in part in the physics of Comptonization, but this is not the only reason. Something in the physics of the accretion flow keeps the Comptonization parameters fairly constant as long as it is in the hard state. ISAFs have been applied with some success in interpreting XRB. They can produce reasonable X-ray spectra and have been used in interpretations of the spectral-state transitions in sources like Cyg X-1 (Esin et al., 1998, and references therein).

An alternative to the ISAF model for the hard state in sources like Cyg X-1 and the black hole X-ray transients is the 'corona' model. A hot corona (Bisnovatyi-Kogan and Ruzmaikin, 1976), heated perhaps by magnetic fields as in the case of the Sun (Galeev et al., 1979) could be the medium that Comptonizes soft photons radiated from the cool disk underneath. The energy balance in such a model produces a Comptonized spectrum within the observed range (Haardt et al., 1994). This model has received further momentum, especially as a model for AGN, with the discovery of broadened X-ray lines interpreted as indicative of the presence of a cool disk close to the last stable orbit around a black hole (Fabian et al., 2002, and references therein). The very rapid X-ray variability seen in some of these sources is interpreted as due magnetic flaring in the corona, like in the solar corona (e.g., Di Matteo et al., 1999).

1.13.2 Transition from cool disk to ISAF: truncated disks

One of the difficulties in applying ISAFs to specific observed systems is the transition from a standard geometrically thin, optically thick disk, which must be the mode of mass transfer at large distances, to an ISAF at closer range. This is shown by Figure 1.4, which illustrates the situation at some distance close to the central object. The standard disk and the optically thin branches are separated from each other for all values of the viscosity parameter. This separation of the optically thin solutions also holds at larger distances. Thus, there is no plausible continuous path from one to the other, and the transition between the two must be due to additional physics that is not included in diagrams such as Figure 1.4.

Circumstantial observational evidence points to the existence of such a transition. The distance from the hole where it is assumed to take place is then called the truncation radius. The extensive datasets from the black hole candidate Cyg X-1 obtained with the Rossi X-ray Timing Explorer (RXTE) have played an important role in the development of the truncated disk model. The X-ray spectrum of Cyg X-1 varies (on time scales of days to years) between softer and harder states, and the characteristic time scale of its fast variability (milliseconds to minutes) correlates closely with these changes. Since all disk time scales decrease with distance from the hole, the fast variability is interpreted as due to a process (still to be identified in detail) that depends on the size of the truncation radius. Variation in time of this radius is then assumed to cause the observed changes. Characteristics of the spectrum that find a natural place in this picture are the slope of the hard X-ray spectrum, the amplitude of the so-called Compton reflection hump, and the behavior of the Fe K_α fluorescence line at 6.7 keV; each of these correlates with the characteristic fast-variability time scale in an interpretable (though still somewhat model-dependent) way (Gilfanov et al., 1999; Revnivtsev et al., 1999). Very similar correlations have been found in X-ray observations of AGN (Zdziarski et al., 1999).

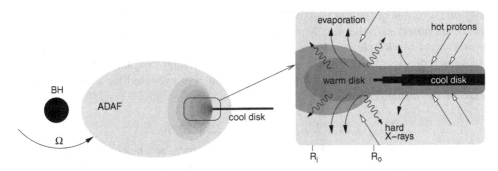

FIGURE 1.5. Interaction between ion-supported accretion flow (ISAF) and cool disk in the truncated disk model for hard X-ray states in X-ray binaries. Accreting black hole on the left, its hot ions illuminating the cool disk create a warm (~100 keV) surface layer producing the hard X-rays, extending inward and heating up to virial temperature by thermal instability ("evaporation") once cooling by soft photons becomes inefficient.

A promising possibility is that the transition takes place through "evaporation." Two distinct mechanisms have been elaborated for such evaporation. In the first (Meyer and Meyer-Hofmeister, 1994; Liu et al., 2002), the evaporation starts at a relatively large distance from the hole, where the virial temperature is of the order of 10^6 to 10^7 K. As in the solar corona, the strong decrease of radiative efficiency of gas with temperature in this range produces a hot optically thin corona in contact with the cool disk below. Exchange of mass can take place through evaporation and condensation, and the process is mediated by electron heat conduction. In this scenario, a corona flow at ~10^7 K at a distance of several hundred Schwarzschild radii transforms into a two-temperature ISAF further in (see Figure 1.5).

Ion illumination

Observations indicate that cool disks can also coexist with a hot, hard X-ray producing plasma quite close to the hole (for references, see Dullemond and Spruit, 2005). At these close distances, evaporation must behave differently from the coronal evaporation model, since the interaction of a two-temperature plasma with a cool disk is very different from that of a plasma at coronal temperatures (Spruit, 1997). The energy in an ISAF is in the ions, and the electron conduction of heat that drives coronal evaporation unimportant. Moreover, the ions penetrate a substantial distance into the cool disk, and the energy they dump is radiated away long before they can heat up the disk to virial temperatures. Nevertheless, evaporation can still take place in this case, since it turns out that the interaction of the ions with the cool disk produces a layer of intermediate temperature (around 100 keV) that becomes thermally unstable in the presence of viscous dissipation and heats up to ISAF temperatures (Deufel and Spruit, 2000; Deufel et al., 2001, 2002; Spruit and Deufel, 2002). The sequence of events is illustrated in Figure 1.5. This model explains both the hard spectra of typical black hole accretors and the coexistence of cool and hot plasma that is indicated by the observations (Dullemond and Spruit, 2005; D'Angelo et al., 2008).

1.13.3 *Quiescent galactic nuclei*

For very low accretion rates, such as inferred for the black hole in the center of our galaxy (identified with the radio source Sgr A*), the broadband spectral energy distribution of an ISAF is predicted to have two humps (Narayan et al., 1995; Quataert et al., 1999). In the X-ray range, the emission is due to bremsstrahlung. In the radio range, the flow emits synchrotron radiation, provided that the magnetic field in the flow has an energy density on the order of the gas pressure ("equipartition"). Synthetic ISAF spectra can be fitted to the observed radio and X-ray emission from Sgr A*. In other galaxies where a massive central black hole is inferred, and the center is populated by an X-ray–emitting gas of known density, ISAFs would also be natural and might explain why the observed luminosities are so low compared with the accretion rate expected for a hole embedded in a gas of the measured density.

1.13.4 ISAF-disk interaction: lithium

One of the strong predictions of ISAF models, whether for black holes or neutron stars, is that the accreting plasma in the inner regions has an ion temperature of 10 to 100 MeV. Nearby is a cool and dense accretion disk feeding this plasma. If only a small fraction of the hot ion plasma gets in contact with the disk, the intense irradiation by ions will produce nuclear reactions (Agaronian and Sunyaev, 1984; Martin et al., 1992). The main effects would be spallation of CNO elements into Li and Be, and the release of neutrons by various reactions. In this context, it is intriguing that the secondaries of both neutron star and black hole accretors have high overabundances of Li compared with other stars of their spectral types (Martin et al., 1992, 1994a). If a fraction of the disk material is carried to the secondary by a disk wind, the observed Li abundances may be accounted for (Martin et al., 1994b).

1.14 Outflows

The energy density in an advection-dominated accretion flow is of the same order as the gravitational binding energy density GM/r, since a significant fraction of that energy went into internal energy of the gas by viscous dissipation, and little of it got lost by radiation. The gas is therefore only marginally bound in the gravitational potential. This suggests that perhaps a part of the accreting gas can escape, producing an outflow or wind. In the case of the ion-supported ISAFs, this wind would be thermally driven by the temperature of the ions, like an "evaporation" from the accretion torus. In the case of the radiation-supported tori, which exist only at a luminosity near the Eddington value, but with much lower temperatures than the ion tori, winds driven by radiation pressure could exist.

The possibility of outflows is enhanced by the viscous energy transport through the disk. In the case of thin accretion disks (not quite appropriate in the present case, but sufficient to demonstrate the effect), the local rate of gravitational energy release (erg cm^{-2}s^{-1}) is $W = \Sigma v_r \partial_r (GM/r)$. The local viscous dissipation rate is $(9/4)\nu\Sigma\Omega^2$. As discussed in Section 1.5.4, they are related by

$$Q_v = 3\left[1 - \left(\frac{r_i}{r}\right)^{1/2}\right] W, \tag{1.115}$$

where r_i is the inner edge of the disk. Part of the gravitational energy released in the inner disk is transported outward by the viscous stresses, so that the energy deposited in the gas is up to three times larger than expected from a local energy balance argument. The temperatures in an ADAF would be correspondingly larger. Blandford and Begelman (1999) have appealed to this effect to argue that in an ADAF, most of the accreting mass of a disk might be expelled through a wind, the energy needed for this being supplied by the viscous energy transport associated with the small amount of mass that actually accretes.

Most outflows from disks such as the observed relativistic jets (and the non-relativistic ones in protostellar objects) are now believed to have a magnetic origin, requiring the presence of a strong, ordered (large-scale) magnetic field anchored in the disk. Three-dimensional numerical MHD simulations of such processes are becoming increasingly realistic (e.g., Moll, 2009; Krolik and Hawley, 2010). The origin of jet-friendly magnetic field configurations, on the other hand, and the reasons for their apparently unpredictable presence are still unknown (not all black holes show jets, and the ones that do don't have them all the time). There is, however, a degree of correlation of their presence with the hard X-ray states in these objects (in the sense that they appear absent in soft states). For a discussion and interpretation of these issues, see Spruit (2010b).

REFERENCES

Abramowicz, M. A., Czerny, B., Lasota, J. P., and Szuszkiewicz, E. 1988. Slim accretion disks. ApJ, **332**(Sept.), 646–658.

Abramowicz, M. A., Kato, S., and Matsumoto, R. 1989. Numerical models of slim accretion disks. PASJ, **41**, 1215–1218.

Agaronian, F. A., and Sunyaev, R. A. 1984. Gamma-ray line emission, nuclear destruction and neutron production in hot astrophysical plasmas – the deuterium boiler as a gamma-ray source. MNRAS, **210**(Sept.), 257–277.

Armitage, P. J. 1998. Turbulence and angular momentum transport in global accretion disk simulation. ApJ, **501**(July), L189+.

Armitage, P. J., and Livio, M. 1998. Hydrodynamics of the stream-disk impact in interacting binaries. ApJ, **493**(Jan.), 898+.

Balbus, S. A. 2003. Enhanced angular momentum transport in accretion disks. ARA&A, **41**, 555–597.

Balbus, S. A., and Hawley, J. F. 1991. A powerful local shear instability in weakly magnetized disks. I – Linear analysis. II – Nonlinear evolution. ApJ, **376**(July), 214–233.

Begelman, M. C., McKee, C. F., and Shields, G. A. 1983. Compton heated winds and coronae above accretion disks. I Dynamics. ApJ, **271**(Aug.), 70–88.

Bisnovatyi-Kogan, G. S. 1993. A self-consistent solution for an accretion disc structure around a rapidly rotating non-magnetized star. A&A, **274**(July), 796+.

Bisnovatyi-Kogan, G. S., and Blinnikov, S. I. 1977. Disk accretion onto a black hole at subcritical luminosity. A&A, **59**(July), 111–125.

Bisnovatyi-Kogan, G. S., and Lamzin, S. A. 1984. Stars with neutron cores – the possibility of the existence of objects with a low neutrino luminosity. Soviet Ast, **28**(Apr.), 187+.

Bisnovatyi-Kogan, G. S., and Ruzmaikin, A. A. 1976. The accretion of matter by a collapsing star in the presence of a magnetic field. II – Selfconsistent stationary picture. Ap&SS, **42**(July), 401–424.

Blaes, O. M., and Hawley, J. F. 1988. Nonaxisymmetric disk instabilities – a linear and nonlinear synthesis. ApJ, **326**(Mar.), 277–291.

Blandford, R. D. 1976. Accretion disc electrodynamics – a model for double radio sources. MNRAS, **176**(Sept.), 465–481.

Blandford, R. D., and Begelman, M. C. 1999. On the fate of gas accreting at a low rate on to a black hole. MNRAS, **303**(Feb.), L1–L5.

Blandford, R. D., and Payne, D. G. 1982. Hydromagnetic flows from accretion discs and the production of radio jets. MNRAS, **199**(June), 883–903.

Bondi, H. 1952. On spherically symmetrical accretion. MNRAS, **112**, 195+.

Bradshaw, P. 1969. A note on reverse transition. *Journal of Fluid Mechanics*, **35**, 387–390.

Brandenburg, A., Nordlund, A., Stein, R. F., and Torkelsson, U. 1995. Dynamo-generated turbulence and large-scale magnetic fields in a Keplerian shear flow. ApJ, **446**(June), 741+.

Cannizzo, J. K., Shafter, A. W., and Wheeler, J. C. 1988. On the outburst recurrence time for the accretion disk limit cycle mechanism in dwarf novae. ApJ, **333**(Oct.), 227–235.

Cannon, R. C., Eggleton, P. P., Zytkow, A. N., and Podsiadlowski, P. 1992. The structure and evolution of Thorne-Zytkow objects. ApJ, **386**(Feb.), 206–214.

Chandrasekhar, C. 1961. *Hydrodynamic and Hydromagnetic Stability*. International Series of Monographs on Physics, Oxford: Clarendon.

Chen, X., Abramowicz, M. A., Lasota, J.-P., Narayan, R., and Yi, I. 1995. Unified description of accretion flows around black holes. ApJ, **443**(Apr.), L61–L64.

Chevalier, R. A. 1993. Neutron star accretion in a stellar envelope. ApJ, **411**(July), L33–L36.

Courant, R., and Friedrichs, K. O. 1948. *Supersonic flow and shock waves.* Pure and Applied Mathematics, New York: Interscience.

D'Angelo, C. R., Giannios, D., Dullemond, C., and Spruit, H. 2008. Soft X-ray components in the hard state of accreting black holes. A&A, **488**(Sept.), 441–450.

D'Angelo, C. R., and Spruit, H. C. 2010. Episodic accretion on to strongly magnetic stars. MNRAS, **406**(Aug.), 1208–1219.

Davis, S. W., Stone, J. M., and Pessah, M. E. 2010. Sustained magnetorotational turbulence in local simulations of stratified disks with zero net magnetic flux. ApJ, **713**(Apr.), 52–65.

De Villiers, J.-P., and Hawley, J. F. 2003. Global general relativistic magnetohydrodynamic simulations of accretion tori. ApJ, **592**(Aug.), 1060–1077.

Deufel, B., and Spruit, H. C. 2000. Comptonization in an accretion disk illuminated by protons. A&A, **362**(Oct.), 1–8.

Deufel, B., Dullemond, C. P., and Spruit, H. C. 2001. X-ray spectra from protons illuminating a neutron star. A&A, **377**(Oct.), 955–963.

Deufel, B., Dullemond, C. P., and Spruit, H. C. 2002. X-ray spectra from accretion disks illuminated by protons. A&A, **387**(June), 907–917.

Di Matteo, T., Celotti, A., and Fabian, A. C. 1999. Magnetic flares in accretion disc coronae and the spectral states of black hole candidates: the case of GX339-4. MNRAS, **304**(Apr.), 809–820.

Dubrulle, B. 1992. A turbulent closure model for thin accretion disks. A&A, **266**(Dec.), 592–604.

Dullemond, C. P., Dominik, C., and Natta, A. 2001. Passive irradiated circumstellar disks with an inner hole. ApJ, **560**(Oct.), 957–969.

Dullemond, C. P., and Spruit, H. C. 2005. Evaporation of ion-irradiated disks. A&A, **434**(May), 415–422.

Emmering, R. T., Blandford, R. D., and Shlosman, I. 1992. Magnetic acceleration of broad emission-line clouds in active galactic nuclei. ApJ, **385**(Feb.), 460–477.

Esin, A. A., Narayan, R., Cui, W., Grove, J. E., and Zhang, S.-N. 1998. Spectral transitions in Cygnus X-1 and other black hole X-ray binaries. ApJ, **505**(Oct.), 854–868.

Fabian, A. C., Vaughan, S., Nandra, K., Iwasawa, K., Ballantyne, D. R., Lee, J. C., De Rosa, A., Turner, A., and Young, A. J. 2002. A long hard look at MCG-6-30-15 with XMM-Newton. MNRAS, **335**(Sept.), L1–L5.

Frank, J., King, A., and Raine, D. J. 2002. *Accretion Power in Astrophysics: Third Edition.* Cambridge, UK: Cambridge University Press.

Fromang, S., Papaloizou, J., Lesur, G., and Heinemann, T. 2007. MHD simulations of the magnetorotational instability in a shearing box with zero net flux. II. The effect of transport coefficients. A&A, **476**(Dec.), 1123–1132.

Galeev, A. A., Rosner, R., and Vaiana, G. S. 1979. Structured coronae of accretion disks. ApJ, **229**(Apr.), 318–326.

Gammie, C. F. 1997. Nonlinear outcome of gravitational instability in optically thick disks. Pages 704+ of: D. T. Wickramasinghe, G. V. Bicknell, and L. Ferrario (eds.), *IAU Colloq. 163: Accretion Phenomena and Related Outflows.* Astronomical Society of the Pacific Conference Series, vol. 121.

Gilfanov, M., Churazov, E., and Revnivtsev, M. 1999. Reflection and noise in Cygnus X-1. A&A, **352**(Dec.), 182–188.

Gilham, S. 1981. Scale-free axisymmetric accretion with weak viscosity. MNRAS, **195**(June), 755–763.

Haardt, F., Maraschi, L., and Ghisellini, G. 1994. A model for the X-ray and ultraviolet emission from Seyfert galaxies and galactic black holes. ApJ, **432**(Sept.), L95–L99.

Hameury, J.-M., Menou, K., Dubus, G., Lasota, J.-P., and Hure, J.-M. 1998. Accretion disc outbursts: a new version of an old model. MNRAS, **298**(Aug.), 1048–1060.

Hawley, J. F. 1991. Three-dimensional simulations of black hole tori. ApJ, **381**(Nov.), 496–507.

Hawley, J. F., Balbus, S. A., and Winters, W. F. 1999. Local hydrodynamic stability of accretion disks. ApJ, **518**(June), 394–404.

Hawley, J. F., Gammie, C. F., and Balbus, S. A. 1995. Local three-dimensional magnetohydrodynamic simulations of accretion disks. ApJ, **440**(Feb.), 742+.

Hirose, S., Krolik, J. H., and Blaes, O. 2009. Radiation-dominated disks are thermally stable. ApJ, **691**(Jan.), 16–31.

Hirose, S., Krolik, J. H., De Villiers, J.-P., and Hawley, J. F. 2004. Magnetically driven accretion flows in the Kerr metric. II. Structure of the magnetic field. ApJ, **606**(May), 1083–1097.

Honma, F., Kato, S., Matsumoto, R., and Abramowicz, M. A. 1991. Stability of slim, transonic accretion disk models. PASJ, **43**(Apr.), 261–273.

Horne, K. 1985. Images of accretion discs. I – The eclipse mapping method. MNRAS, **213**(Mar.), 129–141.

Horne, K. 1993. *Eclipse Mapping of Accretion Disks: The First Decade*. Pages 117–147 of: J. Craig Wheeler (ed.), Accretion Disk In Compact Stellar System. Series: Advanced Series in Astrophysics and Cosmology, World Scientific, vol. 9.

Hōshi, R. 1979. Accretion model for outbursts of dwarf nova. Progress of Theoretical Physics, **61**(May), 1307–1319.

Hurley, K., Boggs, S. E., Smith, D. M., Duncan, R. C., Lin, R., Zoglauer, A., Krucker, S., Hurford, G., Hudson, H., Wigger, C., Hajdas, W., Thompson, C., Mitrofanov, I., Sanin, A., Boynton, W., Fellows, C., von Kienlin, A., Lichti, G., Rau, A., and Cline, T. 2005. An exceptionally bright flare from SGR 1806-20 and the origins of short-duration γ-ray bursts. Nature, **434**(Apr.), 1098–1103.

Ichimaru, S. 1977. Bimodal behavior of accretion disks – theory and application to Cygnus X-1 transitions. ApJ, **214**(June), 840–855.

Illarionov, A. F., and Sunyaev, R. A. 1975. Why the number of galactic X-ray stars is so small? A&A, **39**(Feb.), 185+.

Inogamov, N. A., and Sunyaev, R. A. 1999. Spread of matter over a neutron-star surface during disk accretion. Astronomy Letters, **25**(May), 269–293.

Ji, H., Burin, M., Schartman, E., and Goodman, J. 2006. Hydrodynamic turbulence cannot transport angular momentum effectively in astrophysical disks. Nature, **444**(Nov.), 343–346.

Kato, M. 1997. Optically Thick Winds from Degenerate Dwarfs. I. Classical Novae of Populations I and II. ApJS, **113**(Nov.), 121+.

King, A. R. 1995. *Cataclysmic variable stars*. Edited by B. Warner, Cambridge Astrophysics Series, vol. 28, pp. 419–456.

King, A. R. 2004. Ultraluminous X-ray sources and star formation. MNRAS, **347**(Jan.), L18–L20.

King, A. R., and Begelman, M. C. 1999. Radiatively Driven Outflows and Avoidance of Common-Envelope Evolution in Close Binaries. ApJ, **519**(July), L169–L171.

King, A. R., and Ritter, H. 1999. Cygnus X-2, super-Eddington mass transfer, and pulsar binaries. MNRAS, **309**(Oct.), 253–260.

Kippenhahn, R., and Thomas, H.-C. 1981. Rotation and stellar evolution. Pages 237–254 of: D. Sugimoto, D. Q. Lamb, and D. N. Schramm (eds.), *Fundamental Problems in the Theory of Stellar Evolution*. IAU Symposium, vol. 93.

Klahr, H. H., and Bodenheimer, P. 2003. Turbulence in accretion disks: vorticity generation and angular momentum transport via the global baroclinic instability. ApJ, **582**(Jan.), 869–892.

Kley, W. 1991. On the influence of the viscosity on the structure of the boundary layer of accretion disks. A&A, **247**(July), 95–107.

Kley, W., Papaloizou, J. C. B., and Lin, D. N. C. 1993. On the angular momentum transport associated with convective eddies in accretion disks. ApJ, **416**(Oct.), 679+.

Königl, A. 1989. Self-similar models of magnetized accretion disks. ApJ, **342**(July), 208–223.

Königl, A., and Kartje, J. F. 1994. Disk-driven hydromagnetic winds as a key ingredient of active galactic nuclei unification schemes. ApJ, **434**(Oct.), 446–467.

Krolik, J. H., and Hawley, J. F. 2010 (Mar.). General relativistic MHD jets. Pages 265+ of: T. Belloni (ed.), *Lecture Notes in Physics*, Berlin Springer Verlag. Lecture Notes in Physics, Berlin Springer Verlag, vol. 794.

Kuulkers, E., Howell, S. B., and van Paradijs, J. 1996. SXTs and TOADs: close encounters of the same kind. ApJ, **462**(May), L87+.

Landau, L. D., and Lifshitz, E. M. 1959. *Fluid Mechanics*. Course of theoretical physics, Oxford: Pergamon Press.

Lee, C.-F., Mundy, L. G., Reipurth, B., Ostriker, E. C., and Stone, J. M. 2000. CO outflows from young stars: confronting the jet and wind models. ApJ, **542**(Oct.), 925–945.

Lesur, G., and Longaretti, P.-Y. 2005. On the relevance of subcritical hydrodynamic turbulence to accretion disk transport. A&A, **444**(Dec.), 25–44.

Lesur, G., and Ogilvie, G. I. 2008. On self-sustained dynamo cycles in accretion discs. A&A, **488**(Sept.), 451–461.

Lesur, G., and Papaloizou, J. C. B. 2010. The subcritical baroclinic instability in local accretion disc models. A&A, **513**(Apr.), A60+.

Lewin, W. H. G., van Paradijs, J., and van den Heuvel, E. P. J. 1995. Lewin, Walter H. G.; Van Paradijs, Jan; Van den Heuvel, Edward P. J. (eds.), *X-Ray Binaries*. Cambridge Astrophysics Series, Cambridge, MA: Cambridge University Press.

Lightman, A. P., and Eardley, D. M. 1974. Black holes in binary systems: instability of disk accretion. ApJ, **187**(Jan.), L1+.

Lin, D. N. C., and Taam, R. E. 1984 (May). On the structure, stability and evolution of accretion disks in soft x-ray transient sources. Pages 83–102 of: S. E. Woosley (ed.), *American Institute of Physics Conference Series*. American Institute of Physics Conference Series, vol. 115.

Liu, B. F., Mineshige, S., Meyer, F., Meyer-Hofmeister, E., and Kawaguchi, T. 2002. Two-temperature coronal flow above a thin disk. ApJ, **575**(Aug.), 117–126.

Long, M., Romanova, M. M., and Lovelace, R. V. E. 2008. Three-dimensional simulations of accretion to stars with complex magnetic fields. MNRAS, **386**(May), 1274–1284.

Lovelace, R. V. E. 1976. Dynamo model of double radio sources. Nature, **262**(Aug.), 649–652.

Lüst, R. 1952. Die Entwicklung einer um einen Zentralkörper rotierenden Gasmasse. Z. Naturforsch., **7**.

Lynden-Bell, D., and Pringle, J. E. 1974. The evolution of viscous discs and the origin of the nebular variables. MNRAS, **168**(Sept.), 603–637.

Lyutyi, V. M., and Syunyaev, R. A. 1976a. Nature of the optical variability of the X-ray binary systems CYG X-2 - V1341 CYG and SCO X-1 - V818 SCO. AZh, **53**(June), 511–526.

Lyutyi, V. M., and Syunyaev, R. A. 1976b. Nature of the optical variability in the x-ray binaries Cygnus X-2 and Scorpius X-1. Soviet Ast, **20**(June), 290–298.

Martin, E. L., Rebolo, R., Casares, J., and Charles, P. A. 1992. High lithium abundance in the secondary of the black-hole binary system V404 Cygni. Nature, **358**(July), 129–131.

Martin, E. L., Rebolo, R., Casares, J., and Charles, P. A. 1994a. Li abundances in late-type companions to neutron stars and black hole candidates. ApJ, **435**(Nov.), 791–796.

Martin, E. L., Spruit, H. C., and van Paradijs, J. 1994b. Energy implications of Li production in X-ray transients. A&A, **291**(Nov.), L43–L46.

Massey, B. S. 1968. *Mechanics of Fluids*, Chapman and Hall, London (6th Ed., 1989).

Meyer, F., and Meyer-Hofmeister, E. 1981. On the elusive cause of cataclysmic variable outbursts. A&A, **104**, L10+.

Meyer, F., and Meyer-Hofmeister, E. 1982. Vertical structure of accretion disks. A&A, **106**(Feb.), 34–42.

Meyer, F., and Meyer-Hofmeister, E. 1994. Accretion disk evaporation by a coronal siphon flow. A&A, **288**(Aug.), 175–182.

Meyer-Hofmeister, E., and Meyer, F. 1999. Black hole soft X-ray transients: evolution of the cool disk and mass supply for the ADAF. A&A, **348**(May), 154–160.

Meyer-Hofmeister, E., and Ritter, H. 1993. Accretion disks in close binaries. Pages 143–168 of: J. Sahade, G. E. McCluskey Jr., and Y. Kondo (eds.), *Astrophysics and Space Science Library*. Astrophysics and Space Science Library, vol. 177.

Meyer-Hofmeister, E., Schandl, S., and Meyer, F. 1997. The structure of the accretion disk rim in supersoft X-ray sources. A&A, **321**(May), 245–253.

Migliari, S., and Fender, R. P. 2006. Jets in neutron star X-ray binaries: a comparison with black holes. MNRAS, **366**(Feb.), 79–91.

Miller, K. A., and Stone, J. M. 1997. Magnetohydrodynamic simulations of stellar magnetosphere–accretion disk interaction. ApJ, **489**(Nov.), 890+.

Mineshige, S., and Wheeler, J. C. 1989. Disk-instability model for soft-X-ray transients containing black holes. ApJ, **343**(Aug.), 241–253.

Mineshige, S., and Wood, J. H. 1989. Viscous evolution of accretion discs in the quiescence of dwarf novae. Pages 221+ of: F. Meyer (ed.), *NATO ASIC Proc. 290: Theory of Accretion Disks*.

Moll, R. 2009. Decay of the toroidal field in magnetically driven jets. A&A, **507**(Dec.), 1203–1210.

Nakamura, K. E., Matsumoto, R., Kusunose, M., and Kato, S. 1996. Global structures of advection-dominated two-temperature accretion disks. PASJ, **48**(Oct.), 761–769.

Narayan, R., and Popham, R. 1994. Accretion disk boundary layers. Pages 293+ of: W. J. Duschl, J. Frank, F. Meyer, E. Meyer-Hofmeister, and W. M. Tscharnuter (eds.), *NATO ASIC Proc. 417: Theory of Accretion Disks – 2*.

Narayan, R., and Yi, I. 1994. Advection-dominated accretion: a self-similar solution. ApJ, **428**(June), L13–L16.

Narayan, R., Goldreich, P., and Goodman, J. 1987. Physics of modes in a differentially rotating system – analysis of the shearing sheet. MNRAS, **228**(Sept.), 1–41.

Narayan, R., Kato, S., and Honma, F. 1997. Global structure and dynamics of advection-dominated accretion flows around black holes. ApJ, **476**(Feb.), 49+.

Narayan, R., Yi, I., and Mahadevan, R. 1995. Explaining the spectrum of Sagittarius A* with a model of an accreting black hole. Nature, **374**(Apr.), 623–625.

Nayakshin, S., Rappaport, S., and Melia, F. 1999 (Apr.). Time dependent disk models for the microquasar GRS1915+105. Pages 730+ of: *AAS/High Energy Astrophysics Division #4*. Bulletin of the American Astronomical Society, vol. 31.

O'Donoghue, D., Chen, A., Marang, F., Mittaz, J. P. D., Winkler, H., and Warner, B. 1991. WX CET and the WZ SGE stars. MNRAS, **250**(May), 363–372.

Ogilvie, G. I. 1999. Time-dependent quasi-spherical accretion. MNRAS, **306**(June), L9–L13.

Osaki, Y. 1974. An accretion model for the outbursts of U Geminorum stars. PASJ, **26**, 429–436.

Osaki, Y. 1994. Disk instability model for SU UMa stars: SU UMa/WZ Sge connection. Pages 93+ of: W. J. Duschl, J. Frank, F. Meyer, E. Meyer-Hofmeister, and W. M. Tscharnuter (eds.), *NATO ASIC Proc. 417: Theory of Accretion Disks – 2*.

Ostriker, E. C., Gammie, C. F., and Stone, J. M. 1999. Kinetic and structural evolution of self-gravitating, magnetized clouds: 2.5-dimensional simulations of decaying turbulence. ApJ, **513**(Mar.), 259–274.

Paczyński, B. 1978. A model of self-gravitating accretion disk. Acta Astron, **28**, 91–109.

Paczyński, B. 1991. A polytropic model of an accretion disk, a boundary layer, and a star. ApJ, **370**(Apr.), 597–603.

Paczyński, B. 1998. Are Gamma-Ray Bursts in Star-Forming Regions? ApJ, **494**(Feb.), L45+.

Papaloizou, J. C. B., and Lin, D. N. C. 1995. Theory of accretion disks I: Angular momentum transport processes. ARA&A, **33**, 505–540.

Piran, T. 1978. The role of viscosity and cooling mechanisms in the stability of accretion disks. ApJ, **221**(Apr.), 652–660.

Podolak, M., Hubbard, W. B., and Pollack, J. B. 1993. Gaseous accretion and the formation of giant planets. Pages 1109–1147 of: E. H. Levy and J. I. Lunine (eds.), *Protostars and Planets III*. University of Arizona Press.

Podsiadlowski, P., and Rappaport, S. 2000. Cygnus X-2: the descendant of an intermediate-mass x-ray binary. ApJ, **529**(Feb.), 946–951.

Popham, R. 1997. Boundary layers in cataclysmic variables and pre-main-sequence Stars. Pages 230+ of: D. T. Wickramasinghe, G. V. Bicknell, and L. Ferrario (eds.), *IAU Colloq. 163:*

Accretion Phenomena and Related Outflows. Astronomical Society of the Pacific Conference Series, vol. 121.

Popham, R., and Narayan, R. 1991. Does accretion cease when a star approaches breakup? ApJ, **370**(Apr.), 604–614.

Popham, R., Woosley, S. E., and Fryer, C. 1999. Hyperaccreting black holes and gamma-ray bursts. ApJ, **518**(June), 356–374.

Pringle, J. E. 1981. Accretion discs in astrophysics. ARA&A, **19**, 137–162.

Pringle, J. E., and Savonije, G. J. 1979. X-ray emission from dwarf novae. MNRAS, **187**(June), 777–783.

Quataert, E., Narayan, R., and Reid, M. J. 1999. What is the accretion rate in Sagittarius A*? ApJ, **517**(June), L101–L104.

Rafikov, R. R. 2009. Properties of gravitoturbulent accretion disks. ApJ, **704**(Oct.), 281–291.

Rappaport, S. A., Fregeau, J. M., and Spruit, H. 2004. Accretion onto fast X-ray pulsars. ApJ, **606**(May), 436–443.

Rees, M. J. 1978. Accretion and the quasar phenomenon. Phys Scr, **17**(Mar.), 193–200.

Rees, M. J., Begelman, M. C., Blandford, R. D., and Phinney, E. S. 1982. Ion-supported tori and the origin of radio jets. Nature, **295**(Jan.), 17–21.

Regev, O., and Hougerat, A. A. 1988. Accretion disc boundary layers – geometrically an optically thin case. MNRAS, **232**(May), 81–89.

Revnivtsev, M., Gilfanov, M., and Churazov, E. 1999. The frequency resolved spectroscopy of CYG X-1: fast variability of the Fe K$_\alpha$ line. A&A, **347**(July), L23–L26.

Rincon, F., Ogilvie, G. I., and Cossu, C. 2007. On self-sustaining processes in Rayleigh-stable rotating plane Couette flows and subcritical transition to turbulence in accretion disks. A&A, **463**(Mar.), 817–832.

Różyczka, M., and Spruit, H. C. 1993. Numerical simulations of shock-driven accretion. ApJ, **417**(Nov.), 677+.

Rutten, R. G. M., van Paradijs, J., and Tinbergen, J. 1992. Reconstruction of the accretion disk in six cataclysmic variable stars. A&A, **260**(July), 213–226.

Rybicki, G. B., and Lightman, A. P. 1979. *Radiative Processes in Astrophysics*. New York, Wiley-Interscience, 393 p.

Sakimoto, P. J., and Coroniti, F. V. 1989. Buoyancy-limited magnetic viscosity in quasi-stellar object accretion disk models. ApJ, **342**(July), 49–63.

Sawada, K., Matsuda, T., Inoue, M., and Hachisu, I. 1987. Is the standard accretion disc model invulnerable? MNRAS, **224**(Jan.), 307–322.

Schandl, S., and Meyer, F. 1994. Herculis X-1: Coronal winds producing the tilted shape of the accretion disk. A&A, **289**(Sept.), 149–161.

Schultz, A. L., and Price, R. H. 1985. Pair production in spherical accretion onto black holes. ApJ, **291**(Apr.), 1–7.

Shakura, N. I., and Sunyaev, R. A. 1973. Black holes in binary systems. Observational appearance. A&A, **24**, 337–355.

Shapiro, S. L., Lightman, A. P., and Eardley, D. M. 1976. A two-temperature accretion disk model for Cygnus X-1 – structure and spectrum. ApJ, **204**(Feb.), 187–199.

Shi, J., Krolik, J. H., and Hirose, S. 2010. What is the numerically converged amplitude of magnetohydrodynamics turbulence in stratified shearing boxes? ApJ, **708**(Jan.), 1716–1727.

Siuniaev, R. A., and Shakura, N. I. 1977. Disk reservoirs in binary systems and prospects for observing them. Pis ma Astronomicheskii Zhurnal, **3**(June), 262–266.

Smak, J. 1971. Eruptive binaries. II. U Geminorum. Acta Astron., **21**, 15+.

Smak, J. 1984. Outbursts of dwarf novae. PASP, **96**(Jan.), 5–18.

Spitzer, L. 1965. *Physics of Fully Ionized Gases*. Interscience Tracts on Physics and Astronomy, New York: Interscience Publication, 2nd rev. ed.

Spruit, H. C. 1987. Stationary shocks in accretion disks. A&A, **184**(Oct.), 173–184.

Spruit, H. C. 1996. Cyclic accretion from an accretion disk onto a neutron star magnetosphere. Pages 377–382 of: Kluwer Academic Publishers. *Lives of Neutron Stars*, NATO ASI Series C., Vol. 477.

Spruit, H. C. 1997. X-ray spectrum of a disk illuminated by ions. Pages 67–76 of: E. Meyer-Hofmeister and H. Spruit (ed), *Accretion Disks – New Aspects*. Lecture Notes in Physics, Berlin Springer Verlag, vol. 487.

Spruit, H. C. 2010a. Theories of the solar cycle: a critical view. ArXiv e-prints.

Spruit, H. C. 2010b. Theory of magnetically powered jets. In: *Lecture Notes in Physics*, Berlin Springer Verlag.

Spruit, H. C., and Deufel, B. 2002. The transition from a cool disk to an ion supported flow. A&A, **387**(June), 918–930.

Spruit, H. C., Matsuda, T., Inoue, M., and Sawada, K. 1987. Spiral shocks and accretion in discs. MNRAS, **229**(Dec.), 517–527.

Spruit, H. C., and Rutten, R. G. M. 1998. The stream impact region in the disc of WZ SGE. MNRAS, **299**(Sept.), 768–776.

Spruit, H. C., and Taam, R. E. 1993. An instability associated with a magnetosphere-disk interaction. ApJ, **402**(Jan.), 593–604.

Strom, S. E., Edwards, S., and Skrutskie, M. F. 1993. Evolutionary time scales for circumstellar disks associated with intermediate- and solar-type stars. Pages 837–866 of: E. H. Levy and J. I. Lunine (eds.), *Protostars and Planets III*. University of Arizona Press.

Sunyaev, R. A., and Titarchuk, L. G. 1980. Comptonization of X-rays in plasma clouds – typical radiation spectra. A&A, **86**(June), 121–138.

Syunyaev, R. A., and Shakura, N. I. 1977. Disk reservoirs in binary systems and prospects for observing them. Soviet Astronomy Letters, **3**(June), 138+.

Taam, R. E. 1996. Common-envelope evolution, the formation of CVs, LMXBs, and the fate of HMXBs. Pages 3+ of: J. van Paradijs, E. P. J. van den Heuvel, and E. Kuulkers (eds.), *Compact Stars in Binaries*. IAU Symposium, vol. 165.

Taam, R. E., and Sandquist, E. L. 2000. Common envelope evolution of massive binary stars. ARA&A, **38**, 113–141.

Treves, A., Maraschi, L., and Abramowicz, M. 1988. Basic elements of the theory of accretion. PASP, **100**(Apr.), 427–451.

Turner, N. J. 2004. On the vertical structure of radiation-dominated accretion disks. ApJ, **605**(Apr.), L45–L48.

van Paradijs, J., and McClintock, J. E. 1995. Optical and ultraviolet observations of X-ray binaries. Pages 58–125 of: W. H. G. Lewin, J. van Paradijs, and E. P. J. van den Heuvel (eds.), *X-ray Binaries*. Cambridge University Press.

van Paradijs, J., and Verbunt, F. 1984 (May). A comparison of soft x-ray transients and dwarf novae. Pages 49–62 of: S. E. Woosley (ed), *American Institute of Physics Conference Series*. American Institute of Physics Conference Series, vol. 115.

Velikhov, P. E. 1959. Stability of an ideally conducting liquid flowing between cylinders rotating in a magnetic field. J Expl Theoret Phys (*USSR*), **36**, 1398.

Wang, J.-M., and Zhou, Y.-Y. 1999. Self-similar solution of optically thick advection-dominated flows. ApJ, **516**(May), 420–424.

Warner, B. 1987. Absolute magnitudes of cataclysmic variables. MNRAS, **227**(July), 23–73.

Warner, B. 1995. *Cataclysmic variable stars*. Cambridge Astrophysics Series, vol. 28.

Wood, J. H., Horne, K., Berriman, G., and Wade, R. A. 1989a. Eclipse studies of the dwarf nova OY Carinae in quiescence. ApJ, **341**(June), 974–996.

Wood, J. H., Horne, K., and Vennes, S. 1992. Eclipse studies of the dwarf nova HT Cassiopeiae. II – White dwarf and accretion disk. ApJ, **385**(Jan.), 294–305.

Wood, J. H., Marsh, T. R., Robinson, E. L., Stiening, R. F., Horne, K., Stover, R. J., Schoembs, R., Allen, S. L., Bond, H. E., Jones, D. H. P., Grauer, A. D., and Ciardullo, R. 1989b. The ephemeris and variations of the accretion disc radius in IP Pegasi. MNRAS, **239**(Aug.), 809–824.

Woosley, S. E. 1993. Gamma-ray bursts from stellar mass accretion disks around black holes. ApJ, **405**(Mar.), 273–277.

Yi, I. (1999, Apr.). Advection-dominated accretion flows. Pages 279+ of: J. A. Sellwood and J. Goodman (ed), *Astrophysical Discs – an EC Summer School*. Astronomical Society of the Pacific Conference Series, vol. 160.

Zdziarski, A. A. 1998. Hot accretion discs with thermal Comptonization and advection in luminous black hole sources. MNRAS, **296**(June), L51+.

Zdziarski, A. A., Lubinski, P., and Smith, D. A. 1999. Correlation between Compton reflection and X-ray slope in Seyferts and X-ray binaries. MNRAS, **303**(Feb.), L11–L15.

Zel'dovich, Y. B. 1981. On the friction of fluids between rotating cylinders. Royal Society of London Proceedings Series A, **374**(Feb.), 299–312.

2. The evolution of binary systems
PHILIPP PODSIADLOWSKI

Abstract

One of the most important environments in which accretion disks are found occur in interacting binaries. In this chapter I review the main properties of binary systems and the most important types of binary interactions, stable and unstable mass transfer, the role of mass loss, mass accretion, and, in the most dramatic case, the merging of the two binary components. I particularly emphasize the evolutionary context in which these interactions occur and illustrate this using numerous examples of different types of binaries of current research interest. These include hot subdwarfs; symbiotic binaries; binary supernova progenitors, including the progenitors of type Ia supernovae and potential progenitors of long-duration gamma-ray bursts; low-, intermediate-, and high-mass X-ray binaries, containing both neutron stars and black holes; and their descendants, including binary millisecond pulsars, Thorne-Żytkow objects, and short-duration gamma-ray bursts.

2.1 Introduction

One of the main sites for accretion disks are interacting binary systems. Indeed, the majority of stars are found in binary systems, and in many cases (up to ∼50%), they are close enough that mass flows from one star to the other, in many cases forming an accretion disk. This can happen for a wide variety of different systems: systems containing two normal nondegenerate stars, or one compact star (white dwarf [WD], neutron star [NS], or black hole [BH]), or even two compact stars of various combinations. The purpose of the chapter is to provide an overview of the evolution of binary systems, starting with the fundamentals of binary evolution in Section 2.2, followed by a selection of current topics in binary evolution theory in Section 2.3 and the effects of binary evolution on the final fate of stars and supernovae in Section 2.4. The last two sections, 2.5 and 2.6, discuss low-/intermediate-mass X-ray binaries and high-mass X-ray binaries, respectively.

2.2 Fundamentals of binary evolution
2.2.1 Basic properties

Most stars in the sky are in binary systems or, more generally, in multiple systems (triples, quadruples, quintuplets, ...), where the orbital periods ($P_{\rm orb}$) range all the way from 11 min (for a NS-WD binary) to $\sim 10^6$ yr. Of course, the majority of binaries are in fairly wide systems that do not interact strongly and where both stars evolve essentially as single stars. But there is a large fraction of systems (with $P_{\rm orb} \lesssim 10\,{\rm yr}$) that are close enough that mass is transferred from one star to the other, which changes the structures of both stars and their subsequent evolution. Although the exact numbers are somewhat uncertain, binary surveys suggest that the range of interacting binaries, which are the systems of interest in this chapter, is in the range of 30% to 50% (where the binary fraction is higher for more massive stars; see, e.g., Duquennoy and Mayor, 1991; Kobulnicky and Fryer, 2007). A very approximate period distribution, and very useful rule of thumb, is that the distribution in $\log P_{\rm orb}$ is logarithmically flat (i.e., $f(\log P_{\rm orb}) \simeq$ constant), where each decade of $\log P_{\rm orb}$ contains 10% of systems from 10^{-3} yr to 10^7 yr. The mass-ratio distribution (i.e., $q = M_2/M_1$, where M_1 and M_2 are the initially more massive [the primary] and the initially less massive star [the secondary], respectively) is not very well determined but appears to depend somewhat on the mass range. Whereas massive binaries favor stars of comparable mass (i.e., if the primary is a massive star, the secondary is also likely to be relatively massive), this is less clear for low-mass stars; it is sometimes argued that, for low-mass binaries, the masses may be independently chosen from a standard initial mass function [IMF], although most studies show that there is also some bias, possibly consistent with a flat mass-ratio distribution. These differences clearly reflect differences in the formation processes of low- and high-mass stars that are still extremely poorly understood. Finally, there is generally a large scatter in the distribution of eccentricities $e \equiv \sqrt{1 - b^2/a^2}$, where a and b are the semimajor and semiminor axes, respectively. Close binaries (with $P_{\rm orb} < 10\,{\rm d}$) tend to be circular, but this is the result of tidal interactions that efficiently circularize close eccentric binaries.

2.2.2 Observational classification

One of the main classifications of different types of binary systems is how they appear to an observer. *Visual binaries* are systems where the periodic motion of both components can be seen in the sky (see, e.g., Sirius A and B). If the motion of only one star is observable, the binary is referred to as an *astrometric binary*.

Spectroscopic binaries are systems where the periodic Doppler shifts (due to the orbital motion of the binary components) can be detected in one or more spectral lines. Depending on whether these Doppler shifts can be measured for just one or both binary components, these systems are called *single-lined* or *double-lined* spectroscopic binaries.

Photometric binaries are systems where one can observe a periodic variation of the flux, color, and so forth, of the system. However, this does not necessarily prove the binary nature of a system, as variable stars (e.g., Cepheids, RR Lyrae variables) can show similar periodic variations.

Finally, if at least one star eclipses the other star during part of the orbit, the system is an *eclipsing binary*. These binaries play a particularly important role in determining basic stellar parameters of stars (such as radius and mass).

2.2.3 The binary mass function

Another important concept helping to constrain the masses of the components in a binary is the binary mass function. By equating the gravitational force to the centripetal force of either of the components, using various Newtonian relations and assuming a circular orbit, it is easy to derive the following two relations, referred to as the mass functions for stars 1 and 2, f_1 and f_2:

$$f_1(M_2) = \frac{M_2^3 \sin^3 i}{(M_1 + M_2)^2} = \frac{P(v_1 \sin i)^3}{2\pi G}, \qquad (2.1)$$

and

$$f_2(M_1) = \frac{M_1^3 \sin^3 i}{(M_1 + M_2)^2} = \frac{P(v_2 \sin i)^3}{2\pi G}. \qquad (2.2)$$

Note that the expressions on the right-hand sides of these equations only contain *measurable* quantities, such as the projected radial velocity amplitudes of the two components ($v_{1/2} \sin i$, where i is the orbital inclination of the binary) and the orbital period P (G is the gravitational constant). The terms in the middle contain, besides $\sin i$, the main quantities of interest: the masses of the two components, M_1 and M_2.

Thus, the mass functions directly relate the two masses to observable quantities. For a double-lined spectroscopic binary, where the radial velocity amplitudes of both components can be measured, one can use these relations to determine $M_1 \sin^3 i$ and $M_2 \sin^3 i$. It is often difficult to determine the inclination; however, in cases where this is possible (e.g., for an eclipsing binary with $i \simeq 90°$ or for a visual binary), one obtains the masses of both components. Indeed, this is one of the most important methods for determining stellar masses, including those of compact objects, such as black holes. In the case where one star is much less massive than the other (e.g., $M_1 \ll M_2$), its mass function directly constrains the mass of the other component since, in this case, equation 2.1 simplifies to $f_1(M_2) \simeq M_2 \sin^3 i$.

In the case of doubly eclipsing binaries, one can also determine the radii of both stars. Such systems are the main sources for determining accurate masses and radii of stars, and luminosities, if their distances are also known.

2.2.4 The Roche lobe

One particularly important concept in studying the evolution of binary systems is the *Roche lobe*. Considering the so-called restricted three-body problem, where one follows the motion of a massless test particle in the gravitational field of two orbiting masses M_1 and M_2, one can define an effective potential in a corotating frame that includes the gravitational potential of the two stars and the centrifugal force acting on the test particle (this assumes that the orbit is circular and that the Coriolis force can be neglected, at least initially). This potential has five *Lagrangian points*, where the gradient of the effective potential is zero (i.e., where there is no force in the corotating frame). The three most important ones lie along the line that connects the two stars. Of particular importance is the inner one, referred to as L_1 or the inner Lagrangian point, since the equipotential surface that passes through this point (called the *critical Roche lobe potential*) connects the gravitational spheres of influence of the two stars. This means that, if one star starts to fill its Roche lobe (the part of the critical potential engulfing the star), then matter can flow through the L_1 point into the Roche lobe of the other star. This is the most important way for mass to be transferred from one star to the other and is called *Roche lobe overflow (RLOF)*.

The effective Roche lobe radius R_L depends only on the orbital separation A and the mass-ratio q. For star "1" with mass M_1, it is well approximated by

$$R_\mathrm{L} = \frac{0.49\, q^{-2/3}}{0.6\, q^{-2/3} + \ln(1 + q^{-1/3})} A \qquad (2.3)$$

(Eggleton, 1983), where $q \equiv M_2/M_1$ (and an analogous expression for the effective Roche lobe radius of star '2').

Another useful way of classifying binaries is how the actual radii of the two stars compare to their respective Roche lobe radii. In *detached binaries*, both stars underfill their respective Roche lobes (i.e., have radii smaller than their Roche lobe radii). In this case, no mass transfer via RLOF can take place, and the main gravitational interaction occurs via mutual tides (such as in the Earth–Moon system). In cases where one star has a very strong stellar wind (e.g., for relatively massive stars), a fraction of this wind may be gravitationally accreted by its companion. This alternative but generally much less efficient type of mass transfer is referred to as *wind mass transfer* and can also sometimes be important (e.g., in the case of high-mass X-ray binaries; see Section 2.6).

If one star fills its Roche lobe, the binary is called a *semidetached binary*; these are the systems where mass transfer takes place via RLOF.

Finally, it is also possible that both stars fill or even overfill their Roche lobes. In this case, a common photosphere forms that engulfs both components. These systems are called either *contact binaries* (observationally referred to as W Ursae Majoris stars) or *common-envelope binaries*. For such systems, the other two Lagrangian points, L_2 and L_3, which lie along the axis connecting the two stars but outside their orbit, can become important: if the common envelope reaches either of these two points, then mass can flow through it from the binary to the outside and possibly lead to the formation of a *circumbinary disk* surrounding the whole binary system.

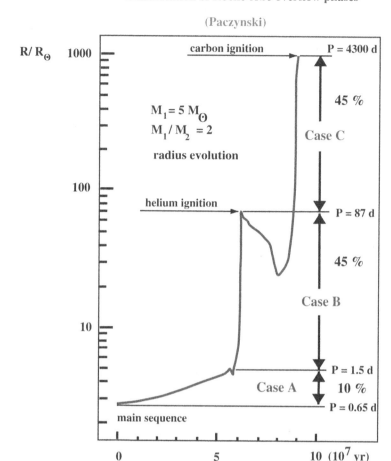

FIGURE 2.1. The evolution of the radius of a $5\,M_\odot$ star as a function of its lifetime to illustrate the ranges in radius and orbital period for the different cases of RLOF phases, as indicated, assuming a $2\,M_\odot$ companion.

2.2.5 Types of binary interactions

Although most stars in the sky are probably in binary system, the only ones we are interested in here are those where at least one of the components transfers mass to the other one by RLOF (typically 30% to 50% of all systems). For the first phase[1] of mass transfer for one of the stars, one distinguishes three cases of mass transfer depending on the nuclear evolutionary state of the star: *case A* (the star is on the main sequence burning hydrogen), *case B* (the star has finished hydrogen burning, but not helium burning in the core), and *case C* (the star has completed core helium burning). Figure 2.1 shows the radius evolution of a $5\,M_\odot$ star as a function of time and indicates the range where the different cases occur. Since the radius of the star expands only very little (a factor of ~ 2) on the main sequence but a factor of more than 10 before helium ignition and again after helium burning, it is much more likely that RLOF starts after the star has continued its main-sequence phase (this assumes a logarithmically flat initial period distribution). On the other hand, because a star spends most of its life on the main sequence, it follows that most binaries observed in the sky have not yet had a strong binary interaction, but

---

[1] If a star experiences more than one mass-transfer phase, the nomenclature quickly becomes complicated, and there is no established standard notation.

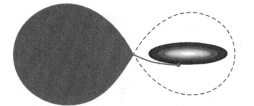

FIGURE 2.2. Cartoon illustrating stable mass transfer.

many of them will do so in the future. This is particularly important when studying the end states of stars and supernovae that probe the late evolutionary phases of a star (see Section 2.4). Note also that quite massive stars ($\gtrsim 20\,\mathrm{M}_\odot$) tend to expand only moderately after helium core burning; hence, for massive stars, case C mass transfer tends to be much less important than case B mass transfer, where most of the expansion occurs.

When RLOF occurs, one has to distinguish between different modes of mass transfer, depending on whether mass transfer is stable or unstable with very different outcomes.

Stable mass transfer

Stable, (quasi-)conservative mass transfer (as illustrated in Fig. 2.2) is the easiest type of mass transfer to understand. In this case, most, but not necessarily all, of the transferred mass is accreted by the companion star, generally leading to a widening of the binary. Mass transfer ends when most of the hydrogen-rich envelope of the donor star has either been transferred to the companion or been lost from the system. The end product will be a hydrogen-exhausted helium star with at most a small hydrogen-rich envelope.[2] Mass accretion will also change the structure of the accreting star. If it is still on the main sequence, the accretor tends to be *rejuvenated* and then behave like a more massive normal main-sequence star. On the other hand, if it has already left the main sequence, its evolution can be drastically altered, and the star may never evolve to become a red supergiant, but explode as a blue supergiant (if it is a massive star; Podsiadlowski and Joss, 1989).

Unstable mass transfer and common-envelope evolution

Mass transfer is unstable when the accreting star cannot accrete all of the material transferred from the donor star. The transferred material then piles up on the accretor and starts to expand, ultimately filling and overfilling the accretor's Roche lobe. This leads to the formation of a common-envelope (CE) system, where the core of the donor and the companion form a binary immersed in the envelope of the donor star (see Fig. 2.3). This typically happens when the donor star is a giant or supergiant with a convective envelope, since a star with a convective envelope tends to *expand* rather than shrink when it loses mass very rapidly (adiabatically), whereas the Roche lobe radius shrinks when mass is transferred from a more massive to a less massive star; this makes the donor overfill its Roche lobe by an ever larger amount and causes runaway mass transfer on a dynamical times scale (so-called dynamical mass transfer).

Once a CE system has formed, friction between the immersed binary and the envelope will make the two components spiral toward each other until enough orbital energy has been released to eject the envelope (Paczyński, 1976). This ends the spiral-in phase and leaves a much closer binary with an orbital period typically between ~ 0.1 and $\sim 10\,\mathrm{d}$, consisting of the core of the giant and a normal-star secondary. In contrast to the stable

---

[2]Stable mass transfer can also occur for an expanding hydrogen-exhausted helium star (so-called case BB mass transfer). In this case, the star is likely to lose a large fraction/most of its helium envelope.

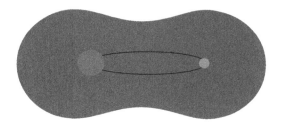

FIGURE 2.3. Cartoon illustrating unstable mass transfer.

RLOF channel, CE evolution tends to produce very short period systems. Indeed, this is believed to be the main mechanism by which an initially wide binary, with an orbital period of possibly many years, can be transformed into a very close binary with an orbital period of hours to days (Paczyński, 1976). Because this spiral-in phase is very short-lived, the immersed companion star will not be able to accrete much matter and will emerge little changed from the CE phase.

Binary mergers

The most dramatic consequence of a CE phase is that the orbital energy that is released in the spiral-in phase is not sufficient to eject the envelope. In this case, the spiral-in process continues until the core of the donor has merged with the companion, producing a *single*, initially rapidly rotating star (such as FK Comae stars).

Binary mergers are one of the least studied phases of binary evolution. Despite this lack of attention, binary mergers are by no means rare events: estimates based on binary population synthesis (BPS) studies suggest that ∼5% to 10% of *all stars* experience a complete merger with a companion star during their evolution. Such binary mergers are likely to be responsible for many eruptive events in the galaxy (e.g., V838 Mon; Tylenda and Soker, 2006).

2.2.6 *Mass-transfer driving mechanisms*

In order to have mass transfer, at least one of the stars has to fill and to continue to fill its Roche lobe. There are two fundamentally different modes of how this is achieved: one is that the donor star tries to expand because of its own internal evolution (this can occur either on a nuclear or a thermal time scale); the other is that the binary system loses angular momentum, causing a shrinking of the orbit.

Expansion of the donor

The simplest driving mechanism is the expansion of the donor star due to its own nuclear evolution. If such a donor transfers mass to a more massive companion, mass transfer will take place on a nuclear time scale, and the mass-transfer rate $\dot{M}$ will be of order $M/t_{\rm nuc}$, where M is the mass of the donor star and $t_{\rm nuc}$ its nuclear time scale.

However, the situation already becomes more complicated if the donor star is initially more massive than the accretor as, in this case, the donor's Roche lobe will shrink initially. To illustrate this, let us consider the case of conservative mass transfer, where all the mass lost by the donor is accreted by the companion, and the total angular momentum of the binary is conserved. The total angular momentum, J, of a binary can be written as

$$J = \frac{M_1 M_2}{M_1 + M_2} \sqrt{G(M_1 + M_2) A} \,. \tag{2.4}$$

If J and $M_1 + M_2$ are constant (conserved), this immediately implies that

$$(M_1 M_2)^2 A = {\rm constant}, \tag{2.5}$$

when $M_1 = M_2$. This implies that, for $M_1 = M_2$, the orbital separation (A) has a minimum. So, if initially $M_1 > M_2$ (assuming star 1 is the donor star), as star 1 transfers mass to star 2 and the masses become more equal, the orbital separation necessarily shrinks. Since the Roche lobe radius scales roughly with the separation (with a relatively weak mass dependence; eq. 2.3), this also implies that the Roche lobe radius of star 1 becomes smaller and that star 1 has to shrink to be just able to fill its Roche lobe. Since this radius will generally be smaller than the star's thermal equilibrium radius, this means that it can no longer remain in thermal equilibrium. On the other hand, a star taken out of thermal equilibrium will try to reestablish thermal equilibrium. In this case, as the equilibrium radius is larger, it will try to expand *against* the shrinking Roche lobe. This expansion will drive more mass across the Roche lobe; indeed, it generally becomes the main mass-transfer driving mechanism. Because the expansion occurs on the thermal time scale of the star, $\dot{M}$ can very roughly be estimated as $M_1/t_{\rm KH}$, where $t_{\rm KH}$ is the thermal (Kelvin-Helmholtz) time scale of the star. This mode of mass transfer is referred to as *thermal time scale mass transfer*.

Once the two masses have equalized and $M_1 < M_2$, the orbital separation and the donor's Roche lobe start to increase and, after another thermal timescale or so, the donor will be able to reestablish thermal equilibrium. Subsequently, mass transfer will occur on its nuclear time scale (*nuclear-driven mass transfer*). Note that the thermal time scale is generally orders of magnitude shorter than the nuclear time scale. This implies that the mass-transfer rate in the initial phase (with $M_1 > M_2$) will be orders of magnitude larger than in the later nuclear-driven phase and that this early phase of *rapid mass transfer* will be concomitantly shorter than the later *slow mass-transfer phase*. This also means that, when one sees a mass-transferring binary, one is always more likely to observe it in the later slow phase.

Angular momentum loss from the system

There are two main causes of angular momentum loss in a binary that cause a shrinking of the orbit and can drive mass transfer: *gravitational radiation* and *magnetic braking*.[3]

Two masses orbiting each other cause a periodic distortion of the space-time continuum around them, that is, they generate a gravitational wave. Since a gravitational wave carries both energy and angular momentum, this means that the binary loses angular momentum. This angular-momentum loss is well described by the standard formula, directly derived from Einstein's theory of general relativity (Landau and Lifshitz, 1959; Faulkner, 1971):

$$\frac{d\ln J_{\rm GR}}{dt} = -\frac{32}{5} G^3 c^5 \frac{M_1 M_2 (M_1 + M_2)}{A^4} , \qquad (2.6)$$

where c is the speed of light. Gravitational radiation as a mass-driving mechanism is only important for fairly close systems, with orbital periods $\lesssim 12\,{\rm h}$ (depending on the masses of the components).

Magnetic braking, on the other hand, is far less well understand. Any star with a convective envelope (like the Sun) loses angular momentum due to the magnetic coupling of its wind to the rotation of the star. For example, in the case of the Sun, the solar wind is in corotation with the Sun's rotation up to about 10 stellar radii. Indeed, the Sun is rotating so slowly because it has been magnetically braked efficiently in its past. The same will be true if such a star is in a binary, except that, for a sufficiently close binary, the rotation of one or both components will be tidally locked to the orbit (i.e., the rotation period is the same as the orbital period). Thus, in this case, the angular momentum

---

[3] Another mechanism in principle is the spiraling in of a binary in a common envelope caused by the friction with the envelope.

carried away in the magnetic wind is ultimately extracted from the binary orbit, not just the rotation of the star, causing the binary system to shrink.

While there is plenty of evidence that magnetic braking is an important angular-momentum loss mechanism in binaries, the details are still quite uncertain. A commonly used formula (Verbunt and Zwaan, 1981) is

$$\frac{dJ_{\rm MB}}{dt} = -3.8 \times 10^{-30} M_2 R^4 \omega^3 \,{\rm dyn\, cm}, \qquad (2.7)$$

where star 2 is the donor star with radius R and ω is its angular rotation speed, assumed to be synchronized with the orbit. Note that generally magnetic braking is included only for stars that have convective envelopes, as stars with radiative envelopes or fully convective stars may not have a magnetic dynamo to produce a magnetically coupled wind.

2.3 Selected current topics in binary evolution

2.3.1 Key uncertainties in modeling binary interactions

Despite the success in modeling a wide variety of different types of binary systems in recent years, there are still major uncertainties in modeling some of the very basic types of binary interactions. In the following, I discuss some of the major uncertainties.

The common-envelope phase

Common-envelope (CE) evolution is undoubtedly the least understood binary interaction (see, e.g., Iben and Livio, 1993; Taam and Sandquist, 2000; Podsiadlowski, 2001). It typically involves the spiraling in of a companion star inside the envelope of a (super-)giant donor star and, in many cases, the ejection of the envelope, transforming an initially wide binary into a very close binary (Paczyński, 1976). Most typically, it occurs when the radius of the mass-losing star expands more rapidly than the radius of its Roche lobe, leading to *dynamical mass transfer*. The conditions for the occurrence of dynamical mass transfer are not very well determined. In BPS simulations, it is still occasionally assumed that mass transfer from a star with a convective envelope is dynamically unstable if the mass ratio q of the mass donor to the mass accretor is larger than a critical value $\simeq 0.7$ (this is the appropriate value for a fully convective polytropic star). However, this does not take into account the stabilizing effect of the compact core of the giant (e.g., Hjellming and Webbink, 1987), and indeed full binary evolution calculations show that a much more typical critical mass ratio is 1.2 (1.1 to 1.3), 70% larger than the commonly used value (see, e.g., Han et al., 2002). Indeed, there is observational evidence, such as from the orbital-period distribution of symbiotic binaries, that a common-envelope phase may lead to drastic mass loss from the system, but without being accompanied by a dramatic spiral-in phase (see Section 2.3.2).

One of the biggest uncertainties in modeling CE evolution is the condition that leads to CE ejection. The most commonly used criterion is that the CE is ejected when the orbital energy times some efficiency factor $\alpha_{\rm CE}$ exceeds the binding energy of the envelope; however, this simple formula involves numerous uncertainties, in particular, whether the binding energy is estimated from a simple analytic expression or from realistic envelope structures obtained from calculated stellar models (e.g., Dewi and Tauris, 2000) and whether the ionization energy should be included in the energy balance (see Han et al., 2002, for discussions). The simplistic application of such a criterion can also lead to the violation of energy conservation (by up to a factor of 10 in some published studies). Moreover, in cases where the spiral-in becomes self-regulated and where all the energy released in the spiral-in can be radiated away at the surface of the common envelope (Meyer and Meyer-Hofmeister, 1979; Podsiadlowski, 2001), an energy criterion is no longer appropriate. Finally, it is also not clear whether this treatment is applicable to CE phases where the donor star has a radiative envelope (as may happen when a star starts to fill its Roche lobe in the Hertzsprung gap). Indeed, it seems more likely that this leads to a frictionally driven wind, at least initially, rather than to a classical CE phase (Podsiadlowski, 2001).

Nonconservative mass transfer

Another major uncertainty in modeling binary evolution is the treatment of nonconservative mass transfer, in particular the amount of specific angular momentum that is lost from the system. Different reasonable prescriptions can give very different evolutionary paths. Depending on how angular momentum is lost from the system, mass transfer can either be stabilizing or destabilizing. Various studies on classes of particular binaries have shown that mass transfer must often be very nonconservative: these include classical Algols (van Rensbergen et al., 2006) and sdB binaries with white-dwarf companions; see Section 2.3.2).

2.3.2 Hot subdwarfs

Hot subdwarfs, or sdB stars, are helium-core-burning stars, typically with a mass $M_{\rm sdB} \simeq 0.5\,M_\odot$, that have lost almost all of their hydrogen-rich envelopes by mass transfer in a binary system (for detailed recent studies, see Han et al., 2002, 2003). In order for the mass donor to be able to ignite helium, the progenitor of the sdB star typically has to fill its Roche lobe near the tip of the first red-giant branch. This can occur either through stable RLOF or in a CE phase. Alternatively, a hot subdwarf can be produced by the merger of two helium white dwarfs if helium is ignited in the merger product (Fig. 2.4 provides an overview of the various channels). This single class of binary system therefore on its own illustrates a large variety of the different types of binary interactions involving a compact component. Since the evolutionary history of sdB stars is so well defined, they are particularly suitable for testing and constraining binary evolution theory. The studies by Han et al. (2002, 2003) have been very successful in reproducing the main observed properties of these systems. Their main conclusions were:

(i) The three major formation channels, involving (a) stable RLOF, (b) CE evolution and (c) binary mergers, are expected to be of comparable importance.
(ii) The orbital period distribution of short-period sdB binaries (with orbital periods $\lesssim 10$ d) is well reproduced if the CE ejection mechanism is very efficient (with $\alpha_{\rm CE} > 0.75$) and a large fraction of the recombination energy can be used in the process.
(iii) In order to reproduce the short-period sdB binaries with white-dwarf companions (in the second, unstable mass-transfer phase; the bottom left channel in Fig. 2.4), the first mass-transfer phase has to be stable and very nonconservative.

Since hot subdwarfs are the dominant source of UV light from an old population in our galaxy, it is only natural to assume that hot subdwarfs in binaries are also an important source of UV light in other old populations, such as early-type galaxies. Indeed, Han et al. (2007) have shown that the same binary model that is so successful in our own galaxy, when applied to early-type galaxies, can reproduce the UV upturn, a long-standing puzzle in the field, without any ad hoc assumptions. It also predicts that the UV upturn should develop after an age of the population of ~ 1 Gyr and only show a fairly weak metallicity dependence (although the latter point is still under investigation). This example illustrates particularly well how important it is to understand the complexities of stellar populations in our galaxy before one can trust the modeling of stellar populations in other galaxies. It would be unreasonable to assume that the stellar populations in external galaxies are much simpler than in the Milky Way, a mistake that is, however, still commonly made.

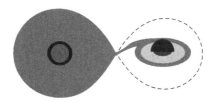
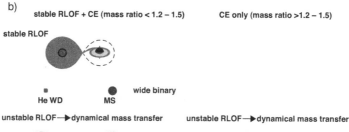

FIGURE 2.4. Binary channels illustrating the formation of hot subdwarfs. *a)* Stable Roche lobe overflow. *b)* CE channels. In addition, a hot subdwarf may form from the merger of two He white dwarfs (or a He and HeCO white dwarf).

2.3.3 Binary mergers

Binary mergers are another topic of major current interest, in particular because present and future transient surveys are likely to detect such mergers in real time (if they have not done so already; see the case of the optical transient in M85; Kulkarni et al., 2007). Since a large fraction of the orbital binding energy is released in a merger, the merger process itself is expected to resemble a faint supernova ("supernova impostor"), such as the outburst of eta Carinae in the 19th century. After the merger, the remnant will be a rapidly rotating supergiant, at least initially rotating near breakup at the equator. This is probably the major channel for producing B[e] supergiants, which are evolved stars rotating near breakup that probably cannot be formed via any other "reasonable" single-star channel (see Podsiadlowski et al., 2006, for further discussions and details).

In general, binary mergers can occur on a large range of time scales. This depends mainly on the structure of the envelope, in particular the density profile, since this determines the friction time scale and hence the spiral-in time scale during the spiral-in phase. This immediately implies that different merger types have to be distinguished depending on whether the envelope is initially convective or radiative.

In the case of *radiative envelopes*, very little mass is contained in the outer low-density envelope (typically less than 1% of the total mass in the outer 50% or the radius of the star), but there is a fairly steep density gradient (see fig. 2 of Podsiadlowski, 2001). This implies that the frictional luminosity, which is proportional to the density, is initially very low, and that there is a long initial "contact" phase without significant spiral-in. This phase may last 100 s to thousands of years. Even a moderate shrinking of the orbit releases enough orbital energy to eject most of the mass surrounding the binary in a frictionally driven wind. If the mass ratio is sufficiently close to 1, this CE phase may be temporary, and the system may survive as a semidetached binary. However, once the immersed companion reaches the high-density layers of the envelope, the spiral-in accelerates and always runs away, ultimately occurring on a dynamical time scale (because of the steep density gradient, envelope expansion can never lead to a self-regulated spiral-in as in the case of convective envelopes).

In contrast, in a *convective envelope* most of the mass of the envelope is contained in the outer parts of the envelope. Hence a convective envelope tends to have a much higher density in the outer parts, but a much shallower density gradient (see fig. 2 of Podsiadlowski, 2001). This implies that the initial spiral-in (after the loss of corotation) is much faster (because of the higher density), but once the envelope has expanded sufficiently the spiral-in can become self-regulated, where the frictional luminosity is low enough to be completely radiated away at the surface rather than drive further expansion of the envelope (Meyer and Meyer-Hofmeister, 1979). The typical time scale for the spiral-in and merging phase in this case is hundreds of years, and we refer to such mergers as *slow mergers* rather than *dynamical mergers*. A third type of merger can occur in dense clusters, where stars can collide directly to merge in *collisional mergers*.

A slow merger model for SN 1987A

Probably one of the most interesting cases of a binary merger is the progenitor of SN 1987A, for which a merger model provides the most likely explanation for the unusual properties of this supernova (Podsiadlowski et al., 2007). However, after the two components in a binary have merged, apart from rapid rotation, there is little direct evidence that the newly formed object once was a binary system. Arguably the best evidence for the former binary nature of the progenitor of SN 1987A stems from the spectacular triple-ring nebula surrounding the supernova, first discovered with the NTT (Wampler et al., 1990) and best imaged with the HST (Burrows et al., 1995). All of the material in the ring nebula was ejected from the progenitor system some ~20,000 yr before the explosion and provides a unique fingerprint of the dramatic events that occurred at that time (see the model by Morris and Podsiadlowski, 2007). Its almost axisymmetric, but very nonspherical structure suggests rotation as an important physical ingredient, but elementary angular-momentum considerations imply that any single massive star that rotated rapidly on the main sequence could not possibly be rotating sufficiently rapidly as a red supergiant to produce the observed asymmetries. A source of angular momentum is required, most likely in the form of orbital angular momentum that was converted into spin angular momentum during a merger event some $\sim 20,000$ yr before the explosion.

The B[e] supergiant R4

The B[e] supergiant R4 in the Small Magellanic Cloud provides perhaps the most convincing evidence that at least some B[e] supergiants are the results of binary mergers. The B[e] component has a luminosity of $L \simeq 10^5 \, L_\odot$ and effective temperature $T_{\text{eff}} \simeq 27,000$ K, and its mass has been estimated to be $12 \, M_\odot$. The B[e] supergiant is a member of a binary system with an orbital period of 21.3 yr, and the companion is an evolved A supergiant (Zickgraf et al., 1996). Moreover, NTT spectra suggest that the system is surrounded by a "cloverleaf" or double bipolar nebula with a characteristic expansion velocity of $v \sim 100 \, \text{km s}^{-1}$ and a size of 2.4 pc, giving the nebula a dynamical age of 10^4 yr (Pasquali et al., 2000). The composition of the nebula is also enriched in nitrogen, suggesting that CNO processed material has been ejected.

The most puzzling feature of the system is, however, the fact that the more evolved A supergiant is much less luminous (roughly by a factor of 10) than the B[e] supergiant, the opposite of what one would expect if the two stars formed at the same time and evolved in isolation. The likely resolution of this Algol-type paradox (Pasquali et al., 2000) is that the system originally consisted of three stars, where the initially most massive component (now the A supergiant) evolved independently from the other two, but where the other two merged after the second most massive component evolved off the main sequence, producing a new object that is more massive and hence more luminous than the original primary, but is still less evolved. The "cloverleaf" nebula could be a combination of an equatorial outflow, associated with an outflow from one of the outer Lagrangian points during the binary contact phase, whereas the perpendicular structure is the result of a bipolar outflow ejected during the merger phase (plus any subsequent wind interaction).

Eta Carinae

One of the most spectacular nebulae in our galaxy is the nebula around η Carinae, which was ejected between 1840 and 1860 in a giant outburst, during which the luminosity of the system reached a luminosity of almost $3 \times 10^7 \, L_\odot$. The mass ejected in the outburst has been estimated to be $\sim 10 \, M_\odot$. Combined with the measured expansion velocities, this gives the nebula a kinetic energy of $\sim 10^{50}$ erg (Smith et al., 2003). This corresponds to about 10% of the energy released in a typical core-collapse supernova, making this a truly remarkable event.

η Car appears to be a member of a relatively wide binary with an orbital period of 5.5 yr in a very eccentric orbit (with an eccentricity $e > 0.6$), leading to periodic X-ray activity when the companion is near periastron. With this relatively long period, it seems unlikely that the binary companion is responsible for the major outburst. There is also evidence for a latitude-dependent wind suggesting that η Car is rapidly rotating.

If it is indeed true that in the major outburst $\sim 10 \, M_\odot$ were ejected with an energy of $\sim 10^{50}$ erg, which would lead to a very dramatic event. Considering that this energy is comparable to the binding energy of a massive early-type supergiant, it seems implausible that this could be caused by an envelope instability, which could at most release the binding energy of the outer envelope, which would be several orders of magnitude less than 10^{50} erg.

All of these facts combined again point in the direction of a merger, quite similar to the case of R4, where two components in a triple system merged to produce the outburst in the mid-1800s. This could:
(i) Provide the energy for the mass ejection; in a dynamical merger, this is expected to be of the order of the orbital energy of the spiraling-in star near the point where it is disrupted. This could easily be as large as $\sim 10^{50}$ erg.
(ii) Cause the spinup of the merger product.
(iii) Provide the excess thermal energy that needs to be radiated away after the merger, driving the posteruption wind with an inferred wind mass-loss rate of $10^{-3}\,M_\odot\,yr^{-1}$.

2.3.4 Symbiotic binaries

A case for quasi-dynamical mass transfer?

Symbiotic binaries, specifically the so-called S-type symbiotics, which contain a giant donor star transferring mass to typically a white-dwarf companion, provide a major challenge to our current understanding of binary interactions. In particular, the orbital-period distribution (~ 100 to $1400\,d$; Mikołajewska, 2007) cannot be explained by present BPS models that only involve the stable and unstable types of mass transfer discussed in Section 2.2. If mass transfer is unstable, leading to a CE and spiral-in phase, one would expect – even with the most optimistic assumptions about the CE ejection process (Han et al., 1995) – much shorter orbital periods than the observed ones. In contrast, if mass transfer is stable, this would generally lead to a widening of the systems. In short, standard BPS simulations predict a gap in the orbital-period distribution where most of the S-type symbiotics are actually observed. This problem was first realized by Webbink (1988), and since then there have been a number of proposals to resolve this problem. Podsiadlowski et al. (1992) proposed a different mode of mass transfer, *quasi-dynamical mass transfer*, which has characteristics of both dynamical and stable mass transfer. The basic idea is that, if the mass ratio is relatively close to 1, the mass-transfer rate is large enough to lead to a common envelope surrounding the binary components (similar to what is seen in contact binaries), but without a significant spiral-in phase. A spiral-in phase is avoided as long as the envelope remains in corotation with the binary since, in this case, there is no friction to drive the spiraling-in. Typically, the envelope remains tidally locked to the binary if the size of the envelope is less than about twice the orbital separation. During this phase, one expects that most of the mass lost from the giant is ultimately lost from the system, mainly through one of the outer Lagrangian points. This may even lead to a circumbinary disk, which itself may tidally couple to the binary and affect the evolution of the binary orbit (e.g., Spruit and Taam, 2001; Frankowski and Jorissen, 2007). Once the mass ratio has reversed sufficiently, the envelope will recede below the critical Roche potential, and the subsequent evolution will resemble the case of stable but most likely still very nonconservative mass transfer.

Wind Roche lobe overflow: a new mode of mass transfer

Recent observations of the symbiotic binary Mira (o Ceti) have provided another example indicating that our present understanding of binary interactions is incomplete. Mira is a so-called D-type symbiotic where the donor star is a Mira variable in a very wide orbit (in the case of Mira, the orbital period has been estimated to be larger than $\sim 1000\,yr$). One would ordinarily not consider such wide binaries as interacting binaries (apart perhaps from some low level of wind accretion). Nevertheless, X-ray observations by Karovska et al. (2005), which were able to resolve the Mira donor star (Mira A), appear to show that Mira A is filling its Roche lobe. Of course, it cannot be the Mira variable itself, as it is a factor of $\gtrsim 10$ smaller than its Roche lobe, but the *slow wind* emanating from it. Mira winds are driven by the pulsations of the dynamically unstable

Mira envelope, but are only accelerated to their terminal speeds at ~5 stellar radii, where dust can form and radiation pressure on the dust can provide the necessary acceleration. If this acceleration region is comparable to the radius of the Roche lobe, the wind flow itself will feel the binary potential and can effectively fill the donor star's Roche lobe. The importance of this new type of *wind Roche lobe overflow* (WRLOF) is that a large fraction of the wind can be transferred to the companion: the mass-transfer rate may exceed the estimate expected from simple Bondi-Hoyle accretion by up to 2 orders of magnitude. This provides an efficient new mechanism for mass transfer in fairly wide binaries (which we have started to refer to as *case D mass transfer*; see Podsiadlowski and Mohamed, 2007). In addition, since any mass that is lost from the system is strongly confined to the orbital plane, producing a disk-like outflow (or even circumbinary disk), this is also likely to have important implications for the shaping of asymmetric planetary nebulae.

Case D mass transfer should also be important for massive stars, as recent calculations (Yoon and Cantiello, 2010) have shown that massive red supergiants can also develop dynamically unstable envelopes – that is, experience the "Mira" phenomenon. Since, in many cases, the consequences of WRLOF are similar to case C mass transfer, this may lead to a dramatic expansion of the period range for which such late phases of mass transfer are important.[4] This may have major implications for various types of supernova progenitors (such as the progenitors of type II-L, IIb supernovae) and even some binary gamma-ray burst progenitor models, which often require such late phases of mass transfer. At the moment, we are only at the beginning of exploring all the consequences of Case D mass transfer.

2.4 Late stellar evolution and supernovae in binaries

2.4.1 Major supernova explosion mechanisms

Generally speaking, a *supernova* is the explosion of a star. For at least a few decades, however, it has been realized that there are (at least) two main supernova explosion mechanisms: *core-collapse supernovae*, involving the final phase in the evolution of a massive star, and *thermonuclear explosions*, most likely related to white dwarfs approaching the Chandrasekhar limit.[5]

Core-collapse supernovae

The evolution of stars and, in particular, massive stars is characterized by an alternation of nuclear burning phases and contraction phases. For a massive star, the evolution ends when it has developed an iron core, surrounded by onion-like structure consisting of shells of increasingly lower mean atomic mass. Since iron is the most stable nucleus (i.e., has the highest nuclear binding energy per baryon), no more energy can be generated by fusing iron with other nuclei. Therefore, if the core exceeds the Chandrasekhar mass for iron, there is no longer a cold hydrostatic equilibrium configuration, and the core has to contract/collapse as it cools and loses its pressure support. Although this contraction may start slowly, it soon accelerates because of a number of instabilities, ultimately reaching free fall. Most of the gravitational energy that is released in the collapse is ultimately converted into neutrinos that, at least initially, freely escape from the core.

This collapse is only stopped once matter reaches nuclear densities ($\rho_{\rm nuc}$) and the strong force becomes important, providing a sudden repulsive force. Because of the initial overcompression of the matter, now mainly composed of neutrons, the core bounces and

---

[4]Note that, for rather massive stars, the orbital period range for case C mass transfer is very narrow and disappears completely for the most massive stars.

[5]The *Chandrasekhar limit* defines the maximum mass at which a zero-temperature, self-gravitating object can be supported by electron degeneracy pressure. For most compositions of interest, this mass is close to $1.4\,M_\odot$.

drives an outward-moving shock into the still infalling outer core. It was once hoped that this shock, which initially carries an energy of $\sim 10^{51}$ erg, could reverse the infall of the outer core into an outflow and drive a *prompt explosion*. But because of the continued photo-disintegration of the infalling material, which requires $\sim 10^{51}$ erg for $0.1\,M_\odot$ of Fe, this energy is quickly consumed, and the shock stalls; it is now believed that this can never drive a successful prompt explosion.

The total energy that is released in the collapse is of the order of the binding energy of the neutron star forming at the center ($GM_{\rm NS}^2/R_{\rm NS} \sim 3\times 10^{53}$ erg $\simeq 0.1 M_{\rm NS} c^2$ for $M_{\rm NS} \simeq 1.4\,M_\odot$ and $R_{\rm NS} \simeq 10$ km). This is several orders of magnitude more than the binding energy of the outer core ($E_{\rm core} \simeq 10^{51}$ erg). However, most of this energy escapes freely in the form of neutrinos that interact only weakly with matter. It has remained one of the most enduring unsolved problems in supernova physics how a fraction ($\sim 1\,\%$) of this energy can be deposited just below the accretion shock and be allowed to accumulate until enough energy is available to drive an explosion. In the presently favored model of *delayed neutrino-driven explosions* (e.g., Janka et al., 2007; Mezzacappa et al., 2007), this may require more than 500 ms, which is extremely long compared to the dynamical time scale of the proto–neutron star (~ 1 ms). If this mechanism fails, matter will continue to fall onto the proto–neutron star and ultimately convert it into a black hole.

Thermonuclear explosions

The second important explosion mechanism has nothing to with massive stars, but is generally believed to occur in accreting CO white dwarfs when their mass approaches the Chandrasekhar mass. When the CO WD mass reaches $\sim 1.37\,M_\odot$, carbon is ignited in or near the center of the white dwarf. Initially this drives convection in the core, transporting the energy outward and radiating it away in the form of neutrinos (this phase of low-level carbon burning, referred to as the *simmering phase*, can last for up to $\sim 10^3$ yr). But, there comes a point when the core is unable to rid itself of the excess nuclear energy, and the burning process becomes explosive. The reason for this nuclear runaway is that the core material is highly degenerate. This means that the core pressure is independent of temperature. Therefore, a rise in central temperature (due to the carbon burning) does not produce an increase in pressure that would limit the increase in temperature (the valve mechanism that keeps burning in ordinary stars, supported by thermal pressure, stable). The further increase in temperature increases the nuclear burning further, producing a runaway process that incinerates a large fraction of the white dwarf and ultimately destroys it completely. In the case of a thermonuclear explosion, unlike the case of core collapse, no remnant is expected, and the energy source is purely nuclear energy ($\sim 10^{51}$ erg). The fact that the energy in the two types of explosion is comparable ($\sim 10^{51}$ erg) is not a coincidence, since, in both cases, the energy scale is set by the binding energy of the core (the CO core in the case of the thermonuclear explosion, and the binding energy of the outer Fe core in the core-collapse case); they are comparable because both are ultimately determined by the same physics of electron degeneracy, which determines the immediate pre-supernova structure.

In the ensuing explosion, a large fraction of the white dwarf is burned, in the inner part completely to *nuclear statistical equilibrium* (NSE), which means mainly to iron-group elements, mostly ^{56}Ni, and incompletely further out, producing mainly intermediate-mass elements, such as ^{28}Si and ^{32}S. The radioactive ^{56}Ni will subsequently decay to ^{56}Co (with a half-life of 7 d), powering the supernova light curve, and ultimately to ^{56}Fe (with a half-life of 77 d). A typical supernova of this type produces $\sim 0.7\,M_\odot$ of ^{56}Fe; hence, these supernovae are believed to be the dominant producers of iron in the universe. Since most of them produce very similar amounts of radioactive ^{56}Ni, the resulting supernova light curves are quite similar, which means that they can be used as standard cosmological distance candles (strictly speaking, "standardizable" distance candles).

Unlike core-collapse supernovae, the physics of thermonuclear explosions is reasonably well understood. One of the lingering uncertainties is how the carbon burning front, which starts as a *deflagration* (i.e., a subsonic burning front), is accelerated into a *detonation* (i.e., a supersonic burning front), which seems to be favored by observations for the majority of thermonuclear explosions.

The main uncertainty, even controversy, is the question of their progenitors, the type of stellar systems in which a CO white dwarf can grow toward the Chandrasekhar mass. I will return to this issue in more detail later.

2.4.2 Supernova classification

The basic classification of supernovae is quite simple: they are classified as type I or type II supernovae, depending on whether they have hydrogen lines in the spectrum (type II) or lack hydrogen lines (type I). For a long time, it was thought that these two observational classes may have a one-to-one relation to the two explosion mechanisms discussed in the last section, core collapse supernovae (type II) and thermonuclear explosions (type I), respectively. However, over the past three decades, it has become clear that this is not the case and that, in principle, both explosion types could come in both observational varieties. As a consequence, the basic classification has become much more complex, requiring the introduction of more and more subtypes.

Main classification scheme

The thermonuclear explosion of a CO white dwarf is now believed to be associated with a *type Ia supernova* (SN Ia). These supernovae have no hydrogen, but strong Si lines. Si and also S are intermediate-mass nuclei that are produced in abundance in the part of an exploding white dwarf that does not burn completely to NSE; hence, this provides a very characteristic signature for a thermonuclear explosion.

In addition to SNe Ia, there are two other subtypes of type I supernovae, type Ib and type Ic. These types are also defined on the basis of their spectroscopic characteristics: both lack hydrogen, but *type Ib supernovae* (SNe Ib) show He lines, whereas *type Ic supernovae* (SNe Ic) lack both H and He lines. Unlike SNe Ia, they produce fairly little ^{56}Ni, are found predominantly in or near star-forming regions, and are therefore believed to be connected with core-collapse supernovae, that is, the explosion of massive stars that have lost their H-rich envelopes and, in the case of SNe Ic, their He-rich layers as well.[6]

There are also several different subtypes of type II supernovae. Unlike SNe I, they are not defined by their spectroscopic properties but by their light curves: their luminosity, measured in a particular waveband (typically B or V) as a function of time. The light curves of *type II-P supernovae* (SNe II-P), where the "P" stands for "plateau," show a long phase, lasting up to ~ 100 d, where the light curve is constant (the *plateau* phase). Their progenitors are most likely massive red supergiants (with a typical mass $\lesssim 20\,\mathrm{M}_\odot$) that experience core collapse. The second, much less common variety, *Type II-L supernovae* (SNe II-L), do not show this plateau, but their luminosity drops off more or less linearly (on a logarithmic scale) after the light-curve peak (hence the letter "L" for "linear"). These are almost certainly also core-collapse supernovae, but in this case, the progenitors must have already lost a large fraction of their H-rich envelopes.

---

[6]It is presently not entirely clear how much He could be present in a SN Ic. Since He is generally nonthermally excited, it requires the presence of a source of energetic photons, such as from the radioactive decay of ^{56}Ni. If the He layer is shielded from this radioactive source, it is possible in principle to hide significant amounts of He.

Complications

Unfortunately, there are many complications going beyond this simple scheme. The progenitor of supernova 1987A (SN 87A) had a large H-rich envelope, but did not have an extended plateau phase, and therefore SN 87A defines a new class of its own. Other supernovae appear to change their type. Supernova 1993J initially looked like a type II supernova but soon transformed into a supernova resembling a SN Ib. As a consequence, this supernova type is now referred to as a *type IIb supernova* (SN IIb). Other subtypes are not directly related to a particular supernova mechanism, but to a supernova-related phenomenon. For example, *type IIn supernovae* stand for supernovae that show narrow H lines (Hα) in emission. This must come from H-rich material in the immediate neighborhood of the supernova, most likely ejected by the progenitor in the not-too-distant past, that was flash-ionized by the first light from the supernova. This is not necessarily related to a particular explosion type; it just implies a particular mass-loss history of the progenitor. In a more extreme version, there may be so much material around the exploding star that the supernova ejecta are rapidly slowed down by the interaction with this material, converting kinetic energy into thermal energy and ultimately radiation. In this case, the light-curve shape itself is determined by this interaction with the circumstellar material. Supernovae that show evidence for such interactions are sometimes referred to as *type IIa supernovae* (SNe IIa), though how this fits into the overall supernova scheme and, in particular, its relation to SNe IIb lack any obvious logic.

Indeed, as this previous discussion shows, the supernova classification scheme has become too complicated and convoluted to be very useful. In fact, sometimes even supernova experts get confused. The problem is that the main scheme is a discrete one, whereas the supernova properties clearly vary in a continuous manner. Indeed, a lot of the diversity of supernova subtypes can be understood as a sequence of increased mass loss.

Thus, the whole sequence

$$\text{SN II-P} \to \text{SN II-L} \to \text{SN IIb} \to \text{SN Ib} \to \text{SN Ic}$$

appears to be a sequence of increased envelope loss, first of the H-rich envelope and then of the He-rich layer. The immediate physical question is what causes this mass loss. While stellar winds may play an important role in some cases, binary interactions are almost certainly even more important, since a large fraction, if not the majority, of all massive stars are in relatively close binaries where the components can interact directly (e.g., by mass transfer causing mass loss and mass accretion and, in the most extreme case, by the complete merger of the binary components, Podsiadlowski et al., 1992). These interactions particularly affect the envelope properties of the massive progenitors and hence help to determine the shapes of the resulting light curves.

2.4.3 The progenitors of type Ia supernovae

Type Ia supernovae (SNe Ia) have been very successfully used as standardizable distance candles and have provided the first indication for an accelerating universe (Riess et al., 1998; Perlmutter et al., 1999). Their use as distance candles relies on the empirical fact that SN Ia light curves appear to form a one-parameter family and that there is a relationship between the supernova peak luminosity and the light-curve width, referred to as the "Phillips relation" (Phillips, 1993), which can be used to infer the peak luminosity and hence the distance. In recent years there has been increasing evidence that not all SNe Ia obey this relation, and there is even some evidence that a subset of SNe Ia have progenitors with a mass exceeding the Chandrasekhar mass (e.g., Howell et al., 2006). Indeed, the nature of the progenitors of SNe Ia is still controversial, and the link between progenitor models and explosion models is presently one of the weakest points in our understanding of SNe Ia.

There is now broad agreement that most SNe Ia are caused by a thermonuclear explosion of a CO white dwarf when its mass approaches the Chandrasekhar mass. At this point, carbon is ignited in the electron-degenerate core. This causes a thermonuclear runaway, leading to the incineration of a large fraction of the white dwarf and ultimately its complete destruction. Unlike the case of a core-collapse supernova, the energy causing the SN Ia explosion ($\sim 10^{51}$ erg) is entirely nuclear energy, and no compact remnant is expected.

What is still uncertain and is indeed controversial is the evolution that produces these Chandrasekhar-mass white dwarfs. The most popular progenitor models fall broadly into two classes, the single-degenerate (SD) model and the double-degenerate (DD) model.

The single-degenerate model

In the SD model, the white dwarf grows in mass by accreting from a nondegenerate companion star (Whelan and Iben, 1973; Nomoto, 1982), where the companion star can be a main-sequence star, a helium star, a subgiant, or even a giant. The main problem with this class of models is that it is generally difficult to increase the mass of a white dwarf by accretion: if the mass-accretion rate is too low, this causes nova explosions and/or helium flashes (Nomoto, 1982) that may eject most of the accreted mass. If the mass-accretion rate is too high, most of the transferred mass must be lost in a disk wind to avoid a merger of the binary, again leading to a low accretion efficiency. There is only a very narrow parameter range where a white dwarf can accrete hydrogen-rich material and burn in a stable manner. This parameter range may be increased if differential rotation affects the accretion process (Yoon and Langer, 2004). One promising channel that has been identified in recent years relates them to supersoft X-ray sources (van den Heuvel et al., 1992). However, it is not clear whether this channel produces a sufficient number of systems to explain the observed SN Ia rate in our galaxy ($\sim 3 \times 10^{-3}$ yr^{-1}; Cappellaro and Turatto, 1997; Han and Podsiadlowski, 2004; Fedorova et al., 2004). On the plus side, a number of binary systems are known that are excellent candidates for SN Ia progenitors: U Sco, RS Oph, and TCrB all contain white dwarfs that are already close to the Chandrasekhar mass, where the latter two systems are symbiotic binaries containing a giant companion (see Hachisu et al., 1999, for a discussion of this channel). However, in none of these cases is it clear whether the massive white dwarf is a CO or an ONeMg white dwarf (the latter is not expected to produce a SN Ia).

The double-degenerate model

In contrast to the SD model, the DD model (Iben and Tutukov, 1984; Webbink, 1984) involves the merger of two CO white dwarfs with a combined mass in excess of the Chandrasekhar mass. This model has the advantage that the theoretically predicted merger rate is quite high (see, e.g., Yungelson et al., 1994; Han et al., 1995; Nelemans et al., 2001), consistent with the observed SN Ia rate and probably the observed number of DD systems discovered by the SPY survey (Napiwotzki et al., 2002). The main problem with this scenario is that it seems more likely that the disruption of the lighter white dwarf and the accretion of its debris onto the more massive one leads to the transformation of the surviving CO white dwarf into an ONeMg white dwarf that subsequently collapses to form a neutron star (i.e., undergoes accretion-induced collapse) rather than experiencing a thermonuclear explosion (e.g., Nomoto and Iben, 1985).

Recently, Yoon et al. (2007) modeled the expected postmerger evolution after a double-degenerate CO merger. They found that the immediate postmerger product was a low-entropy core, the initially more massive CO white dwarf, surrounded by a high-entropy envelope and an accretion disk (from the disrupted lower-mass object). Following the thermal and angular-momentum evolution of the merger product, they showed that the evolution is controlled by neutrino losses at the bottom of the envelope and that, despite the very high core accretion rate, carbon ignition could be avoided under some circumstances. More generally, they concluded that the merger could lead to a thermonuclear explosion if two main conditions are satisfied: (1) carbon ignition must be avoided during the merging process and (2) the mass accretion rate from the surrounding disk must be less than $\sim 10^{-5}\,M_\odot\,\mathrm{yr}^{-1}$.

Thus, there may be some parameter range where the conversion into an ONeMg white dwarf can be avoided (also see Piersanti et al., 2003). This raises the possibility that more than one channel may lead to a SN Ia, perhaps explaining part of the observed diversity. Interestingly, in the context of supersoft sources, the time that elapses between the merger of two CO white dwarfs and the actual supernova is $\sim 10^5$ yr, and, during this phase, the merged object would look like a supersoft source without a companion star (with $T \simeq 0.5\text{--}1 \times 10^6\,\mathrm{K}$ and $L_\mathrm{X} \simeq 10^{37}\,\mathrm{erg\,s}^{-1}$), which could provide a potential test for this channel (Voss and Nelemans, 2008).

2.4.4 Observational tests of SN Ia progenitor models

The detection of circumstellar material

Marietta et al. (2000) simulated the interaction of the supernova ejecta with the companion star and showed that a significant fraction of the envelope of the companion ($\sim$10% to 20% for a main-sequence star or subgiant) is stripped from the companion and mixed with the supernova ejecta. Since this material is likely to be dominated by hydrogen (at least in the classical supersoft channel), this should then lead to easily detectable hydrogen emission lines in the nebular phase of the supernova. To date, with the exception of some extremely unusual supernovae (e.g., SN 2002ic; Hamuy et al., 2003), no hydrogen has ever been detected in a normal SN Ia. Indeed, the present lowest upper limits (less than $\sim 0.01\text{--}0.02\,M_\odot$; Leonard, 2007) now provide a strong constraint on the supersoft model, since these limits are not consistent with the Marietta predictions. However, it now seems that the amount of stripping in the original Marietta calculations may have been significantly overestimated and that it is substantially smaller if realistic stellar models for the companion are employed, marginally compatible with the observational limits (Pakmor et al., 2008; also see Meng et al., 2007).

More encouragingly, Patat et al. (2007) recently found some direct evidence for circumstellar material in a normal SN Ia, SN 2006X. They observed a variation of Na lines immediately after the supernova which they interpreted as arising from the ionization and subsequent recombination of Na in circumstellar material. This strongly favors a SD progenitor for this supernova, but at present only about 10% to 20% of SNe Ia show this behavior, and the interpretation is not unambiguous.

Detecting surviving SN Ia companions

One of the most direct ways of confirming the SD model would be the discovery of the surviving companion that is now a runaway star moving away from the center of a supernova remnant. Indeed, Ruiz-Lapuente et al. (2004) have claimed to have identified such a companion in the Tycho supernova remnant, a G2IV star with a high peculiar velocity. Although such a star is consistent with the SD model, this claim is still quite controversial. Most importantly, the rapid rotation predicted by this model, assuming that the companion is corotating with the orbit at the time of the explosion, is not observed (Kerzendorf et al., 2009). This does not yet rule out that this star is the surviving companion, as there is an alternative model in which the star is a stripped, slowly rotating red giant, but this is not a priori very likely.

2.4.5 Binary evolution and the final fate of massive stars

Although it has been clear for many years that binary interactions strongly affect the structures of stellar envelopes, both by mass loss and by mass accretion, and hence are likely to be a major cause for the observed diversity of supernova subtypes, it has only recently become clear that they can also alter the core evolution and, in fact, the final fate of a star. Generically, one expects that, if mass loss/accretion occurs during an early evolutionary phase, the core continues to evolve subsequently like a less or more massive star. However, this is not true if mass loss occurs after the main-sequence phase.

Black hole or neutron star?

If a star loses its envelope after hydrogen core burning, but before helium ignition (or early during helium core burning; i.e., experiences case B mass transfer), the evolution of the core can be drastically altered. Because of the lack of a H-burning shell, the convective core does not grow during helium core burning, and stars end up with much smaller CO and ultimately iron cores (Brown et al., 1999, 2001). Indeed, because of this, such H-deficient stars formed in binaries as a result of case B mass transfer, with initial masses as high as $50/60\,M_\odot$ (the exact limit depends on the Wolf-Rayet mass-loss rate), are expected to end their evolution as neutron stars rather than as black holes (Brown et al., 2001), the expected fate for their single-star counterparts. Single stars become Wolf-Rayet stars only if their initial mass is larger than ~ 25 to $35\,M_\odot$ (again dependent on the exact mass-loss rate). This is larger than the initial mass where single stars are believed to produce black holes (~ 20–$25\,M_\odot$; e.g., Fryer and Kalogera, 2001). Since the formation of a slowly rotating black hole is not a priori expected to be associated with a bright supernova (as the whole star can just collapse into a black hole), this has the important implication that all normal H-deficient core-collapse supernovae (SNe Ib/Ic) may require a close binary companion.

Electron-capture supernovae

Another mass range where binary interactions can drastically change the final fate of a massive star is near the minimum mass for stars to explode as supernovae (around $7\,M_\odot$, where the exact value depends on the amount of convective overshooting and the metallicity of the star). Single stars in this mass range experience a second dredge-up phase when they ascend the asymptotic giant branch (AGB), where a large fraction of the H-exhausted core is dredged up and mixed with the envelope. This reduces the core mass at the end of the AGB phase; as a consequence, single stars as massive as $10/11\,M_\odot$ probably produce ONeMg white dwarfs rather than a supernova. In contrast, if such stars lose their H-rich envelopes due to a binary interaction before reaching the AGB, they end up with much larger He cores and are likely to produce an electron-capture (e-capture) supernova.

An e-capture supernova occurs in a very degenerate ONeMg core, long before an iron core has developed, and is triggered by the sudden capture of electrons onto Ne nuclei, taking away the hydrostatic support provided by the degenerate electrons (Nomoto, 1984). This occurs at a characteristic density ($\sim 4.5 \times 10^9\,\mathrm{g\,cm^{-3}}$; Podsiadlowski et al., 2005), which can be related to a critical precollapse mass for the ONeMg core of $\sim 1.37\,M_\odot$.

A dichotomous scenario for neutron-star formation and neutron-star kicks

The postcollapse (i.e., NS) mass of an e-capture supernova depends on the equation of state for matter at nuclear density, but has been estimated to be close to $1.25\,M_\odot$ (Podsiadlowski et al., 2005). This is significantly lower than the NS mass from iron core collapse ($\sim 1.35\,M_\odot$). Indeed, Schwab et al. (2010) analyzed the NS mass distribution of pulsars with very well-determined masses and found a bimodal NS mass distribution with a sharp peak at the e-capture mass of $1.25\,M_\odot$ and a broader distribution around $1.35\,M_\odot$, lending further support for two NS formation channels.

This also has important implications for neutron-star birth kicks. It has long been known that young pulsars have rather high space velocities. The interpretation of these high velocities is that they must have received a large kick when they were born in the supernova because of an asymmetry in the explosion mechanism. In the most recent study of pulsar birth kicks, Hobbs et al. (2005) found that the natal kick distribution is well approximated by a Gaussian kick distribution with a velocity dispersion of $265\,\mathrm{km\,s^{-1}}$, with no evidence for a low-velocity tail in the distribution. On the other hand, there has been mounting evidence that not all neutron stars receive large kicks at birth (e.g., the eccentricity distribution of Be X-ray binaries [Pfahl et al., 2002b]; the problem of pulsar retention in globular clusters [Pfahl et al., 2002a, and references therein]; and the properties of the double pulsar, J0737–3039 [Podsiadlowski et al., 2005]). It is tempting to associate the two NS formation channels with different pulsar kick distributions (Podsiadlowski et al., 2004a; also see van den Heuvel, 2004). Since, in an e-capture supernova, the whole core collapses, it is relatively easy to eject the rest of the loosely bound envelope. This probably leads to a "fast" supernova explosion, where the instabilities in the accretion shock that are presently the best candidates for the origin of supernova kicks (Blondin and Mezzacappa, 2006, 2007; Foglizzo et al., 2007) did not have time to develop. Indeed, because e-capture supernovae are expected to occur mainly in binary systems (since, as discussed earlier, a binary may be required to prevent the second dredgeup; Podsiadlowski et al., 2004a), this would also naturally explain why there is no evidence for low-kick neutron stars in the single pulsar population.

2.4.6 Hypernovae and gamma-ray bursts

In the late 1990s it was realized that, in addition to the normal core-collapse and thermonuclear explosions, there are more energetic supernovae with an energy output $\gtrsim 10^{52}$ erg, that is, they are at least 10 times as energetic as a normal supernova. These are now often referred to as *hypernovae*, with the prototype being SN 1998bw (Iwamoto et al., 1998). Interestingly, at least some of the hypernovae are associated with long-duration gamma-ray bursts (LGRBs), the most powerful explosive events known in the universe. On the other hand, the nature of their progenitors is almost completely unknown. It is clear that LGRBs are relatively rare events (see Podsiadlowski et al., 2004b, and references therein). After correcting for beaming, the LGRB rate in a typical galaxy like ours is $\sim 10^{-5}$ yr^{-1}, where the estimate has an uncertainty of about an order of magnitude. This rate is in fact comparable to the hypernova rate (Podsiadlowski et al., 2004b). This rate implies that fewer than about 1 in 1,000 core-collapse supernovae produce a LGRB and that the production of a LGRB requires some special circumstances: that is, the progenitors cannot just be more massive single stars, but stars that are unusual in some respects, such as, because of a combination of low metallicity and rapid rotation (Yoon and Langer, 2005; Woosley and Heger, 2006) or because of binary evolution effects, as I now discuss.

In the presently favored collapsar model (Woosley, 1993; MacFadyen and Woosley, 1999), a LGRB is triggered by the collapse of a rapidly rotating massive core. In order for the collapse to proceed via a disk phase, the specific angular momentum in the core has to be larger than a few 10^{16} cm^2 s^{-1}. In the case of a single star, this requires that the star for some reason must not have been spun down during its evolution, the normal fate for most massive stars. Alternatively, in a binary scenario it may have been spun up by various binary processes.

Tidal spinup models

In many respects the simplest binary process that can produce a rapidly rotating helium star is tidal spinup since, in a tidally locked binary, a star can be spun up (or down) until its spin angular velocity is equal to the orbital angular velocity (e.g., Izzard et al.,

2004). Simple angular-momentum estimates suggest that this requires an orbital period shorter than ~10 hr (Podsiadlowski et al., 2004b). In practice, this means that the companion is most likely a compact object (a neutron star or a black hole). Such systems are indeed observed; for example, the X-ray binary Cygnus X-3 contains a Wolf-Rayet star in orbit with a neutron star or black hole (van Kerkwijk et al., 1992). In the case of Cygnus X-3, it is not clear whether the Wolf-Rayet star will ultimately collapse to form a black hole and produce a LGRB. Nevertheless, similar systems that produce a black hole are likely to exist, indeed with a rate compatible with the LGRB rate.

Detmers et al. (2008) have recently modeled the evolution of such systems and, in particular, the spinup evolution of the companion star. Indeed, they found that tidal spinup of the core is possible. However, they also showed that, at solar metallicity, the expected strong wind from the Wolf-Rayet star leads to a significant widening of the binary and the ultimate *spindown* of the companion. As a consequence, this channel is only likely to work at low metallicity when the wind mass-loss rate is expected to be much lower.

In cases where the Wolf-Rayet companion filled its Roche lobe, Detmers et al. (2008) found that it was then likely that the Wolf-Rayet star would merge completely with the compact companion, quite similar to another LGRB model proposed originally by Fryer and Woosley (1998).

Merger models

Most binary models for LGRBs proposed to date involve the merger of two stars. This is a particularly efficient way for converting *orbital* angular momentum into *spin* angular momentum. A variety of different types of binary mergers can be distinguished depending on the nature of the components and the cause of the merging.

The most widely discussed merger models consider the merging of two compact cores inside a common envelope (e.g., Fryer and Woosley, 1998; Fryer and Heger, 2005), where one of the cores can already be a compact star (e.g., a neutron star or a black hole). One of the most interesting cases involves the merger of the nondegenerate cores of two massive stars. This occurs when the initial masses of the binary components are very close (typically within ~5% to 10%) and both stars already have a compact core at the time of the binary interaction, leading to a *double-core* common-envelope phase (as first discussed by Brown, 1995; also see Dewi et al., 2006). Statistically, it is more likely that the initially more massive star has already developed a CO core while the less massive star has a less evolved He core. When the two cores merge, this will lead to a rapidly rotating object consisting of a CO core with a helium envelope. Since the merging process is driven by friction within the common hydrogen-rich envelope, it is not entirely clear how the merger can proceed to its conclusion and still eject the hydrogen-rich envelope completely at the same time.

Explosive common-envelope ejection

A rather different route to a LGRB was discovered by N. Ivanova (Ivanova, 2002; Podsiadlowski et al., 2010) when studying the slow merger of two massive stars after helium core burning (i.e., involving case C mass transfer). This evolution occurs when mass transfer from a red supergiant to a less massive companion is unstable. This leads to a common-envelope phase where the secondary spirals in are inside the envelope of the original supergiant. At some stage during this spiral-in, the immersed companion itself will fill its Roche lobe and start to transfer mass to the core of the supergiant (as illustrated in Fig. 2.5). Most importantly, the stream emanating from the secondary, which initially is mainly composed of hydrogen-rich material, can penetrate deep into the helium core of the supergiant, eroding it in the process (Ivanova and Podsiadlowski, 2003).

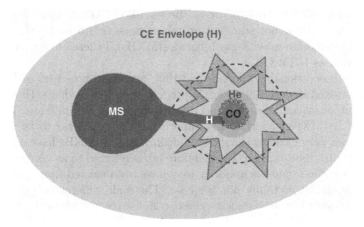

FIGURE 2.5. Schematic illustration of the process of explosive common-envelope ejection. The H-rich stream from the Roche lobe–filling immersed companion penetrates deep into the core of the primary, mixing hydrogen into the helium-burning shell. This leads to a thermonuclear runaway ejecting the helium shell and the hydrogen-rich envelope. (From Podsiadlowski et al., 2010.)

If the initial mass ratio of the binary is relatively large, it can happen that, at some point during the merging process, the H-rich material from the secondary is mixed into the hot helium-burning shell (with a temperature of a few 10^8 K). This leads to a nuclear runaway and the rapid expansion and ultimately ejection of the He-rich shell, and with it of the total H-rich envelope. This mechanism of "explosive" merging provides a new mechanism for ejecting a common envelope. Unlike the standard case (Paczyński, 1976), where the ejection energy is orbital energy, the energy source here is nuclear energy. In order to eject the envelope in a typical case, only a few percent of a solar mass of H-rich material has to be burned, less than found in the actual calculations. Explosive common-envelope ejection provides a new mechanism for CE ejection that can operate even when the orbital energy is insufficient to eject the envelope otherwise. The stream penetrates particularly deep into the core when the entropy of the secondary is low. This favors relatively low-mass companions with masses less than $\sim 3\,{\rm M}_\odot$.

In the context of LGRBs, one implication is that this process *predicts* that both the hydrogen envelope and the helium-rich layer are ejected and that the final product is a pure CO core, consistent with the constraint that all LGRB-related supernovae to date are of type Ic. Furthermore, the CO core is moderately spun up in the phase where the stream interacts with the helium core (i.e., *before* the explosive phase). In our calculations, the final specific angular momentum of the core was $\sim 10^{16}$ cm^2 s^{-1}, consistent with the angular-momentum requirement in the collapsar model. Simple estimates for the rate of this channel suggest a rate of $\sim 10^{-6}$ yr^{-1}, which would be somewhat too low to explain the total local LGRB rate; however, this rate is expected to be higher at lower metallicity, where there is a larger orbital-period range for case C mass transfer, and it could be much higher if case D mass, as discussed in Section 2.3.4, also contributes to this channel.

2.5 Low- and intermediate-mass X-ray binaries and the formation of millisecond pulsars

Generically, X-ray binaries are binary systems where a star transfers matter either by Roche lobe overflow or a stellar wind to a compact companion star, which typically is a neutron star or a black hole. Traditionally, X-ray binaries are divided into two classes: low-mass X-ray binaries (LMXBs, with donor masses $\lesssim 1.5\,{\rm M}_\odot$), the topic of this section; and high-mass X-ray binaries (HMXBs, with donor masses $\gtrsim 10\,{\rm M}_\odot$), the topic of the next section. As I discuss, many of the so-called LMXBs most likely descended from systems where the donor star had an initial mass of 1.5 to 4.5 $M_\odot$, that is, they really involve intermediate-mass X-ray binaries (IMXBs); I therefore refer to these in this section collectively as L/IMXBs.

The detailed observational properties and their accretion processes of L/IMXBs and HMXBs are discussed in detail in other chapters of these proceedings. Here I would just like to comment on a common misconception considering the different galactic space distributions of L/IMXBs and HMXBs. Superficially, they appear to be very different: HMXBs are found very close to the galactic disk, while L/IMXBs have a much broader galactic latitude distribution; they are therefore often referred to as "bulge sources," and it has even been argued that the majority may have been ejected from globular clusters. However, this is almost certainly not the case. The main difference between these two classes is the difference in kick velocity these systems receive when the neutron star is born and the subsequent lifetime of the system. As discussed in Section 2.4, most neutron stars are believed to receive a large kick when they are born in a supernova. In fact, in many cases, the binary system may become unbound as a consequence. In cases where the system remains bound, a binary with a low-mass companion will receive a larger *systemic* kick than a binary with a high-mass companion (assuming the same NS kick momentum). In addition, because of the longer lifetime, the X-ray binary with the low-mass companion will live longer and hence be able to travel farther away from the galactic plane than a system with a high-mass companion and a much shorter lifetime. Brandt and Podsiadlowski (1995) modeled the galactic space distribution of LMXBs and HMXBs and showed that, assuming they are all born in the galactic plane and adopting the same NS kick distribution for both cases, they could perfectly reproduce the observed space distributions of both classes without the need to invoke different initial populations.

2.5.1 *L/IMXBs formation scenarios*

L/IMXBs are typically rather bright X-ray sources (with $L_X > 10^{36} \mathrm{erg\,s}^{-1}$) and can be seen throughout the galaxy. There are only about 200 bright L/IMXBs; this already indicates that they must be rather rare objects. Indeed, the formation of an L/IMXB requires a couple of rather improbable steps. Since the typical orbital separation in L/IMXBs is 0.1 to 10 $R_\odot$, much smaller than the size of the NS progenitor, this implies that their formation requires a CE and spiral-in phase. However, in the case of an LMXB, this implies that the orbital energy released in the spiral-in of the low-mass companion inside the envelope of the NS progenitor has to be enough to eject a rather massive envelope. This is energetically challenging and implies that, in most cases, such binaries are likely to merge completely. Only if the NS progenitor is a very extended red supergiant at the beginning of mass transfer (with a low envelope binding energy) can a low-mass star eject the supergiant's envelope. Even if the system has passed this first hurdle, it still has to survive the supernova in which the neutron star is born. Since generally more than half the total mass of the system is ejected in the supernova, most systems with low-mass companions are likely to be disrupted in the supernova. On the other hand, systems that receive a supernova kick in the right direction (against their orbital motion) have a higher probability of remaining bound (e.g., Brandt and Podsiadlowski, 1995). These constraints are less severe for intermediate-mass companions, strongly favoring them in BPS studies (Pfahl *et al.*, 2003).

L/IMXBs in globular clusters

It has long been known that the number of L/IMXBs per unit mass is much larger in globular clusters (GCs) than in the galaxy, implying a formation rate (per unit stellar mass) that is a factor of ∼20 larger in the GC population and suggesting different formation channels. Indeed, since the stellar densities in GCs are so high, this suggests *dynamical* formation channels as the explanation for the higher rate.

The dynamical formation process that has been studied the longest involves the *tidal capture* of a star by a passing neutron star (Fabian et al., 1975). If a neutron star comes close to a normal star, it induces tidal oscillations in its envelope. The energy of these oscillations is taken from the relative orbital energy of the two stars. Therefore, if the encounter is close enough (the neutron star typically has to come within ∼3 stellar radii), enough energy can be taken out of the orbit to change the relative orbit from an unbound orbit to a bound orbit, that is, produce a bound NS binary.

The encounter of a neutron star with a preexisting binary in a globular cluster could be an even more efficient process for producing a NS binary (see e.g, Davies, 1995). This leads to a complicated three-body interaction, the most likely outcome of which is the ejection of the lightest component of the unstable triple, typically one of the stars in the initial binary, leaving the NS with a new companion. Even though primordial binary systems in globular clusters may be relatively rare, the cross section for a three-body interaction is the orbital separation of the initial binary, which can be much larger than the cross section required for a tidal capture. Note, however, that this process is likely to produce relatively wide NS binaries. Whether tidal capture or three-body (and possibly even four-body) interactions are the dominant formation for L/IMXBs in globular clusters is still the source of much current debate.

2.5.2 *The origin of millisecond pulsars*

Among the population of radio pulsars, there are ∼200 radio pulsars with very short spin periods (as short as 1.4 ms) and relatively weak magnetic fields ($\lesssim 10^9$ G). Unlike the bulk of the normal radio pulsar population, the majority of these are found in binary systems. It is generally believed that these are recycled pulsars that achieved their short spin periods by accretion of mass and angular momentum from a companion star (see, e.g., Bhattacharya and van den Heuvel, 1991).

However, it has also now been established for more than two decades (Ruderman et al., 1989) that the standard model, where their evolution is considered similar to the evolution of cataclysmic variables (CVs; systems similar to LMXBs but where the compact object is a white dwarf), cannot explain some of the main observed characteristics of L/IMXBs.

Problems with the standard model

Two of the main problems with the standard model for LMXBs (e.g., Ruderman et al., 1989; Bhattacharya and van den Heuvel, 1991) are that it cannot explain their distributions of orbital periods (which is qualitatively different from the CV distribution) and X-ray luminosities. The typical luminosities, and hence mass-transfer rates, appear to be about an order of magnitude larger than the standard model predicts. A further problem with the standard model, referred to as the "birthrate" problem for ms pulsars, is that the birthrate of LMXBs appears to be a factor of 10 to 100 lower than the birthrate of ms pulsars both in the galactic disk (Kulkarni and Narayan, 1988; Johnston and Bailes, 1991) and in globular clusters (Fruchter and Goss, 1990; Kulkarni et al., 1990).

2.5.3 *The case of Cyg X-2: the importance of intermediate-mass X-ray binaries*

Until about a decade ago, intermediate-mass X-ray binaries (IMXBs) had received fairly little attention (see, however, Pylyser and Savonije, 1988, 1989). This changed with several new developments. First Davies and Hansen (1998), studying dynamical interactions in globular clusters, found that IMXBs are much easier to form dynamically than LMXBs and speculated that these IMXBs, which do not exist in globular clusters at the present epoch, might be the progenitors of the observed ms pulsars rather than the presently observed LMXBs.

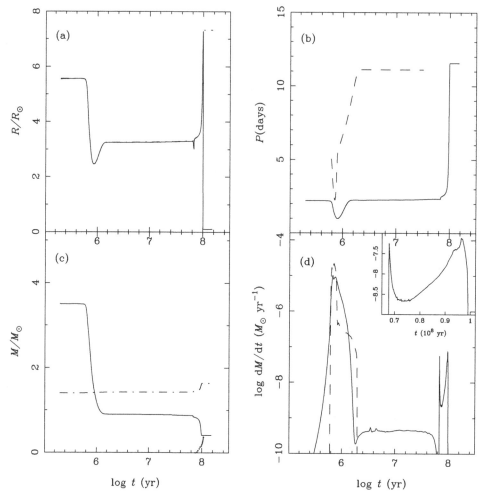

FIGURE 2.6. Key binary parameters as a function of time (with arbitrary offset) in a binary calculation illustrating the possible evolution of Cygnus X-2. The calculation assumes that mass transfer started when the secondary was near the end of the main sequence (case AB mass transfer). a) Radius (solid curve) and Roche lobe radius (dot-dashed curve) of the secondary; b) the orbital period (solid curve); c) the mass of the secondary (solid curve) and the primary (dot-dashed curve); d) the mass-loss rate from the secondary (solid curve); the inset shows a blowup of the second slow mass-transfer phase (hydrogen shell burning). The dashed curves in b) and d) show the orbital period and mass-transfer rate for a case B calculation. (From Podsiadlowski and Rappaport, 2000, Reproduced by permission of the AAS.)

The second development was a re-assessment of the evolutionary status of the X-ray binary Cyg X-2. Spectroscopic observations of Cyg X-2 by Casares et al. (1998) combined with the modeling of the ellipsoidal light curve (Orosz and Kuulkers, 1999) showed that the secondary in Cyg X-2 was a low-mass star of $\sim 0.6 \pm 0.13\,M_\odot$ that was much hotter and almost a factor of 10 too luminous to be consistent with a low-mass subgiant with an orbital period of 9.84 days (see Podsiadlowski and Rappaport, 2000). The explanation for this paradox was found independently by King and Ritter (1999) and Podsiadlowski and Rappaport (2000) (also see Kolb et al., 2000; Tauris et al., 2000), who showed that the characteristics of Cyg X-2 can best be understood if the system was the descendant of an IMXB where the secondary initially had a mass of $\sim 3.5\,M_\odot$ and lost most of its mass in very nonconservative case AB or case B mass transfer (see Fig. 2.6). Thus, Cyg X-2 provides observational proof that IMXBs can eject most of the mass that is

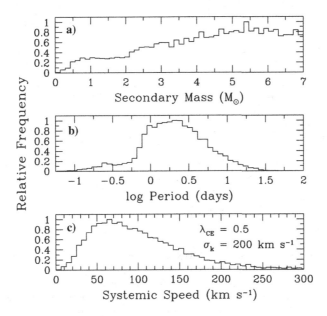

FIGURE 2.7. The initial properties of L/IMXBs at the beginning of the mass-transfer phase. From *a)* to *c)*: initial secondary mass distribution, orbital period distribution and system velocity. (From Pfahl *et al.*, 2003, Reproduced by permission of the AAS.)

being transferred from the secondary (perhaps in the form of an equatorial outflow, as is observed in SS 433; Blundell *et al.*, 2001) and subsequently resemble classical LMXBs. This immediately suggests that a large fraction of so-called LMXBs may in reality be IMXBs or their descendants.

2.5.4 Modeling the L/IMXB population

In order to assess the importance of IMXBs, Podsiadlowski *et al.* (2002) [PRP] and Pfahl *et al.* (2003) carried out a comprehensive study consisting of two parts. The first involved a series of ∼150 binary evolution calculations using a realistic binary evolution code (PRP). These calculations covered the mass range of 0.8 to 7 $M_\odot$ and all evolutionary phases from early case A to late case B mass transfer. The second part involved the integration of these binary sequences into a state-of-the-art Monte Carlo BPS code that allowed a detailed comparison of the calculations with observations (Pfahl *et al.*, 2003).

Figure 2.7 shows the initial properties of systems that become X-ray binaries. As the top panel shows, the majority of X-ray binaries initially have intermediate-mass companions, simply because these are much easier to form. However, because the initial mass-transfer phase for IMXBs is very rapid (see Fig. 2.6), systems are most likely to be observed after the secondaries have transferred/lost a large fraction of their mass and essentially look like standard "LMXBs" (although they may show evidence for CNO processing and He enrichment in their envelopes). The panels of Figure 2.8 show the distributions of the secondary mass, orbital period, and neutron-star mass-accretion rate during the X-ray binary phase at the current epoch. Unlike the initial mass distribution, the mass distribution at the current epoch is dominated by relatively low-mass systems, and there are hardly any systems above ∼2 $M_\odot$ because of the initial high mass-transfer rate for IMXBs. The orbital-period distribution shows no period gap and extends to very short periods. Finally, the luminosity distribution displays a fairly strong peak around $5 \times 10^{-11}\,M_\odot\,\mathrm{yr}^{-1}$ and has a sharp cutoff at ∼$10^{-11}\,M_\odot\,\mathrm{yr}^{-1}$.

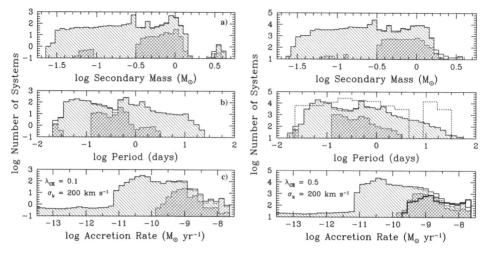

FIGURE 2.8. The present properties of L/IMXBs. From a) to c): secondary (donor) mass distribution, orbital-period distribution, and NS mass-accretion rate. The panels on the left assume a lower (but probably more realistic) efficiency for the CE ejection process than the panels on the right (parametrized using a lower envelope structure parameter λ_{CE}). The double-hatched regions indicate persistent X-ray sources, the single-hatched regions transients. The thick, solid distribution on the bottom right panel illustrates how the inclusion of X-ray irradiation effects may affect the theoretical $\dot{M}$ distribution (From Pfahl et al., 2003, Reproduced by permission of the AAS.)

While these distributions show many of the characteristics of the observed distributions of "LMXBs," in fact more so than a model that only includes CV-like systems, there are still some fairly obvious discrepancies. First, there are too many short-period systems to be consistent with the observed period distribution (e.g., Ritter and Kolb, 1998). Second, although the distribution of mass-accretion rate (and hence X-ray luminosity) has a sharp cutoff at $\sim 10^{-11}\,M_\odot\,\mathrm{yr}^{-1}$, as is desirable, the peak in the distribution is probably too low by about an order of magnitude.

Irradiation-driven mass-transfer cycles

Our binary evolution calculations at the moment do not account for irradiation effects of the secondary that can dramatically change the evolution of the system either by irradiation-driven winds (Ruderman et al., 1989) or irradiation-driven expansion of the secondary (Podsiadlowski, 1991). Podsiadlowski (1991) showed that if a star with a convective envelope is irradiated by a sufficiently high X-ray flux (so that hydrogen is being ionized), it will try to expand by a factor of 2 to 4. If such a star is already filling its Roche lobe, this will drive mass transfer at a highly enhanced rate on a time scale determined by the thermal time scale of the convective envelope. These early calculations assumed spherical illumination; this is unrealistic in a binary situation because, in this case, the energy that drives the expansion can, in principle, be redirected to the unilluminated side, where it can be radiated away. But even then one expects a moderate expansion that can lead to mass-transfer cycles (Hameury et al., 1993), characterized by relatively short phases of enhanced mass transfer and long detached phases. These would not only increase the mass-transfer rates during the X-ray–active portion of the cycles but also reduce the duration of the X-ray–active lifetime of these systems by a proportionate amount. This could provide a simultaneous solution of the X-ray luminosity and the birthrate problem.

To simulate this in our BPS simulations, we have taken the binary sequences calculated with our standard assumptions, but assumed that the mass-transfer rate was a factor of 10 larger during X-ray–active phases than in the standard calculation and was interrupted

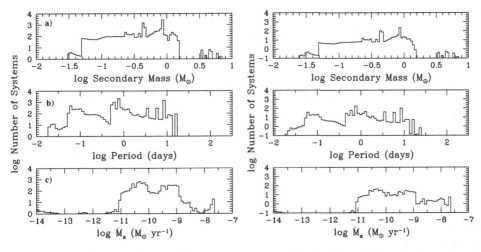

FIGURE 2.9. a) The distributions of secondary mass, b) orbital period, and c) neutron-star mass-accretion rate at the current epoch for a standard reference model (left panels) and a model simulating irradiation-driven mass-transfer cycles. In the latter, it is assumed that during phases of irradiation-driven mass transfer, the mass transfer is a factor of 10 larger than the secular mass-transfer rate with correspondingly shorter lifetimes for the X-ray–active phase. The standard model on the left predicts ∼2000 observable X-ray binaries in the galaxy, whereas the model with irradiation cycles predicts ∼200 systems, more consistent with the observed sample. (From Podsiadlowski et al., 2003b.)

by long X-ray–quiet phases, which we calculated in such a way that the secular evolution of the systems (e.g., the orbital period, secondary mass as a function of time) were the same as in the standard sequences; this procedure is consistent with the results of Hameury et al. (1993) but in some sense assumes that irradiation effects are relatively modest.

The panels on the right in Figure 2.9 show the results of such a simulation. As one would expect, there are now many systems with higher mass-transfer rates, consistent with observations, and the number of observable systems is reduced by a factor of 10 to ∼200, again in better agreement with the actual numbers.

2.6 High-mass X-ray binaries

2.6.1 Systems with neutron stars

The mass-transfer process in high-mass X-ray binaries (HMXBs) is in some respects quite different from those in L/IMXBs. If the compact object is a neutron star, Roche lobe overflow will always become unstable because of the large mass ratio. In general, one distinguishes two main modes of mass transfer: *atmospheric Roche lobe overflow* and *wind mass transfer*.

Atmospheric Roche lobe overflow

Because there is an atmosphere above the photosphere of a star, mass transfer generally starts somewhat before the photosphere reaches the Roche lobe. Since, for low-mass main-sequence stars (but not for giants!), the radial extent of this atmosphere is very small, this initial phase is not usually very important. However, for intermediate-mass stars, and even more so for high-mass stars, the extent can be substantial, and a large amount of mass can be transferred in this phase of atmospheric Roche lobe overflow (see fig. 12 in PRP). In particular, if mass transfer ultimately becomes unstable, this is the only phase in which such a system can be observed as an X-ray source. The mass-transfer rate, in this phase, can be orders of magnitude larger than the Eddington mass-accretion rate.[7] Therefore, these systems typically appear as very bright X-ray sources.

Wind mass transfer

In addition, massive stars have strong stellar winds, and part of this wind can be accreted by the companion. Because of gravitational focusing, the accretion cross section is generally much larger than the geometric cross section of the accretor; it is given by the *Bondi-Hoyle accretion radius*

$$R_{\rm BH} = \frac{2GM}{v^2}, \qquad (2.8)$$

where M is the mass of the accretor and v the relative velocity between the wind and the accreting object (assumed to be supersonic here). For a spherically symmetric wind emanating from the donor with mass-loss rate $\dot{M}_{\rm wind}$, it is easy to show that the fraction of the wind accreted by the companion is given by

$$\frac{\dot{M}_{\rm acc}}{\dot{M}_{\rm wind}} = \left(\frac{v_{\rm orb}}{v_{\rm wind}}\right)^4 \left(\frac{M_{\rm acc}}{M_{\rm acc} + M_{\rm donor}}\right)^2, \qquad (2.9)$$

where $v_{\rm orb}$ is the orbital velocity and the subscripts "acc" and "donor" refer to the donor and accretor. For an HMXB, the last factor in this expression is generally a small number, and hence only a small fraction of the wind can be accreted in most cases. As a consequence, wind-accreting HMXBs tend to be fainter than those that accrete by atmospheric Roche lobe overflow.

Be X-ray binaries

A third class of HMXBs are *Be X-ray binaries*, where the mass donor is a Be star. Be stars are rapidly rotating stars, rotating at a speed close to breakup. Be X-ray binaries are believed to be systems in which the first mass-transfer phase (in the phase before the formation of the neutron star) was stable and the companion star was able to accrete a substantial amount of mass, being spun up to breakup in the process. Because of their large rotation rates, Be stars tend to have substantial winds and eject matter episodically. This ejection is strongly concentrated toward their equatorial planes. Once a neutron star has formed in a supernova, the neutron star can accrete part of this wind and appear as an X-ray source. In particular, if the orbit of the neutron star is eccentric (e.g., due to a supernova kick), these X-ray outbursts tend to be very transient. Be X-ray binaries probably form the largest subgroup of HMXBs, but because of their transient nature, their total number is somewhat uncertain.

2.6.2 The final fate of HMXBs containing neutron stars

As already mentioned, mass transfer in an HMXB is generally expected to become unstable because of the large mass ratio. The reason is not only that the mass ratio is above some critical value, but also that there is an additional instability, the *Darwin instability*. The neutron star orbiting the massive star induces a tide in the massive star that will try to spin it up, so that ultimately it would spin in corotation with the neutron star's orbit. However, for a sufficiently large mass ratio, there is not enough angular momentum in the

---

[7] The Eddington mass-accretion rate for an accreting object is the rate at which it has to accrete so that the accretion luminosity equals the *Eddington limit* at which radiation pressure stops accretion: $\dot{M}_{\rm Edd} = 4\pi cR/\kappa$, where R is the radius of the accreting object, and κ is the opacity of the material being accreted. For a neutron star of $1.4\,{\rm M}_\odot$, this accretion rate is $\simeq 2 \times 10^{-8}\,{\rm M}_\odot\,{\rm yr}^{-1}$.

neutron star's orbit to bring the massive star into corotation.[8] The transfer of angular momentum from the orbit to the star nevertheless makes the orbit shrink, forcing the neutron star to merge with the massive star and form a common envelope.

The formation of Thorne-Żytkow objects

The fate of the CE phase depends on whether the orbital energy released in the spiral-in is enough to eject the envelope or not. If the initial orbital period is relatively short ($\lesssim 1\,\mathrm{yr}$; Terman et al., 1995), such systems are expected to merge completely; this means that the neutron star sinks to the center, replacing or disrupting whatever there was before. Such objects with neutron cores are known as TŻOs (Thorne and Zytkow, 1975, 1977). These objects will appear as very cool red supergiants. Because these are difficult to distinguish from normal red supergiants, it is presently not clear whether they actually exist. Since this is the possible fate for the majority of known HMXBs, their birthrate in the galaxy is expected to be quite high ($\sim 2 \times 10^{-4}\,\mathrm{yr}^{-1}$; Podsiadlowski et al., 1995). Depending on the uncertain lifetime of this phase (limited, e.g., by the wind mass-loss rate), a few percent to 10% of all red supergiants with a luminosity comparable to or above the Eddington limit for a neutron star could harbor neutron cores. One way of distinguishing them from normal red supergiants is through their anomalously large abundances of proton-rich elements, in particular molybdenum (Biehle, 1991; Cannon, 1993).

One criticism that has been raised against the very existence of TŻOs (see, e.g., Chevalier, 1993) is that, during the initial spiral-in phase, the accretion rate onto the neutron star may occur in the neutrino-dominated regime where all the accretion energy is radiated away in the form of neutrinos and becomes hypercritical (i.e., can exceed the photon Eddington limit by an arbitrary amount). If this were the case, one would expect the neutron star to accrete enough matter to be converted into a black hole. The resulting object would presumably be a black hole surrounded by a massive disk.[9]

The formation of double neutron star binaries

If the orbital period of the HMXB is relatively long ($\gtrsim 1\,\mathrm{yr}$), the orbital energy released by the spiraling-in neutron star is expected to be sufficient to eject the common envelope. The post-CE system will be a much closer binary consisting of the neutron star (assuming that it did not experience supercritical accretion and was converted into a black hole) in orbit with the hydrogen-exhausted core of the massive star, that is, a helium star (see the left panel of Fig. 2.10). If the helium star is sufficiently massive ($\gtrsim 4\,\mathrm{M}_\odot$), it will appear as a *Wolf-Rayet star* with a very powerful, optically thick wind (with $\dot{M}_{\mathrm{wind}} \gtrsim 10^{-6}\,\mathrm{M}_\odot\,\mathrm{yr}^{-1}$). If even a small fraction of this wind is accreted by the neutron star, the system will again appear as a bright X-ray source. Cyg X-3 with an orbital period of 4.8 hr provides a prototypical example for this type of system.[10]

Eventually, the helium star will explode in a supernova (of type Ib/Ic) and itself produce a neutron star. Because of the natal NS supernova kick, there is a high probability that the system becomes disrupted in this second supernova. If this is the case, both neutron stars (one a young pulsar, the other a relatively old neutron star) will move apart as *runaway neutron stars* with velocities comparable to their final orbital velocities

---

[8]The exact criterion for the Darwin instability is that a binary system is unstable if the moment of inertia of the star being spun up (assuming solid-body rotation) is larger than one-third of the momentum inertia of the orbit, that is, $I_{\mathrm{star}} > 1/3\,\mu A^2$, where μ is the reduced mass of the binary and A the orbital separation.

[9]If this disk is sufficiently massive, it is conceivable that self-gravitating objects, even low-mass stars, could form in such a disk because of gravitational instabilities (Podsiadlowski et al., 1995).

[10]Note, however, that, in the case of Cyg X-3, it is not clear whether the compact object is a neutron star or a black hole.

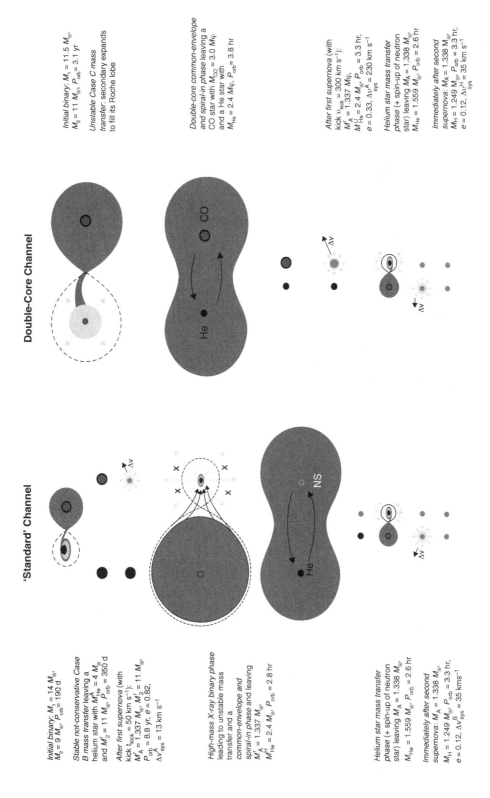

FIGURE 2.10. The standard and the double-core channel for the formation of the double pulsar, PSR J0737−3039. (From Podsiadlowski et al., 1995.)

in the disrupted binary (typically a few $100\,\mathrm{km\,s^{-1}}$). On the other hand, if the system remains bound, the surviving system is a binary containing two neutron stars. The first double neutron star (DNS) system discovered was the Hulse-Taylor pulsar, PSR 1913+16 (Hulse and Taylor, 1975), with an orbital period of about 8 hr and a spin period of 59 ms (the pulsar was mildly spun up [recycled] by accretion from the helium star in the previous He-star binary phase). Since the original discovery, half a dozen more DNS systems have been found (see, e.g., the list in Schwab et al., 2010). These systems have become extremely important probes of fundamental physics. The orbital evolution of these systems is entirely driven by gravitational radiation. Since the neutron stars are too small to interact in any way (even tidally), the resulting measured decrease of the orbital period in these systems provides a direct and very accurate test of Einstein's theory of general relativity (for this achievement, Hulse and Taylor were awarded the Nobel prize in 1993).

In addition, if the orbital period of a DNS system is short enough ($\lesssim 10$ hr), gravitational radiation will bring the system together in a Hubble time, making the two neutron stars merge in a final cataclysmic event. Such merger events are accompanied by a major burst of gravitational waves, which could be directly detectable with current and future gravitational wave experiments (e.g., Advanced LIGO). If such a merger occurs in a relatively nearby galaxy, such mergers should be detectable within the next few years.

In addition, such mergers are likely to produce a burst of gamma rays, and mergers of two neutron stars or a neutron star and a black hole are presently the best candidates for short-duration GRBs.[11] Finally, DNS mergers may also be the source of some unusual nucleosynthesis: in particular, because of the overabundance of neutrons, they are a potential source for all the neutron-rich, r-process elements in the universe, for which no source has yet been identified unambiguously.[12]

As mentioned before, it is not clear whether a neutron star spiraling in inside a massive envelope will survive as a neutron star or be converted into a black hole by hypercritical accretion. If this were the case, the preceding scenario could not produce a DNS system. An alternative scenario to produce DNSs was proposed by Brown (1995), which is shown in the right panel of Figure 2.10. If the masses of the initial binary are sufficiently close (typically within 4%) and mass transfer occurs when the primary has already finished helium core burning (case C mass transfer), the secondary will already have finished its hydrogen-core–burning phase and developed a helium core. In this case, one may expect a common-envelope phase, where the common envelope contains the envelopes of both stars, and the embedded binary consists of the He-exhausted core of the primary and the H-exhausted core of the secondary (so-called double-core evolution). Once the common envelope is ejected, the system has become a close binary with two H-exhausted stars, one most likely containing a CO core, the other a He core. After two supernovae, the system will end up as a DNS system. Even though this evolution requires rather special circumstances, Dewi et al. (2006) found that, within the substantial uncertainties of this channel, this channel could account for a large fraction of DNS systems and potentially for all.

The double pulsar: PSR J0737−3039

A particularly important recent discovery is the binary pulsar PSR J0737−3039 (Burgay et al., 2003), which consists of two pulsars: one old, recycled pulsar with a spin period of 22.7 ms and one younger pulsar with a spin period of 2.77 s (Fig. 2.10 illustrates two possible evolutionary histories for the double pulsar). Because of the short

---

[11] Unlike long-duration GRBs, the average duration of a short-duration GRB is less than 1 s.

[12] Explosive nucleosynthesis in supernovae is often considered a possible source for r-process elements, but it is still unclear whether the conditions for the r-process are right for a sufficiently long time during the explosive supernova phase.

orbital period of this system (2.4 hr), general relativistic effects are much more important than in the Hulse-Taylor pulsar, and this system has now become our most important laboratory for testing general relativity (see Kramer and Stairs, 2008). Note also that the second-born pulsar (pulsar B) has a mass of 1.249 $M_\odot$, very close to the mass expected for an e-capture supernova. Indeed, there is strong evidence that the second-born neutron star did not receive a large kick: (i) the orbit is almost circular, (ii) the system space velocity is small, and (iii) the spin of the recycled pulsar (pulsar A) is aligned with the orbit (see Podsiadlowski et al., 2005, for further discussion).

2.6.3 X-ray binaries containing black holes

Interlude: do black holes exist?

These days we often take the existence of black holes for granted, but we should ask whether it has actually been proven. In the case of stellar-mass black holes, the argument is usually just based on the mass of the compact object, since the maximum mass of a neutron star is believed to be less than $\sim 3\,M_\odot$. The inferred masses for the compact objects in some of the "black-hole" binaries, as largely determined by their mass function (eq. 2.1), exceeds $10\,M_\odot$ in some of the best cases (e.g., GRS 1915+105; Greiner et al., 2001). Although this is well in excess of the maximum neutron-star mass, it does not prove that the compact object is a black hole: the equation of state of matter at these densities is very poorly understood, and it has been postulated that other states of matter could exist (e.g., involving strange matter, Q-balls, etc.) that do not have a maximum mass limit.

What one needs to show the existence of a black hole is the existence of an event horizon, the defining feature of a black hole. This is possible in principle, if one has a system where one knows the accretion rate onto a compact object, but one does not see the accretion luminosity that would be associated with this accretion rate, because all the mass–energy disappears below the event horizon without trace (in contrast, for objects with a surface, most of that energy has to be radiated away). This is possible in principle but has not yet been demonstrated convincingly (at least to this author), despite some claims in the literature.

2.6.4 The origin of black-hole binaries with low-mass companions

A large fraction of the black-hole (BH) binaries known to date appear to contain low-mass donor stars, in many cases resembling LMXBs except that they often tend to be transient rather than persistent X-ray sources (see, e.g., Lee et al., 2002). This poses an immediate problem concerning their formation. As already discussed in the context of the formation of LMXBs, it is challenging for a low-mass star to eject the massive envelope of an NS progenitor in the CE phase. This becomes even harder for a BH progenitor that is necessarily more massive. Indeed, if one uses realistic envelope structures, the maximum orbital energy available from the spiral-in of a low-mass star falls short, by about a factor of 5 to 10, of what is needed to eject the envelope of a BH progenitor (see Podsiadlowski et al., 2003a, for details). This problem has long been known (even though it is often ignored!), and a number of solutions have been proposed:

(i) the models of the envelopes of massive red supergiants may be wrong (in particular, due to uncertainties in the wind mass loss);

(ii) the modeling of the CE phase may be in error (this would not be surprising considering that this is a very poorly understood phase; however, energy conservation should not be violated);

(iii) there are alternative exotic formation scenarios, involving triple systems (Eggleton and Verbunt, 1986) or the formation of a low-mass star in the debris disk of, for example, a Thorne-Żytkow object (Podsiadlowski et al., 1995, 2003a);

(iv) the companions of these systems may have descended from intermediate-mass objects (similar to the case of most LMXBs; Podsiadlowski et al., 2003a; Justham et al., 2006); or

(v) the energy source for the ejection of the common envelope is nuclear energy rather than orbital energy ("explosive common-envelope ejection"; Podsiadlowski et al., 2010, and Section 2.4). At present, there is no consensus on the resolution of this conundrum.

2.6.5 The evolution of black-hole binaries

In X-ray binaries where the accreting compact source is a black hole, the mass-transfer process and the overall evolution differ significantly in some respects from the case of neutron-star binaries.

If the donor star is a low-mass star, the main processes are similar to the case of LMXBs, except that, because of the larger mass of the black hole, the mass-transfer rate tends to be lower. One consequence of this is that most black-hole binaries with low-mass companions are X-ray transients with alternating phases of high and low accretion rates depending on the state of the accretion disk (see Chapters 4, 6 and 7 in these proceedings).

For black-hole binaries with massive companions, the main difference from NS systems is that standard Roche lobe overflow is not necessarily unstable for mass ratios as high as ~ 2. Therefore, these systems may experience long phases of stable mass transfer as bright X-ray sources.

Podsiadlowski et al. (2003a) have systematically explored the evolution of black-hole binaries. Figure 2.11 shows some of the key results from this study. It shows the evolution of initially unevolved main-sequence stars (ranging from 2 to $17\,M_\odot$ initially), transferring mass to a black hole with an initial mass of $10\,M_\odot$. In all of these sequences, mass transfer is stable at all times (even a model with $20\,M_\odot$ was only marginally unstable). In systems where the initial donor mass exceeds the black-hole mass, mass transfer initially occurs on a thermal time scale, leading to very high mass-transfer rates and the spikes in the $\dot{M}$ distribution in Figure 2.11. After the mass ratio has been reversed, mass transfer continues to be driven by the nuclear evolution of the donor star. Since this phase is much longer-lived than the thermal time scale phase, BH X-ray binaries are most likely to be observed in this phase. As the donor ascends the giant branch, the mass-transfer rate goes up again, producing another spike in the $\dot{M}$ evolution.

The main results of this study were as follows:

(i) RLOF in BH X-ray binaries is stable for mass ratios as high as 2, and, because of the mass loss, the X-ray active phase can be much longer than the lifetime of a single star of the same initial mass.

(ii) Even if mass accretion onto the black hole is Eddington limited, the black holes can accrete substantial amounts of mass (see the bottom left panel of Fig. 2.11); hence, the present observed BH mass is not necessarily a good indicator of the initial postcollapse BH mass.

(iii) Black holes can also accrete substantial amounts of angular momentum (top left panel of Fig. 2.11) and be spun up in the process to spin parameters $a \sim 0.4$–0.9 (assuming that the black holes were initially nonrotating).

2.6.6 The nature of ultraluminous X-ray sources

Ultraluminous X-ray sources (ULXs; Fabbiano, 1989) are typically defined as X-ray sources that have a luminosity exceeding $L_X = 10^{39}\,\mathrm{erg\,s^{-1}}$, which is roughly the Eddington luminosity for a $10\,M_\odot$ black hole (the exact definition varies somewhat from author to author). The particular interest of these systems is that they may harbor intermediate-mass black holes with masses of 10^2 to $10^3\,M_\odot$ (Colbert and Mushotzky, 1999), which

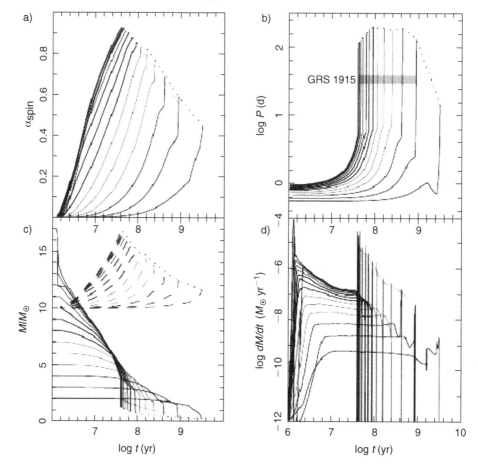

FIGURE 2.11. Selected properties of black-hole binary sequences as a function of time since the beginning of mass transfer: *a)* black-hole spin parameter; *b)* orbital period; *c)* black-hole mass and secondary mass (dashed and solid curves); *d)* mass-transfer rate. In all sequences, the black hole has an initial mass of $10\,M_\odot$ and is initially nonrotating. The secondaries (mass donors) range from 2 to $17\,M_\odot$ and are initially unevolved (the larger the initial mass of the secondary, the shorter the duration of the mass-transfer phase). The shaded regions in each panel indicate the period range of 30 to 40 d (similar to the orbital period of GRS 1915+105 with $P_{\rm orb} = 33.5$ d). (From Podsiadlowski *et al.*, 2003a.)

could possibly represent the missing link between stellar-mass black holes and supermassive black holes at the centers of galaxies. Indeed, they could be the key building block for supermassive black holes. As discussed in detail in Chapter 5 by Fabbiano in these proceedings, it is now clear that probably the vast majority of ULXs are physically associated with regions of very active massive star formation (such as in the Antennae interacting galaxies and the Cartwheel galaxy). This strongly suggests that most of them (but not necessarily all!) are linked to massive BH binary populations. Indeed, it is no problem to feed black holes at a mass-transfer rate at which they could appear as ULXs. This is shown in Figure 2.12, in which the right panel shows the potential X-ray luminosity of the evolutionary sequences in Figure 2.11 (for comparison, the left panel shows the luminosity in these sequences if accretion were Eddington limited). The potential X-ray luminosity is defined as the accretion luminosity one would observe if all the mass lost from the donor star were accreted by the black hole. Note that the potential luminosities can reach values as high as $\sim 10^{41}\,\rm erg\,s^{-1}$, albeit for only very short periods of time.

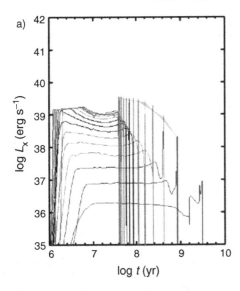
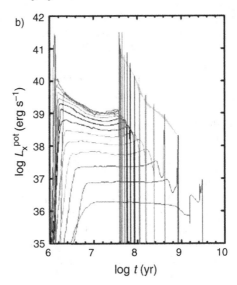

FIGURE 2.12. *a)* X-ray luminosities, assuming Eddington-limited accretion, and *b)* potential X-ray luminosities, assuming non-Eddington-limited accretion, for the binary evolution sequences in Figure 2.11 as a function of time since the beginning of mass transfer. (From Podsiadlowski et al., 2003a.)

In this context, I would like to clear up a common misconception in the literature. One often hears that stellar BH X-ray binaries may look like ULXs if they are in a thermal time-scale mass-transfer phase. However, as Figure 2.12 shows, these systems reach potential ULX luminosities for the more massive companions during most of the phases where mass transfer is driven by the nuclear evolution of the donor.[13] Because this phase is many orders of magnitude longer-lived, it is much more likely that the observed ULXs are in this phase rather than the thermal time-scale phase (the very short initial spikes in Fig. 2.12). (For recent detailed studies of massive BH binaries and their relation to ULXs, see Rappaport et al., 2005; Madhusudhan et al., 2006, 2008).

The main issue with ULXs being stellar-mass BH binaries is that they have to accrete at a rate that exceeds the Eddington accretion rate by a significant factor (for some of the more luminous systems, by a factor up to ~ 20). It has been suggested that, in magnetic accretion disk, such high accretion rates can be reached, but the exact physical mechanism remains unclear and may involve a photon bubble instability (Begelman, 2002, 2006; Ruszkowski and Begelman, 2003); emission from a hot, optically thin corona magnetically coupled to the accretion flow (Socrates and Davis, 2006); or something else. Until we understand the physics of magnetic accretion disks better, this issue remains unresolved (also see Chapter 8 by Hawley in these proceedings).

REFERENCES

Begelman, M. C. 2002. Super-Eddington fluxes from thin accretion disks? ApJ, **568**(Apr.), L97–L100.

Begelman, M. C. 2006. Photon bubbles and the vertical structure of accretion disks. ApJ, **643**(June), 1065–1080.

Bhattacharya, D., and van den Heuvel, E. P. J. 1991. Formation and evolution of binary and millisecond radio pulsars. Phys. Rep., **203**, 1–124.

---

[13] In addition, beaming, both geometric (King et al., 2001) and relativistic (Körding et al., 2002), may increase the apparent luminosities of these systems further if they are observed along the direction where the radiation is beamed.

Biehle, G. T. 1991. High-mass stars with degenerate neutron cores. ApJ, **380**(Oct.), 167–184.

Blondin, J. M., and Mezzacappa, A. 2006. The spherical accretion shock instability in the linear regime. ApJ, **642**(May), 401–409.

Blondin, J. M., and Mezzacappa, A. 2007. Pulsar spins from an instability in the accretion shock of supernovae. Nature, **445**(Jan.), 58–60.

Blundell, K. M., Mioduszewski, A. J., Muxlow, T. W. B., Podsiadlowski, P., and Rupen, M. P. 2001. Images of an equatorial outflow in SS 433. ApJ, **562**(Nov.), L79–L82.

Brandt, N., and Podsiadlowski, P. 1995. The effects of high-velocity supernova kicks on the orbital properties and sky distributions of neutron-star binaries. MNRAS, **274**(May), 461–484.

Brown, G. E. 1995. Neutron star accretion and binary pulsar formation. ApJ, **440**(Feb.), 270–279.

Brown, G. E., Heger, A., Langer, N., Lee, C.-H., Wellstein, S., and Bethe, H. A. 2001. Formation of high mass X-ray black hole binaries. New A, **6**(Oct.), 457–470.

Brown, G. E., Lee, C.-H., and Bethe, H. A. 1999. The formation of high-mass black holes in low-mass X-ray binaries. New A, **4**(July), 313–323.

Burgay, M., D'Amico, N., Possenti, A., Manchester, R. N., Lyne, A. G., Joshi, B. C., McLaughlin, M. A., Kramer, M., Sarkissian, J. M., Camilo, F., Kalogera, V., Kim, C., and Lorimer, D. R. 2003. An increased estimate of the merger rate of double neutron stars from observations of a highly relativistic system. Nature, **426**(Dec.), 531–533.

Burrows, C. J., Krist, J., Hester, J. J., Sahai, R., Trauger, J. T., Stapelfeldt, K. R., Gallagher, J. S. III, Ballester, G. E., Casertano, S., Clarke, J. T., Crisp, D., Evans, R. W., Griffiths, R. E., Hoessel, J. G., Holtzman, J. A., Mould, J. R., Scowen, P. A., Watson, A. M., and Westphal, J. A. 1995. Hubble Space Telescope observations of the SN 1987A triple ring nebula. ApJ, **452**(Oct.), 680+.

Cannon, R. C. 1993. Massive Thorne-Żytkow objects – structure and nucleosynthesis. MNRAS, **263**(Aug.), 817+.

Cappellaro, E., and Turatto, M. 1997. The rate of supernovae: biases and uncertainties. Pages 77+ of: P. Ruiz-Lapuente, R. Canal, and J. Isern (eds.), *NATO ASIC Proc. 486: Thermonuclear Supernovae*, Proceedings of the NATO Advanced Study Institute, held in Begur, Girona, Spain, June 20–30, 1995, Dordrecht: Kluwer Academic Publishers.

Casares, J., Charles, P. A., and Kuulkers, E. 1998. The mass of the neutron star in Cygnus X-2 (V1341 Cygni). ApJ, **493**(Jan.), L39+.

Chevalier, R. A. 1993. Neutron star accretion in a stellar envelope. ApJ, **411**(July), L33–L36.

Colbert, E. J. M., and Mushotzky, R. F. 1999. The nature of accreting black holes in nearby galaxy nuclei. ApJ, **519**(July), 89–107.

Davies, M. B. 1995. The binary zoo: the calculation of production rates of binaries through 2+1 encounters in globular clusters. MNRAS, **276**(Oct.), 887–905.

Davies, M. B., and Hansen, B. M. S. 1998. Neutron star retention and millisecond pulsar production in globular clusters. MNRAS, **301**(Nov.), 15–24.

Detmers, R. G., Langer, N., Podsiadlowski, P., and Izzard, R. G. 2008. Gamma-ray bursts from tidally spun-up Wolf-Rayet stars? A&A, **484**(June), 831–839.

Dewi, J. D. M., Podsiadlowski, P., and Sena, A. 2006. Double-core evolution and the formation of neutron star binaries with compact companions. MNRAS, **368**(June), 1742–1748.

Dewi, J. D. M., and Tauris, T. M. 2000. On the energy equation and efficiency parameter of the common envelope evolution. A&A, **360**(Aug.), 1043–1051.

Duquennoy, A., and Mayor, M. 1991. Multiplicity among solar-type stars in the solar neighbourhood. II – Distribution of the orbital elements in an unbiased sample. A&A, **248**(Aug.), 485–524.

Eggleton, P. P. 1983. Approximations to the radii of Roche lobes. ApJ, **268**(May), 368+.

Eggleton, P. P., and Verbunt, F. 1986. Triple star evolution and the formation of short-period, low mass X-ray binaries. MNRAS, **220**(May), 13P–18P.

Fabbiano, G. 1989. X rays from normal galaxies. ARA&A, **27**, 87–138.

Fabian, A. C., Pringle, J. E., and Rees, M. J. 1975. Tidal capture formation of binary systems and X-ray sources in globular clusters. MNRAS, **172**(Aug.), 15P+.

Faulkner, J. 1971. Ultrashort-period binaries, gravitational radiation, and mass transfer. I. The standard model, with applications to WZ Sagittae and Z Camelopardalis. ApJ, **170**(Dec.), L99+.

Fedorova, A. V., Tutukov, A. V., and Yungelson, L. R. 2004. Type-Ia supernovae in semidetached binaries. Astronomy Letters, **30**(Feb.), 73–85.

Foglizzo, T., Galletti, P., Scheck, L., and Janka, H.-T. 2007. Instability of a stalled accretion shock: evidence for the advective-acoustic cycle. ApJ, **654**(Jan.), 1006–1021.

Frankowski, A., and Jorissen, A. 2007. Binary life after the AGB – towards a unified picture. Baltic Astronomy, **16**, 104–111.

Fruchter, A. S., and Goss, W. M. 1990. The integrated flux density of pulsars in globular clusters. ApJ, **365**(Dec.), L63–L66.

Fryer, C. L., and Heger, A. 2005. Binary merger progenitors for gamma-ray bursts and hypernovae. ApJ, **623**(Apr.), 302–313.

Fryer, C. L., and Kalogera, V. 2001. Theoretical black hole mass distributions. ApJ, **554**(June), 548–560.

Fryer, C. L., and Woosley, S. E. 1998. Helium star/black hole mergers: a new gamma-ray burst model. ApJ, **502**(July), L9+.

Greiner, J., Cuby, J. G., and McCaughrean, M. J. 2001. An unusually massive stellar black hole in the Galaxy. Nature, **414**(Nov.), 522–525.

Hachisu, I., Kato, M., Nomoto, K., and Umeda, H. 1999. A new evolutionary path to type IA supernovae: A helium-rich supersoft X-ray source channel. ApJ, **519**(July), 314–323.

Hameury, J. M., King, A. R., Lasota, J. P., and Raison, F. 1993. Structure and evolution of X-ray heated compact binaries. A&A, **277**(Sept.), 81+.

Hamuy, M., Phillips, M. M., Suntzeff, N. B., Maza, J., González, L. E., Roth, M., Krisciunas, K., Morrell, N., Green, E. M., Persson, S. E., and McCarthy, P. J. 2003. An asymptotic-giant-branch star in the progenitor system of a type Ia supernova. Nature, **424**(Aug.), 651–654.

Han, Z., and Podsiadlowski, P. 2004. The single-degenerate channel for the progenitors of Type Ia supernovae. MNRAS, **350**(June), 1301–1309.

Han, Z., Podsiadlowski, P., and Eggleton, P. P. 1995. The formation of bipolar planetary nebulae and close white dwarf binaries. MNRAS, **272**(Feb.), 800–820.

Han, Z., Podsiadlowski, P., and Lynas-Gray, A. E. 2007. A binary model for the UV-upturn of elliptical galaxies. MNRAS, **380**(Sept.), 1098–1118.

Han, Z., Podsiadlowski, P., Maxted, P. F. L., and Marsh, T. R. 2003. The origin of subdwarf B stars – II. MNRAS, **341**(May), 669–691.

Han, Z., Podsiadlowski, P., Maxted, P. F. L., Marsh, T. R., and Ivanova, N. 2002. The origin of subdwarf B stars – I. The formation channels. MNRAS, **336**(Oct.), 449–466.

Hjellming, M. S., and Webbink, R. F. 1987. Thresholds for rapid mass transfer in binary systems. I – Polytropic models. ApJ, **318**(July), 794–808.

Hobbs, G., Lorimer, D. R., Lyne, A. G., and Kramer, M. 2005. A statistical study of 233 pulsar proper motions. MNRAS, **360**(July), 974–992.

Howell, D. A., Sullivan, M., Nugent, P. E., Ellis, R. S., Conley, A. J., Le Borgne, D., Carlberg, R. G., Guy, J., Balam, D., Basa, S., Fouchez, D., Hook, I. M., Hsiao, E. Y., Neill, J. D., Pain, R., Perrett, K. M., and Pritchet, C. J. 2006. The type Ia supernova SNLS-03D3bb from a super-Chandrasekhar-mass white dwarf star. Nature, **443**(Sept.), 308–311.

Hulse, R. A., and Taylor, J. H. 1975. Discovery of a pulsar in a binary system. ApJ, **195**(Jan.), L51–L53.

Iben, I., Jr., and Livio, M. 1993. Common envelopes in binary star evolution. PASP, **105**(Dec.), 1373–1406.

Iben, I., Jr., and Tutukov, A. V. 1984. Supernovae of type I as end products of the evolution of binaries with components of moderate initial mass (M not greater than about 9 solar masses). ApJS, **54**(Feb.), 335–372.

Ivanova, N. 2002. Slow mergers of massive stars. Ph.D. thesis, Balliol College, Oxford.

Ivanova, N., and Podsiadlowski, P. 2003. The Slow Merger of Massive Stars. Pages 19+ of: W. Hillebrandt & B. Leibundgut (ed), *From Twilight to Highlight: The Physics of Supernovae*, Springer-Verlag.

Iwamoto, K., Mazzali, P. A., Nomoto, K., Umeda, H., Nakamura, T., Patat, F., Danziger, I. J., Young, T. R., Suzuki, T., Shigeyama, T., Augusteijn, T., Doublier, V., Gonzalez, J.-F., Boehnhardt, H., Brewer, J., Hainaut, O. R., Lidman, C., Leibundgut, B., Cappellaro, E., Turatto, M., Galama, T. J., Vreeswijk, P. M., Kouveliotou, C., van Paradijs, J., Pian, E., Palazzi, E., and Frontera, F. 1998. A hypernova model for the supernova associated with the γ-ray burst of 25 April 1998. Nature, **395**(Oct.), 672–674.

Izzard, R. G., Ramirez-Ruiz, E., and Tout, C. A. 2004. Formation rates of core-collapse supernovae and gamma-ray bursts. MNRAS, **348**(Mar.), 1215–1228.

Janka, H.-T., Langanke, K., Marek, A., Martínez-Pinedo, G., and Müller, B. 2007. Theory of core-collapse supernovae. Phys. Rep., **442**(Apr.), 38–74.

Johnston, S., and Bailes, M. 1991. New limits on the population of millisecond pulsars in the galactic plane. MNRAS, **252**(Sept.), 277–281.

Justham, S., Rappaport, S., and Podsiadlowski, P. 2006. Magnetic braking of Ap/Bp stars: application to compact black-hole X-ray binaries. MNRAS, **366**(Mar.), 1415–1423.

Karovska, M., Schlegel, E., Hack, W., Raymond, J. C., and Wood, B. E. 2005. A large X-ray outburst in Mira A. ApJ, **623**(Apr.), L137–L140.

Kerzendorf, W. E., Schmidt, B. P., Asplund, M., Nomoto, K., Podsiadlowski, P., Frebel, A., Fesen, R. A., and Yong, D. 2009. Subaru high-resolution spectroscopy of Star G in the Tycho supernova remnant. ApJ, **701**(Aug.), 1665–1672.

King, A. R., Davies, M. B., Ward, M. J., Fabbiano, G., and Elvis, M. 2001. Ultraluminous X-ray sources in external galaxies. ApJ, **552**(May), L109–L112.

King, A. R., and Ritter, H. 1999. Cygnus X-2, super-Eddington mass transfer, and pulsar binaries. MNRAS, **309**(Oct.), 253–260.

Kobulnicky, H. A., and Fryer, C. L. 2007. A new look at the binary characteristics of massive stars. ApJ, **670**(Nov.), 747–765.

Kolb, U., Davies, M. B., King, A., and Ritter, H. 2000. The violent past of Cygnus X-2. MNRAS, **317**(Sept.), 438–446.

Körding, E., Falcke, H., and Markoff, S. 2002. Population X: are the super-Eddington X-ray sources beamed jets in microblazars or intermediate mass black holes? A&A, **382**(Jan.), L13–L16.

Kramer, M., and Stairs, I. H. 2008. The double pulsar. ARA&A, **46**(Sept.), 541–572.

Kulkarni, S. R., and Narayan, R. 1988. Birthrates of low-mass binary pulsars and low-mass X-ray binaries. ApJ, **335**(Dec.), 755–768.

Kulkarni, S. R., Narayan, R., and Romani, R. W. 1990. The pulsar content of globular clusters. ApJ, **356**(June), 174–183.

Kulkarni, S. R., Ofek, E. O., Rau, A., Cenko, S. B., Soderberg, A. M., Fox, D. B., Gal-Yam, A., Capak, P. L., Moon, D. S., Li, W., Filippenko, A. V., Egami, E., Kartaltepe, J., and Sanders, D. B. 2007. An unusually brilliant transient in the galaxy M85. Nature, **447**(May), 458–460.

Landau, L. D., and Lifshitz, E. M. 1959. *The Classical Theory of Fields*. 2nd ed., vol. 2, Butterworth-Heinemann.

Lee, C.-H., Brown, G. E., and Wijers, R. A. M. J. 2002. Discovery of a black hole mass-period correlation in soft X-ray transients and its implication for gamma-ray burst and hypernova mechanisms. ApJ, **575**(Aug.), 996–1006.

Leonard, D. C. 2007. Constraining the type Ia supernova progenitor: the search for hydrogen in nebular spectra. ApJ, **670**(Dec.), 1275–1282.

MacFadyen, A. I., and Woosley, S. E. 1999. Collapsars: gamma-ray bursts and explosions in "failed supernovae." ApJ, **524**(Oct.), 262–289.

Madhusudhan, N., Justham, S., Nelson, L., Paxton, B., Pfahl, E., Podsiadlowski, P., and Rappaport, S. 2006. Models of ultraluminous X-ray sources with intermediate-mass black holes. ApJ, **640**(Apr.), 918–922.

Madhusudhan, N., Rappaport, S., Podsiadlowski, P., and Nelson, L. 2008. Models for the observable system parameters of ultraluminous X-ray sources. ApJ, **688**(Dec.), 1235–1249.

Marietta, E., Burrows, A., and Fryxell, B. 2000. Type IA supernova explosions in binary systems: the impact on the secondary star and its consequences. ApJS, **128**(June), 615–650.

Meng, X., Chen, X., and Han, Z. 2007. The impact of Type Ia supernova explosions on the companions in a binary system. PASJ, **59**(Aug.), 835–840.

Meyer, F., and Meyer-Hofmeister, E. 1979. Formation of cataclysmic binaries through common envelope evolution. A&A, **78**(Sept.), 167–176.

Mezzacappa, A., Bruenn, S. W., Blondin, J. M., Hix, W. R., and Bronson Messer, O. E. 2007 (Aug.). Ascertaining the core collapse supernova mechanism: an emerging picture? Pages 234–242 of: T. di Salvo, G. L. Israel, L. Piersant, L. Burderi, G. Matt, A. Tornambe, & M. T. Menna (eds.), *The Multicolored Landscape of Compact Objects and Their Explosive Origins*. American Institute of Physics Conference Series, vol. 924.

Mikołajewska, J. 2007. Symbiotic stars: continually embarrassing binaries. Baltic Astronomy, **16**, 1–9.

Morris, T., and Podsiadlowski, P. 2007. The triple-ring nebula around SN 1987A: fingerprint of a binary merger. Science, **315**(Feb.), 1103+.

Napiwotzki, R., Koester, D., Nelemans, G., Yungelson, L., Christlieb, N., Renzini, A., Reimers, D., Drechsel, H., and Leibundgut, B. 2002. Binaries discovered by the SPY project. II. HE 1414-0848: a double degenerate with a mass close to the Chandrasekhar limit. A&A, **386**(May), 957–963.

Nelemans, G., Yungelson, L. R., Portegies Zwart, S. F., and Verbunt, F. 2001. Population synthesis for double white dwarfs . I. Close detached systems. A&A, **365**(Jan.), 491–507.

Nomoto, K. 1982. Accreting white dwarf models for type I supernovae. I – Presupernova evolution and triggering mechanisms. ApJ, **253**(Feb.), 798–810.

Nomoto, K. 1984. Evolution of 8-10 solar mass stars toward electron capture supernovae. I – Formation of electron-degenerate O + NE + MG cores. ApJ, **277**(Feb.), 791–805.

Nomoto, K., and Iben, I., Jr., 1985. Carbon ignition in a rapidly accreting degenerate dwarf – a clue to the nature of the merging process in close binaries. ApJ, **297**(Oct.), 531–537.

Orosz, J. A., and Kuulkers, E. 1999. The optical light curves of Cygnus X-2 (V1341 Cyg) and the mass of its neutron star. MNRAS, **305**(May), 132–142.

Paczyński, B. 1976. Common envelope binaries. Pages 75+ of: P. Eggleton, S. Mitton, & J. Whelan (eds.), *Structure and Evolution of Close Binary Systems*. IAU Symposium, vol. 73.

Pakmor, R., Röpke, F. K., Weiss, A., and Hillebrandt, W. 2008. The impact of type Ia supernovae on main sequence binary companions. A&A, **489**(Oct.), 943–951.

Pasquali, A., Nota, A., Langer, N., Schulte-Ladbeck, R. E., and Clampin, M. 2000. R4 and its circumstellar nebula: evidence for a binary merger? AJ, **119**(Mar.), 1352–1358.

Patat, F., Chandra, P., Chevalier, R., Justham, S., Podsiadlowski, P., Wolf, C., Gal-Yam, A., Pasquini, L., Crawford, I. A., Mazzali, P. A., Pauldrach, A. W. A., Nomoto, K., Benetti, S., Cappellaro, E., Elias-Rosa, N., Hillebrandt, W., Leonard, D. C., Pastorello, A., Renzini, A., Sabbadin, F., Simon, J. D., and Turatto, M. 2007. Detection of circumstellar material in a normal type Ia supernova. Science, **317**(Aug.), 924+.

Perlmutter, S., Aldering, G., Goldhaber, G., Knop, R. A., Nugent, P., Castro, P. G., Deustua, S., Fabbro, S., Goobar, A., Groom, D. E., Hook, I. M., Kim, A. G., Kim, M. Y., Lee, J. C., Nunes, N. J., Pain, R., Pennypacker, C. R., Quimby, R., Lidman, C., Ellis, R. S., Irwin, M., McMahon, R. G., Ruiz-Lapuente, P., Walton, N., Schaefer, B., Boyle, B. J., Filippenko, A. V., Matheson, T., Fruchter, A. S., Panagia, N., Newberg, H. J. M., Couch, W. J., and The Supernova Cosmology Project. 1999. Measurements of omega and lambda from 42 high-redshift supernovae. ApJ, **517**(June), 565–586.

Pfahl, E., Rappaport, S., and Podsiadlowski, P. 2002a. A comprehensive study of neutron star retention in globular clusters. ApJ, **573**(July) 283–305.

Pfahl, E., Rappaport, S., and Podsiadlowski, P. 2003. The galactic population of low- and intermediate-mass X-ray binaries. ApJ, **597**(Nov.), 1036–1048.

Pfahl, E., Rappaport, S., Podsiadlowski, P., and Spruit, H. 2002b. A new class of high-mass X-ray binaries: implications for core collapse and neutron star recoil. ApJ, **574**(July), 364–376.

Phillips, M. M. 1993. The absolute magnitudes of Type IA supernovae. ApJ, **413**(Aug.), L105–L108.

Piersanti, L., Gagliardi, S., Iben, I., Jr., and Tornambé, A. 2003. Carbon-oxygen white dwarf accreting CO-rich matter. II. Self-regulating accretion process up to the explosive stage. ApJ, **598**(Dec.), 1229–1238.

Podsiadlowski, P. 1991. Irradiation-driven mass transfer low-mass X-ray binaries. Nature, **350**(Mar.), 136–138.

Podsiadlowski, P. 2001. Common-envelope evolution and stellar mergers. Pages 239+ of: P. Podsiadlowski, S. Rappaport, A. R. King, F. D'Antona, and L. Burderi (eds.), *Evolution of Binary and Multiple Star Systems*. Astronomical Society of the Pacific Conference Series, vol. 229.

Podsiadlowski, P., Cannon, R. C., and Rees, M. J. 1995. The evolution and final fate of massive Thorne-Zytkow objects. MNRAS, **274**(May), 485–490.

Podsiadlowski, P., Dewi, J. D. M., Lesaffre, P., Miller, J. C., Newton, W. G., and Stone, J. R. 2005. The double pulsar J0737-3039: testing the neutron star equation of state. MNRAS, **361**(Aug.), 1243–1249.

Podsiadlowski, P., Ivanova, N., Justham, S., and Rappaport, S. 2010. Explosive common-envelope ejection: implications for gamma-ray bursts and low-mass black-hole binaries. MNRAS, **406**(Aug.), 840–847.

Podsiadlowski, P., and Joss, P. C. 1989. An alternative binary model for SN1987A. Nature, **338**(Mar.), 401–403.

Podsiadlowski, P., Joss, P. C., and Hsu, J. J. L. 1992. Presupernova evolution in massive interacting binaries. ApJ, **391**(May), 246–264.

Podsiadlowski, P., Langer, N., Poelarends, A. J. T., Rappaport, S., Heger, A., and Pfahl, E. 2004a. The effects of binary evolution on the dynamics of core collapse and neutron star kicks. ApJ, **612**(Sept.), 1044–1051.

Podsiadlowski, P., Mazzali, P. A., Nomoto, K., Lazzati, D., and Cappellaro, E. 2004b. The rates of hypernovae and gamma-ray bursts: implications for their progenitors. ApJ, **607**(May), L17–L20.

Podsiadlowski, P., and Mohamed, S. 2007. The origin and evolution of symbiotic binaries. Baltic Astronomy, **16**, 26–33.

Podsiadlowski, P., Morris, T. S., and Ivanova, N. 2006 (Dec.). Massive binary mergers: a unique scenario for the sgB[e] phenomenon? Pages 259+ of: M. Kraus and A. S. Miroshnichenko (eds.), *Stars with the B[e] Phenomenon*. Astronomical Society of the Pacific Conference Series, vol. 355.

Podsiadlowski, P., Morris, T. S., and Ivanova, N. 2007. The progenitor of SN 1987A. In: *Supernova 1987A: 20 Years After: Supernovae and Gamma-Ray Bursters*, American Institute of Physics Conference Series, **937**, 125–133.

Podsiadlowski, P., and Rappaport, S. 2000. Cygnus X-2: The descendant of an intermediate-mass X-ray binary. ApJ, **529**(Feb.), 946–951.

Podsiadlowski, P., Rappaport, S., and Han, Z. 2003a. On the formation and evolution of black hole binaries. MNRAS, **341**(May), 385–404.

Podsiadlowski, P., Rappaport, S., and Pfahl, E. D. 2003b. X-ray binaries and the origin of binary millisecond pulsars. Pages 283+ of: M. Bailes, D. J. Nice, and S. E. Thorsett (eds.), *Radio Pulsars*. Astronomical Society of the Pacific Conference Series, vol. 302.

Podsiadlowski, P., Rappaport, S., and Pfahl, E. D. 2002. Evolutionary sequences for low- and intermediate-mass X-ray binaries. ApJ, **565**(Feb.), 1107–1133.

Pylyser, E., and Savonije, G. J. 1988. Evolution of low-mass close binary sytems with a compact mass accreting component. A&A, **191**(Feb.), 57–70.

Pylyser, E. H. P., and Savonije, G. J. 1989. The evolution of low-mass close binary systems with a compact component. II – Systems captured by angular momentum losses. A&A, **208**(Jan.), 52–62.

Rappaport, S. A., Podsiadlowski, P., and Pfahl, E. 2005. Stellar-mass black hole binaries as ultraluminous X-ray sources. MNRAS, **356**(Jan.), 401–414.

Riess, A. G., Filippenko, A. V., Challis, P., Clocchiatti, A., Diercks, A., Garnavich, P. M., Gilliland, R. L., Hogan, C. J., Jha, S., Kirshner, R. P., Leibundgut, B., Phillips, M. M., Reiss, D., Schmidt, B. P., Schommer, R. A., Smith, R. C., Spyromilio, J., Stubbs, C., Suntzeff, N. B., and Tonry, J. 1998. Observational evidence from supernovae for an accelerating universe and a cosmological constant. AJ, **116**(Sept.), 1009–1038.

Ritter, H., and Kolb, U. 1998. Catalogue of cataclysmic binaries, low-mass X-ray binaries and related objects (Sixth edition). A&AS, **129**(Apr.), 83–85.

Ruderman, M., Shaham, J., and Tavani, M. 1989. Accretion turnoff and rapid evaporation of very light secondaries in low-mass X-ray binaries. ApJ, **336**(Jan.), 507–518.

Ruiz-Lapuente, P., Comeron, F., Méndez, J., Canal, R., Smartt, S. J., Filippenko, A. V., Kurucz, R. L., Chornock, R., Foley, R. J., Stanishev, V., and Ibata, R. 2004. The binary progenitor of Tycho Brahe's 1572 supernova. Nature, **431**(Oct.), 1069–1072.

Ruszkowski, M., and Begelman, M. C. 2003. Eddington limit and radiative transfer in highly inhomogeneous atmospheres. ApJ, **586**(Mar.), 384–388.

Schwab, J., Podsiadlowski, P., and Rappaport, S. 2010. Further evidence for the bimodal distribution of neutron-star masses. ApJ, **719**(Aug.), 722–727.

Smith, N., Davidson, K., Gull, T. R., Ishibashi, K., and Hillier, D. J. 2003. Latitude-dependent effects in the stellar wind of η Carinae. ApJ, **586**(Mar.), 432–450.

Socrates, A., and Davis, S. W. 2006. Ultraluminous X-ray sources powered by radiatively efficient two-phase super-Eddington accretion onto stellar-mass black holes. ApJ, **651**(Nov.), 1049–1058.

Spruit, H. C., and Taam, R. E. 2001. Circumbinary disks and cataclysmic variable evolution. ApJ, **548**(Feb.), 900–907.

Taam, R. E., and Sandquist, E. L. 2000. Common envelope evolution of massive binary stars. ARA&A, **38**, 113–141.

Tauris, T. M., van den Heuvel, E. P. J., and Savonije, G. J. 2000. Formation of millisecond pulsars with heavy white dwarf companions: extreme mass transfer on subthermal timescales. ApJ, **530**(Feb.), L93–L96.

Terman, J. L., Taam, R. E., and Hernquist, L. 1995. Double core evolution. 7: The infall of a neutron star through the envelope of its massive star companion. ApJ, **445**(May), 367–376.

Thorne, K. S., and Zytkow, A. N. 1975. Red giants and supergiants with degenerate neutron cores. ApJ, **199**(July), L19–L24.

Thorne, K. S., and Zytkow, A. N. 1977. Stars with degenerate neutron cores. I – Structure of equilibrium models. ApJ, **212**(Mar.), 832–858.

Tylenda, R., and Soker, N. 2006. Eruptions of the V838 Mon type: stellar merger versus nuclear outburst models. A&A, **451**(May), 223–236.

van den Heuvel, E. P. J. 2004 (Oct.). X-ray binaries and their descendants: binary radio pulsars; evidence for three classes of neutron stars? Pages 185+ of: V. Schoenfelder, G. Lichti, and C. Winkler (eds.), *5th INTEGRAL Workshop on the INTEGRAL Universe*. ESA Special Publication, vol. 552.

van den Heuvel, E. P. J., Bhattacharya, D., Nomoto, K., and Rappaport, S. A. 1992. Accreting white dwarf models for CAL 83, CAL 87 and other ultrasoft X-ray sources in the LMC. A&A, **262**(Aug.), 97–105.

van Kerkwijk, M. H., Charles, P. A., Geballe, T. R., King, D. L., Miley, G. K., Molnar, L. A., van den Heuvel, E. P. J., van der Klis, M., and van Paradijs, J. 1992. Infrared helium emission lines from Cygnus X-3 suggesting a Wolf-Rayet star companion. Nature, **355**(Feb.), 703–705.

van Rensbergen, W., De Loore, C., and Jansen, K. 2006. Evolution of interacting binaries with a B type primary at birth. A&A, **446**(Feb.), 1071–1079.

Verbunt, F., and Zwaan, C. 1981. Magnetic braking in low-mass X-ray binaries. A&A, **100**(July), L7–L9.

Voss, R., and Nelemans, G. 2008. Discovery of the progenitor of the type Ia supernova 2007on. Nature, **451**(Feb.), 802–804.

Wampler, E. J., Wang, L., Baade, D., Banse, K., D'Odorico, S., Gouiffes, C., and Tarenghi, M. 1990. Observations of the nebulosities near SN 1987A. ApJ, **362**(Oct.), L13–L16.

Webbink, R. F. 1984. Double white dwarfs as progenitors of R Coronae Borealis stars and type I supernovae. ApJ, **277**(Feb.), 355–360.

Webbink, R. F. 1988. Late stages of close binary systems – clues to common envelope evolution. Pages 403–446 of: K.-C. Leung (ed), *Critical Observations versus Physical Models for Close Binary Systems*, Gordon and Breach, New York.

Whelan, J., and Iben, I., Jr., 1973. Binaries and supernovae of type I. ApJ, **186**(Dec.), 1007–1014.

Woosley, S. E. 1993. Gamma-ray bursts from stellar mass accretion disks around black holes. ApJ, **405**(Mar.), 273–277.

Woosley, S. E., and Heger, A. 2006. The progenitor stars of gamma-ray bursts. ApJ, **637**(Feb.), 914–921.

Yoon, S.-C., and Cantiello, M. 2010. Evolution of massive stars with pulsation-driven superwinds during the red supergiant phase. ApJ, **717**(July), L62–L65.

Yoon, S.-C., and Langer, N. 2004. Presupernova evolution of accreting white dwarfs with rotation. A&A, **419**(May), 623–644.

Yoon, S.-C., and Langer, N. 2005. Evolution of rapidly rotating metal-poor massive stars towards gamma-ray bursts. A&A, **443**(Nov.), 643–648.

Yoon, S.-C., Podsiadlowski, P., and Rosswog, S. 2007. Remnant evolution after a carbon-oxygen white dwarf merger. MNRAS, **380**(Sept.), 933–948.

Yungelson, L. R., Livio, M., Tutukov, A. V., and Saffer, R. A. 1994. Are the observed frequencies of double degenerates and SN IA contradictory? ApJ, **420**(Jan.), 336–340.

Zickgraf, F.-J., Kovacs, J., Wolf, B., Stahl, O., Kaufer, A., and Appenzeller, I. 1996. R4 in the Small Magellanic Cloud: a spectroscopic binary with a B[e]/LBV-type component. A&A, **309**(May), 505–514.

3. Accretion onto white dwarfs

BRIAN WARNER

3.1 Accretion from the ISM and winds

3.1.1 Introductory remarks on white dwarfs

The spectra of white dwarfs (WD) are classified according to the scheme devised by Sion et al. (1983), of which we need here to use only the types DA (with strong H lines), DB (with He I lines and no H), and DZ (metallic lines, e.g., Ca, but excluding C, subdivided into DAZ and DBZ). In addition, magnetic fields in WDs play important roles in accretion processes. Their occurrence in isolated form (or as members of noninteracting binaries) is observed by Zeeman splitting or polarization, and the distribution of field strengths appears bimodal: Wickramasinghe and Ferrario (2000, 2005) conclude that $\sim 16\%$ of WDs have strong fields (≥ 0.5 MG); a much smaller fraction have lower fields, but there are indications of a rise of up to 25% at the kG level.

3.1.2 Accretion from the ISM

Most isolated WDs are of type DA or DB, but a small fraction at the cool end of the WD sequence are of type DZ (Fig. 3.1). The reason for ignoring carbon in this spectral type is because it can be dredged up from the interior, whereas the other metals must have a different origin. Levitation by radiation pressure is not strong enough to keep metals in the atmospheres of such stars (for T < 40,000 K), and gravitational settling time scales are short compared with the cooling time scale, so the metals must have been delivered from outside the star – such as from the interstellar medium (ISM).

Accretion onto a star in general is an alliance of its environment and (particularly for a WD) its intrinsic magnetic field. The lowest rates of accretion come from the ISM and are estimated from the standard Bondi and Hoyle (1944) formula (which uses the gravitational cross section – much larger than the physical size of a star because of convergent flow lines). The equation usually used is a modification of earlier formulae (see review by Edgar, 2004):

$$\dot{M} = 4\pi \frac{G^2 M^2 \rho_\infty}{(c_\infty^2 + v_\infty^2)^{3/2}} \quad (3.1)$$

where ρ_∞ and c_∞ are the interstellar gas density and sound speed, and v_∞ is the speed of the star through the gas. Typical values for WDs are $\sim 10^{-21}$ to 10^{-16} $M_\odot$ yr^{-1}, with the lowest value arising from characteristic interstellar gas densities and the higher from interstellar gas clouds (through which WDs typically are passing for about 2% of their life: a million years in every 50 million years (Wesemael, 1979). Clearly such interstellar accretion does not noticeably increase the mass of the WD, and the heating effect of accretion impact at these rates is negligible. But it can significantly affect the atmospheric composition of a WD atmosphere, depending on the diffusion time – the time scale on which heavier elements settle below the photosphere (typically $\sim 5 \times 10^5$ yr). This process explains the presence of metal lines (e.g., Ca, Na, Mg, Si) in the spectra of some cooler (T $\leq$ 15,000 K) DA and DB white dwarfs and obviously requires the accretion time scale to be of the order of, or shorter than, the diffusion time scale. The details are quite complicated – metals are largely accreted as interstellar grains, and the observed ratios of metal abundances is in moderate agreement with ratios of the individual time scales (Dupuis et al., 1993), but, from the absence of H lines in the DBZ stars, some mechanism evidently prevents the hydrogen gas from being captured at the same rate. This turns out to be weak magnetic fields.

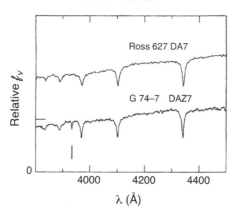

FIGURE 3.1. Spectra of two cool DA white dwarfs, one showing additional 3933 Å Ca II absorption. From Wesemael et al. (1993).

Whereas accreting dust grains are not affected by weak fields, hydrogen is ionized at a sufficient distance that it becomes attached to the field lines and from even a slowly rotating WD is centrifuged away – the "propeller mechanism" of Wesemael and Truran (1982). Thus, the M/H ratio of the accreted material is enhanced by many orders of magnitude, and H is not added to DB atmospheres. Despite the low accretion rates and the intervention of a weak magnetic field, the observational result is visibly dramatic: strong metallic lines in WDs.

There is also a circumstellar source of metals in WDs – the accretion of dust rings that have formed around them as collision debris from asteroids that survived the red giant phase. This also does not accrete any H and may be the dominant process that generates the DZ stars. A number of WDs with infrared excesses indicative of such dust rings are known (e.g., Gänsicke et al., 2008).

3.1.3 Accretion from stellar winds

Accretion at much higher rates occurs if the WD is immersed in the wind from a giant companion. Such combinations of giant plus low-luminosity companions are probably quite common, but usually difficult to detect because of the large difference in luminosities. However, there is one that is sufficiently near to us ($\sim$130 pc) to enable a spatially resolved study to be made of wind accretion; it is Mira (Karovska et al., 1997). Mira is the prototype of long-period variable stars; it has a pulsation period of 332 d and a companion, Mira B, at a separation of 0.4″, with an orbital period $\sim$400 yr. Mira A has a low-velocity ($\sim$6 km s^{-1}) wind with $\dot{M} \sim 4 \times 10^{-7}$ M$_\odot$ yr^{-1} through which Mira B moves with a velocity $\sim$7 km s^{-1}. The wind density near Mira B is $\sim$9 $\times$ 10^{-19} g cm^{-3}, which from equation 3.1 gives an accretion rate onto Mira B of $\sim$8 $\times$ 10^{-10} M$_\odot$ yr^{-1} (Reimers and Cassatella, 1985). Unlike accretion from the ISM, here the accretion luminosity, $L = GM\dot{M}/R$, is highly significant. Evidence for accretion from a turbulent wind was first found from the short–time-scale ($\sim$min) variations of brightness of Mira B observed in the U band when Mira A was at an unusually low minimum (Warner, 1972). Later spectra showed the emission lines typical of optically thin accreting gas (Reimers and Cassatella, 1985). The observed luminosity of Mira B is $\sim$0.2 L$_\odot$, which could be that of a lower main sequence radius rather than that of a WD (e.g., Ireland et al., 2007), though recent work favors the WD solution (Sanad et al., 2009). Furthermore, determining the nature of Mira B is complicated – not least because of interaction of strong winds from both Mira A and Mira B, the existence of the latter itself suggesting a compact accretor.

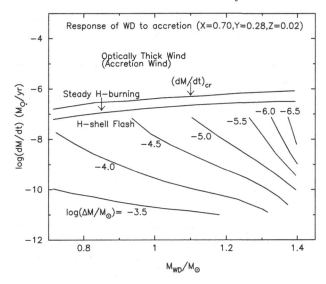

FIGURE 3.2. Response of a white dwarf to mass accretion (Kato, 2010).

Before investigating higher $\dot{M}$, we need to look at how a WD deals with the mass falling onto it.

3.1.4 White dwarf response to accretion

A hydrogen-accreting WD spends its life searching for an (unattainable) equilibrium. The outcome depends sensitively on the mass of the WD and on $\dot{M}$. For lower $\dot{M}$, the WD is, for most of its time, accumulating a surface layer and has a luminosity largely generated by impact and compressional heating as the accreting gas falls onto it. This layer has a critical mass, determined by conditions at its base (where the gas is partially degenerate), that leads to thermonuclear runaway when exceeded, changing the star into a hot giant structure with a nuclear shell source. But this has only a short existence because it quickly exhausts the available nuclear fuel, and so it again reverts to the low-luminosity accretion state. This is the repetitive cycle between nova eruptions and their "quiescent" state (Fujimoto, 1982) and has been followed in great detail in time-dependent stellar structure calculations (e.g., Prialnik, 1986; Prialnik and Kovetz, 1995; Epelstain et al., 2007). The nova eruption ejects the $\sim 10^{-4}$ $M_\odot$ that has accumulated over the previous $\sim 10^4$ yr.

For large $\dot{M}$ ($>2 \times 10^{-7}$ $M_\odot$ yr^{-1}), *the accreting hydrogen burns continuously* near the surface, so there is no eruption, and the star becomes a supersoft X-ray source (Kato, 2010), which we describe later. At even higher $\dot{M}$ the WD accretes at the equator and expels most of the mass in optically thick winds in other directions (Fig. 3.2).

Nova eruptions largely involve explosive mass loss, nonequilibrium nuclear reactions, and strong winds and so are of direct relevance here, but the long-lived phases of accretion and the rich phenomenology and variety of the accretion structures between eruptions leads to the "zoo" of cataclysmic variable stars (CVs) – the main focus of this article, as seen in Section 3.3.

3.1.5 Symbiotic stars and supersoft sources

Symbiotic stars (SySs) are a mixture of different binary types; here we are interested only in those that contain WDs orbiting in very strong winds. Mira is classified as a weak SyS. Most SySs have orbital periods of a few years to several decades and contain cool supergiants with wind $\dot{M} \sim (1\text{--}100) \cdot 10^{-7}$ $M_\odot$ yr^{-1}, of which $\sim (1\text{--}10) \cdot 10^{-8}$ $M_\odot$ yr^{-1} accretes onto the companion star, which can be a dwarf, a hot subdwarf, a WD, or a neutron star (Kenyon, 1988; Iben, 2003). To some extent, the accreting component can be identified according to the nature of the X-rays emitted: the majority of SySs have detectable X-rays. As mentioned previously, the supersoft sources (peak energies below 0.4 keV) have steady H burning, but there are other "soft" sources where the X-rays are strong from 0.1 to 10 keV but peak at 0.8 keV. The hot regions are thought to arise from collision between the wind from the giant and the wind from the WD. Note that because of interstellar absorption, we know only a small fraction of the supersofts in our galaxy (they must be closer than $\sim$2 kpc to be detected: only about 4 out of an estimated galactic population of $\sim$200 are observable).

The SySs provide examples of higher $\dot{M}$ onto WDs; where thermonuclear runaways occur, they produce symbiotic novae, which are slow-motion equivalents of classical novae, remaining in eruption for up to several decades and having amplitudes of $\sim$10 mag. Examples are Z And, V1016 Cyg, HM Sge, and RR Tel. An advantage that study of symbiotic novae has over classical novae is that the components are so widely separated that there is no common envelope phase (which can complicate matters with, e.g., exchange of angular momentum between the gas ejected and the orbit).

Examples of supersoft sources are CAL 87 and SMG 13; two supersofts that have strong winds are V Sge and RX J0513-69.

3.2 Roche lobe overflow

The physics of Roche lobe overflow is dealt with in Henk Spruit's chapter. Such overflow results in a variety of remarkable phenomena that define the various types and subgroups of CVs. These take their name from the nova eruptions and dwarf nova outbursts that draw on nuclear energy and gravitational potential energy sources respectively.

We begin with WD primaries that have negligible intrinsic magnetic fields and then examine the effects of strong, medium, and weak fields (comprehensive reviews of CVs are given in Warner, 1995a, and Hellier, 2001).

3.2.1 Low (effectively zero) field accretion

In the early studies of CVs (before 1977), little thought was given to the effects of accretion onto a WD with an intrinsic magnetic field. The basic physics then considered was that of stream flow, stream impact at the outer rim of an accretion disk, mass and angular momentum transfer through the disk, and the boundary layer where the disk meets the surface of the WD. This is still applicable to "nonmagnetic" CVs and is adequate to account for most of their outburst properties. We examine first the range of time scales of dwarf novae (DNe), remembering that a DN outburst results from intermittent rapid accretion of gas that has accumulated in the disk.

Outburst light curves of dwarf novae

The outburst light curves (Fig. 3.3) of DNe are characterized by four time scales: rise time scale τ_r; fall time scale τ_d (usually rate of final descent from maximum or plateau); duration Δ (mid-rise to mid-fall); and recurrence time T_R (interval between outbursts). τ_d is determined by the velocity of the cooling front as it runs from the outer rim of the disk to the inner edge; the radius of the disk is largely dependent on size of the orbit and hence the orbital period P_{orb}. The observed relationship, covering all P_{orb} up to 5 days, is $\tau_d \sim 0.53\, P_{orb}(h)^{0.84}$ days per magnitude, which can often be used to estimate P_{orb}. The rise time scale is not simply a function of P_{orb}, as can be seen by the presence of both slow and fast rises in some individual DNe. This is because a heating front propagating outward in the disk moves much slower than an inward-propagating wave, so in/out outbursts take longer to switch a disk to the high state than do out/in outbursts. For short P_{orb}, typically $\tau_r \sim 0.3$ to $0.5\, \tau_d$, and for very long P_{orb} (1 to 5 days), the rise and

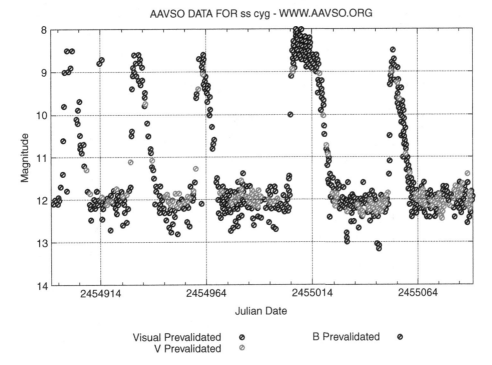

FIGURE 3.3. Light curve of SS Cygni (from AAVSO archives).

fall rates are almost equal. The duration Δ depends on the mass of gas that accumulates before the critical condition is reached and is largely dependent on the size of the disk: $\Delta \sim 0.90\,\mathrm{P}_{orb}(\mathrm{h})^{0.80}$ days. The recurrence time depends both on the amount of gas that the disk can hold and the rate $\dot{M}$ at which it is being filled; it has a very wide range, from a few days to several decades.

These empirical time scales are quantitatively reproduced by general analytical and detailed computer models of accretion disk instabilities (reviewed in Warner, 1995a), which is part of the support for this interpretation of disk outbursts. There is additional evidence from other directions. For example, in some DNe it is possible to measure variations of disk radius during outburst by using eclipse timings of the bright spot. Figure 3.4 shows the radius r_d of the disk in Z Cha, in fractions of the orbital separation a, as a function of the time between outbursts (O'Donoghue, 1986). The disk expands to $r_d/a \sim 0.38$ at the start of outburst and shrinks back to ~ 0.25 before the next outburst. This is a direct result of conservation of angular momentum: the gas accreting onto the WD must lose angular momentum, which it does by giving it to gas that then moves further out from the WD. The shrinkage after outburst is caused by mixing of the low specific angular momentum gas, arriving from the companion star, with the gas in the outer regions of the disk.

A small subset of the DNe are called Z Cam stars, after the prototype. These spend most of their time acting like DNe but can get stuck in outburst (though ~ 0.8 mag below maximum brightness) for days to years (Fig. 3.5). They arise in systems where the mean $\dot{M}$ is very close to the critical value above which the disk is in a stable high-viscosity state, and below which the disk is unstable, so small variations have large consequences.

For $\dot{M}$ steadily above the critical value, the disk is permanently in a high-viscosity, high-luminosity state, and the system is called a *nova-like variable*. In such stars, the radiation is dominated by the disk (except for those with large P_{orb}, in which case there may be significant contribution from the large secondary). These are ideal for studying the properties of accretion disks – in particular their almost two-dimensionality, which results

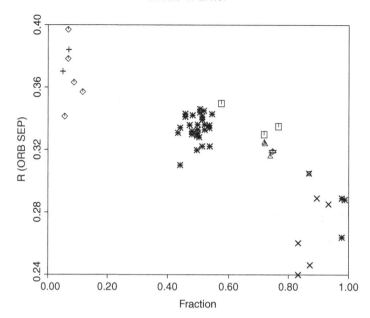

FIGURE 3.4. Disk radius of Z Cha between outbursts (from O'Donoghue, 1986).

in great sensitivity of *apparent luminosity* to *inclination*, i. For example, an optically thick disk seen at $i = 0°$ is 3.5 mag brighter than when viewed at $i = 85°$, and the fainter disk continuum when seen side-on allows its emission lines to achieve greater prominence (Fig. 3.6).

There is another subclass of DNe that shows additional phenomena connected with the accretion flow through disks. These are the SU UMa stars, in which superoutbursts occur in addition to normal outbursts (Fig. 3.7). They are a property of short-period systems ($P_{orb} \lesssim 3$ h), last typically for 10 to 14 d (i.e., 3 to 5 times the length of normal outbursts), and at maximum are about twice as bright as normal outbursts. The accretion energy released in these is $\sim 10^{39}$ erg for normal and $\sim 10^{40}$ erg for superoutbursts, which is an indicator that normal outbursts drain only about 10% of the disk, whereas superoutbursts drain out almost all of the gas.

The underlying physics of a superoutburst is a resonance effect in the orbits of particles in the outer part of the disk: if an orbit in the disk has a period that is a simple rational fraction of P_{orb}, then periodic perturbations from the secondary will accumulate and distort the orbit, causing collisions, equivalent to a local increase of viscosity in the gas. The increased viscosity maintains the disk in a high state until almost all the gas has drained out. Between superoutbursts (i.e., during a supercycle), successive normal outbursts fail to drain all the mass gained from the secondary, so the disk grows in mass and radius until an outburst starts in which the outer parts of the disk finally are pushed out to the 3:1 resonance radius.

The large range of $\dot{M}$ from secondary stars results in two extreme kinds of SU UMa stars: for very low $\dot{M}$ the intervals between outbursts and/or superoutbursts are a year or more, with few normal outbursts in a supercyle. The longest known interval is in WZ Sge, which has no normal outbursts, other than the one that triggers a superoutburst every 20 to 30 years. In some of these "WZ Sge" stars, there is a chain of short-lived normal outbursts, termed echo outbursts, immediately after the infrequent superoutburst – probably the result of irradiation of the secondary (and consequently enhanced $\dot{M}$) while the WD is still cooling down after being heated by the mass accretion during the superoutburst.

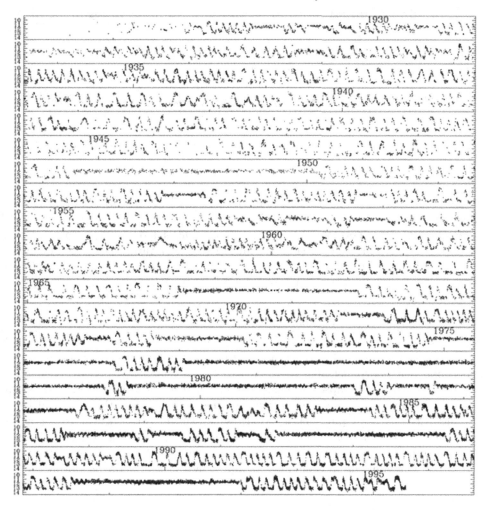

FIGURE 3.5. Light curve of Z Cam (from AAVSO archives).

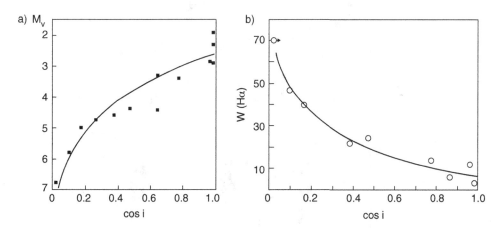

FIGURE 3.6. *a)* Dependence of absolute magnitude and *b)* Hα equivalent width on disk inclination in nova remnants (Warner, 1986b, 1987, respectively).

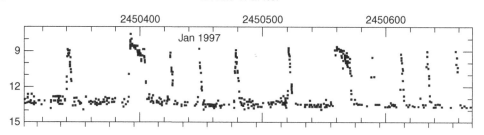

FIGURE 3.7. One year of the light curve of VW Hyi (from RAS New Zealand data – Hellier, 2001).

At the other extreme of the SU UMa stars are the ER UMa (or RZ LMi) stars, where $\dot{M}$ is so high that it is close to the critical rate for a stable disk. Only about six of these are known. They never reach a quiescent state, always descending from outburst or rising to the next one, and as such are among the most variable objects in the sky. The intervals between superoutbursts are ~20 to 50 days, and there is time for only two or three normal outbursts per supercycle (Fig. 3.8).

Brightness variations on orbital time scales

Before moving to the main part of this review, we briefly look at several powerful analysis techniques that are used in many areas of interacting binary stars. Some examples appear among the later illustrations.

(i) *Eclipse mapping:* Eclipses in binary systems offer opportunities to map the intensity distribution of emitting regions. In the simplest case, this can lead to an intensity and color (and hence radial temperature) distribution across an accretion disk. Because of intrinsic flickering in mass-transferring systems, observed eclipse profiles are noisy; the result is that a unique deconvolution is not possible. A commonly used technique, originated by Horne (1985), transforms the 1-D

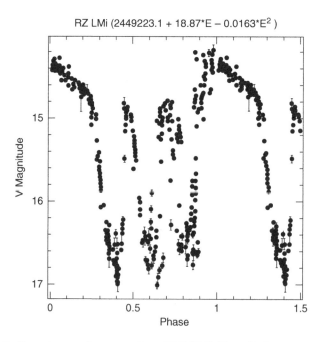

FIGURE 3.8. Superimposed supercycles of RZ LMi (from Robertson *et al.*, 1995).

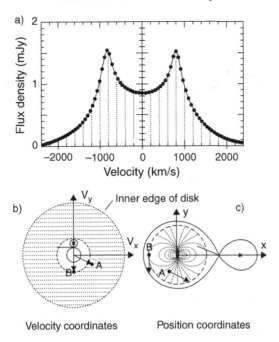

FIGURE 3.9. *a)* Line profile in velocity units. *b)* Lines of constant radial velocity (Vy) (horizontal) and *c)* lines of equal Doppler shift for a disk on Keplerian rotation. The points A and B are equivalent regions in the space and velocity planes (Marsh and Horne, 1988).

eclipse profile to a 2-D intensity distribution in a way analogous to X-ray tomography, adding the requirement of maximum entropy of the process (i.e., a minimum of constraints – though maximal axial symmetry of the distribution often has to be imposed).

(ii) *Doppler tomography:* This technique uses the Doppler-broadened emission lines that arise in optically thin parts of a mass transfer process and transforms their profiles into 2-D velocity space (Marsh and Horne, 1988). The principles are illustrated in Figure 3.9; it is assumed that all the motions lie in the orbital plane. In order to avoid the assumption of Keplerian motion, the transformation is left in velocity space. The theoretical velocity components of the surface of the secondary star and the interstar gas stream are often superimposed in the velocity diagram.

(iii) *Orbital modulations:* Among the most prominent variations on orbital time scales are those generated by revolution in orbit – for example, eclipses and systematic brightness modulations due to varying aspects of the bright spot region. Eclipse mapping (see Section 3.2.1) leads to radial temperature $T(r)$ determinations that, for DNe in outburst and for nova-likes, are in agreement with the $T(r) \sim r^{-3/4}$ expectation of accretion disk theory, and for DNe in quiescence show an almost flat distribution of $T(r)$ (Mineshige and Wood, 1989). For high-inclination disks, the effect of the upturn ("flaring") at the outer edge of the disk must be taken into account; otherwise, false $T(r)$ are obtained (Smak, 1994). The technique of eclipse mapping can also be applied to spectrum lines (Rutten *et al.*, 1994; Baptista *et al.*, 1998), in which a change can be seen from an optically thick spectrum with absorption lines in the inner disk to optically thin with emission lines in the outer disk.

Light curves of CVs do not normally show secondary eclipses in the optical – the secondary star contributes so little to the total luminosity that its eclipse by the accretion disk (seen almost side-on) has no effect. But recently Nova Puppis 2007 (V597 Pup: $P_{orb} = 2.67$ h) has proved an exception to this rule: the orbital

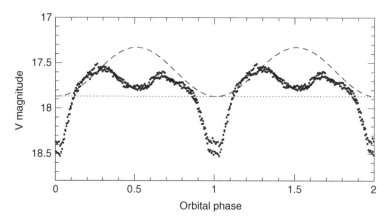

FIGURE 3.10. Light curve of the nova remnant V597 Pup (Warner and Woudt, 2009).

light curve obtained only 4 months after its eruption shows substantial secondary eclipses because the hemisphere of the companion star was brightly illuminated by radiation from the still very hot white dwarf (Fig. 3.10).

(iv) *Superhumps:* During superoutbursts of DNe, a photometric modulation is seen that has a period P_s typically a few percent longer than P_{orb} (Vogt, 1974; Warner, 1975), the excess $\varepsilon = P_s - P_{orb}/P_{orb}$ being correlated with P_{orb} (Fig. 3.11; Patterson, 1998). The explanation is that the 3:1 resonance (see Section 3.2.1, *Outburst light curves of dwarf novae*) excites the outer parts of the accretion disk into an elliptical shape, which then precesses with a period P_b ($\sim$2 d) due to the tidal influence of the secondary star. As a result, the passage of the secondary in orbit stresses and heats the outer parts of the disk with period P_s, where $1/P_s = 1/P_{orb} - 1/P_b$. The severe distortion of the disk is confirmed by the observed strongly asymmetric absorption line profiles of the disk, which change their asymmetries periodically at the period P_b.

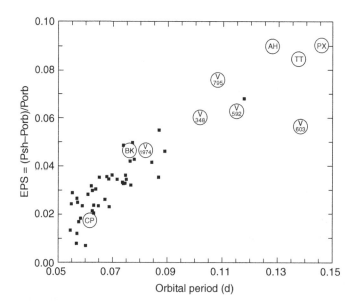

FIGURE 3.11. Correlation between superhump excess ε and orbital period (Patterson, 1998). Reproduced by permission of the AAS.

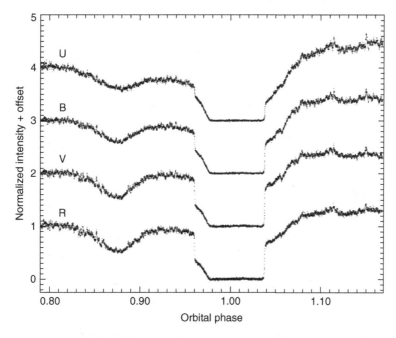

FIGURE 3.12. Optical light curve of the polar HU Aqr (Harrop-Allin et al., 1999).

Before leaving the low-field CVs, we mention briefly that there is another group that are helium-transferring systems, with 5 min $\lesssim P_{orb} \lesssim$ 1 h, known collectively as AM CVn stars. These have behaviors very similar to those of the H-rich CVs, showing DN outbursts, superoutbursts, superhumps, and nova-like light curves (e.g., Warner, 1995b).

3.2.2 High field accretion (polars)

We begin (cf. Hellier, 2001) with the CV containing the largest known magnetic field, AR UMa. The field is ∼230 MG, which is strong enough for the WD's field lines to join onto those of the secondary star. A consequence of this is that the WD rotates synchronously in orbit – if it started out asynchronously the two rotating magnets would quickly align themselves. There is therefore only one "clock" in such a system – P_{orb}. Another consequence is that ionized gas leaving the secondary at the L1 point immediately is threaded onto field lines and falls along them to the surface of the WD. There is therefore no accretion disk in this accreting binary.

More generally, CVs with sufficiently strong fields ($\gtrsim$10 MG) have synchronous rotation of their primaries, but the stream from the secondary is able to penetrate into the magnetosphere of the WD before being stopped by the pressure of the field at the threading (or coupling, or stagnation) region, from where the gas accretes along field lines. Near the surface of the WD, the gas is heated by a stand-off shock (typically at a height of a few percent of the WD radius), below which it forms an accretion column, is slowed by pressure, and radiates its thermal energy so that it falls gently onto the surface. The height of the shock is set by these requirements.

Radiation from the column is bremsstrahlung and cyclotron. The latter gives rise to strong linear and circular polarization, a characteristic that results in these systems being called polars (also AM Her stars, after the first one discovered). The shock region has a temperature ∼10^8 K, producing strong X-ray emission. There are ∼100 polars known, many of them discovered from their X-rays. The optical light curve of HU Aqr (Fig. 3.12: Harrop-Allin et al., 1999), made with 0.2 s time resolution, shows a number of

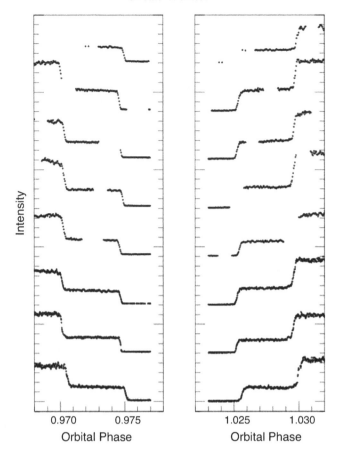

FIGURE 3.13. Ingress and egress light curves of SDSS J0155+0028, obtained with the SALT 11-m telescope (O'Donoghue et al., 2006).

characteristic features of an eclipsing polar: the wide dip before the eclipse is caused by the stream partially obscuring the accreting WD; and the eclipse shows rapid ingress of the accretion region on the WD followed by a slower ingress of the illuminated stream, followed by egress of those components in reverse order. In the polar SDSS J0155+0028, eclipses of two bright accretion zones are seen, ingresses and egresses last only 1 to 2 s, and there is no detectable radiation from the surface of the WD itself (Fig. 3.13). The accretion spots cover only $\sim 1\%$ of the surface of the WD.

As an example of Doppler tomography applied to a polar, we show results for HU Aqr in Figure 3.14 (Schwope et al., 1997; Heerlein et al., 1999). The time-resolved spectrum of the He II emission line is show in the left panel, and the derived Doppler tomogram is on the right. The latter shows emission from the surface of the secondary (which is exposed to the X-rays emitted by the accretion zone on the WD) and the emitting stream, which initially terminates at the threading zone (three possible positions along the trajectory are illustrated) and then reappears (weakly) as gas falling along field lines to the WD.

Figure 3.15 shows another way of analyzing the same star – from Roche tomography (Watson et al., 2003). This is a special case of Doppler tomography, using the emission lines from the secondary star, which are emitted over its Roche surface and so their positions and velocities are known.

Emission from the cyclotron component of the accretion zone exhibits three characteristics: linear and circular broadband polarization that varies around orbit as the angle to the observer changes, cyclotron harmonic humps in the spectral energy distribution

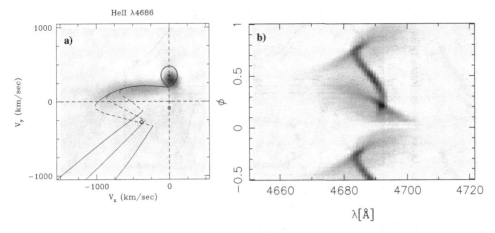

FIGURE 3.14. a) Doppler and b) Spectra tomogram of HU Aqr (from Schwope et al. 1997; Heerlein et al. 1999, respectively).

(also varying in amplitude and peak position around orbit), and Zeeman splitting of absorption lines arising in the atmosphere of the WD (or the cooler gas on the outside of the infalling flow).

Figure 3.16 shows broadband ("white light") polarimetric observations (total intensity, linear polarization percentage and direction, percentage circular polarization) for V834 Cen around orbit. These and Doppler tomograms lead to the derived field geometry and accretion zone intensity distribution (Potter et al., 2004). The tomograms show the emission to be asymmetric, which is attributed to shadowing of the surface (by the infalling stream) of X-rays from the WD.

Cyclotron emission is spread over a number of harmonics, the separation between the harmonic peaks gives a direct measurement of the field strength (they are equally spaced in frequency) and varies around orbit as the projection of the field changes. An example is shown in Figure 3.17a (Schwope et al., 1993); Zeeman components in the same star, but in a low state of $\dot{M}$, when the WD continuum was visible, are shown in Figure 3.17b.

The field strengths in polars are so large that simple quadratic Zeeman-effect calculations are inadequate. At the upper end, the calculations can be made simply by considering the electrostatic field of the nucleus to be a small perturbation on the states of electrons spirally around the field lines. For intermediate strengths, full quantum mechanical calculations are required and have been made for only a few atoms and ions. Figure 3.18 shows how the field strength is deduced for AM Her itself.

The overall energy distribution of a polar has three components: hard bremsstrahlung (∼20 keV) from the accretion shock; soft X-ray emission (∼20 eV) from reprocessing of

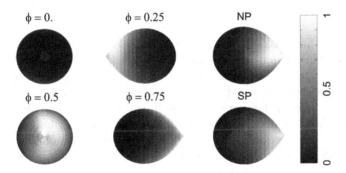

FIGURE 3.15. Brightness distribution on the secondary of HU Aqr (Watson et al., 2003).

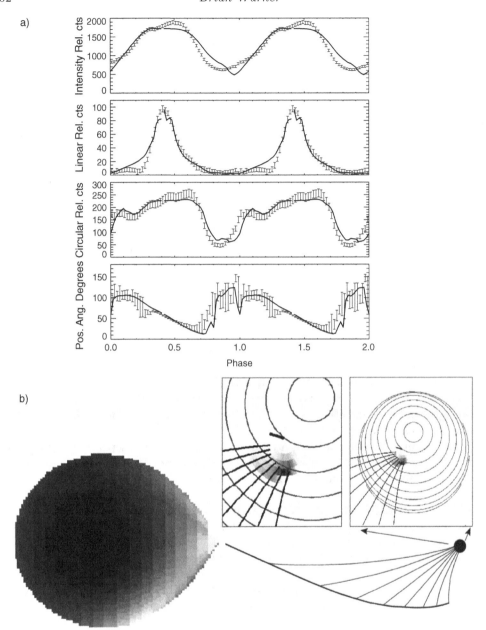

FIGURE 3.16. *a)* Polarimetric observations and *b)* derived field geometry of the polar V834 Cen (Potter *et al.*, 2004).

downward-emitted hard X-rays reaching the surface of the WD and/or from accreting large blobs that bury themselves in the surface of the WD; and cyclotron emission, typically at infrared to ultraviolet wavelengths. The buried blobs produce shocks at large optical depths, the energy emitted being thermalized before emerging through the WD surface in a patch around the accretion zone. The threading region shatters the stream into blobs with a range of sizes, with the result that different distributions of blobs can be delivered to opposite accretion zones, in some cases leading to predominantly hard-X-ray emission from one zone and predominantly soft X-rays from the other zone: then soft and hard X-rays are $\sim$180° out of phase in the light curves (Figure 3.19; Heise *et al.*, 1985).

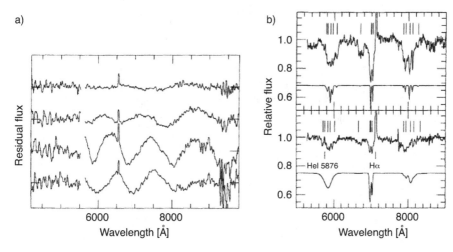

FIGURE 3.17. a) Cyclotron harmonic bumps in the spectral distribution of MR Ser, at orbital phases 0.41, 0.50, 0.58, 0.67 from bottom to top respectively. b) Zeeman components of Hα in MR Ser in its low state. Both from Schwope et al. 1993.

3.2.3 Intermediate field accretion (intermediate polars)

Reducing the strength of the WD's magnetic moment lowers the coupling with the secondary, so the primary no longer rotates synchronously. There are now two "clocks" in the binary – P_{orb} and the rotation (or spin) period P_{rot} of the WD – either or both of which can appear in optical or X-ray light curves and in spectroscopic observations. These systems, of intermediate field strengths (∼1–10 MG), are known as intermediate polars (IPs: sometimes called DQ Herculis systems, after the first one to be identified). There are two types: a rare variety in which the magnetospheric radius of the WD is too large to allow the gas stream to pass around the WD, preventing formation of an accretion disk (leading to "diskless accretion"), and the majority kind in which a disk and impact bright spot occur just as in nonmagnetic systems, but in which the inner disk is absent below a radius where the field strength is sufficient to generate magnetically controlled accretion. There are about 34 confirmed IPs known at present – the site http://asd.gsfc.nasa.gov/Koji.Mukai/iphome/iphome.html provides updates of the list. Note that accreting neutron star binaries resemble IPs, not polars – in the neutron star systems, the magnetic moments ($\mu = BR^3$), despite the huge field strengths, are smaller than in polars, so synchronous rotation does not occur.

Historically, the first comprehensive study of the physics of accretion from a disk onto a rotating primary led to the structure shown in Figure 3.20 (from Ghosh and Lamb, 1978), which is the simplest situation, where the field is dipolar and aligned along the rotation axis of the WD (assumed to be orthogonal to the orbital plane). Field lines thread through the slowly rotating outer parts of the disk and produce a slow-down torque on the primary. The inner parts produce a spinup torque. The boundary layer is the region where field starts to dominate the dynamics of the accretion flow, and within that, gas flows along the field lines. The boundary layer is near the region where the Keplerian angular velocity in the disk equals the angular velocity of the primary (known as the corotation radius, r_{co}) that is, where the gas can most easily thread onto the rotating field lines (beyond r_{co} gas attached to the field lines will be centrifuged outward; only gas threaded within r_{co}, will accrete easily).

In reality, the axis of the field is tilted relative to the pole of rotation, with the result that two accretion curtains channel gas onto the primary, one above and one below the orbital plane, as in Figure 3.21. These play an important role in the optical and X-ray light curves and spectroscopic variations: the accretion footprints are long arcs like

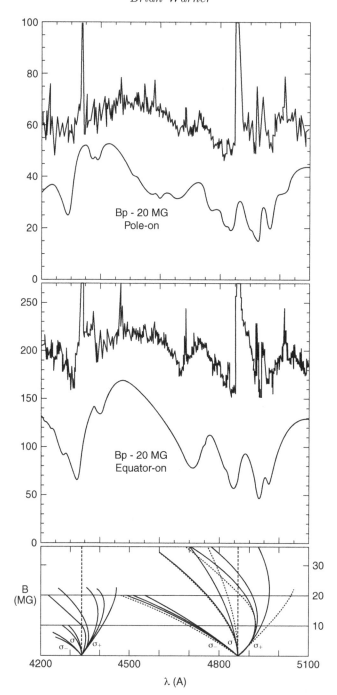

FIGURE 3.18. Spectrum of AM Her in a low state, compared with theoretical spectrum for a dipole field of 20 MG. From Latham *et al.* (1981). Reproduced by permission of the AAS.

the terrestrial aurorae; X-ray emissions from these accretion zones are modulated by the optical thickness of the accretion curtains (with variations according to energy and path thickness); the curtains themselves, illuminated by the nearby X-ray emission, emit radiation that pulsates at the period P_{rot}.

The optical and infrared luminosity of the truncated disk, being unpolarized, dilutes the cyclotron emission from the accretion zones, so only about five of the IPs have

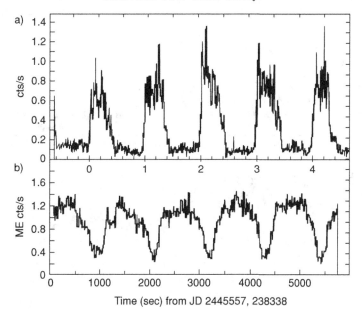

FIGURE 3.19. *a)* Soft X-ray light curve and *b)* hard X-ray light curve of AM Her (Heise *et al.*, 1985).

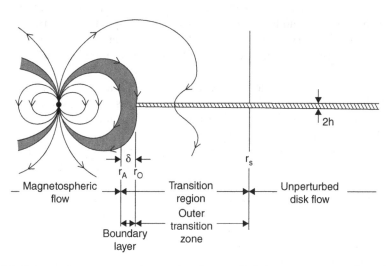

FIGURE 3.20. Schematic structure of accretion from a disk into a magnetosphere (Ghosh and Lamb, 1978). Reproduced by permission of the AAS.

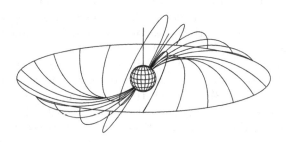

FIGURE 3.21. Accretion curtains (Hellier, 2001).

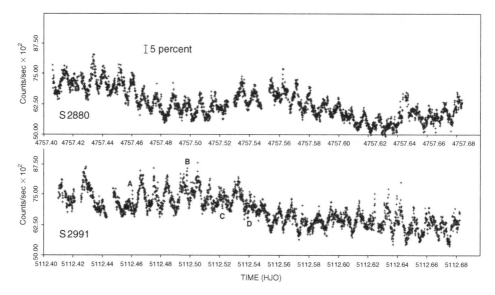

FIGURE 3.22. Light curves of V1223 Sgr (Warner and Cropper, 1984).

detectable polarization. An example light curve is shown in Figure 3.22 – that of V1223 Sgr, which has $P_{orb} = 3.37$ h and a faster modulation, P_b, at 794 s (= 13.2 min), both of which are easily visible in the light curve (Warner and Cropper, 1984). However, Fourier transforms of the light curve and the X-ray light curve both show that there is also a period 745 s (= 12.4 min), which, because the X-ray period must be that of the accretion zones, is P_{rot}.

Where does the additional "clock," P_b, originate? The answer comes from noticing that (to very high accuracy) $1/P_{rot} - 1/P_b = 1/P_{orb}$. This relationship (similar to that connecting sidereal and synodic periods of planets), commonly seen among IPs, implies that the high-energy radiation from the accretion zones, rotating at P_{rot}, sweeps across and is reprocessed by something revolving at P_{orb} – for example, the hemisphere of the secondary facing the primary, or the enlarged region of the disk where the stream impacts and heats it. This generates the beat period, P_b. The ratios of the amplitudes of the signals vary from object to object, and even in a given object, each can vary with time. In the optical, both modulations arise from reprocessing, that is, they are derived from the high-energy beam intersecting the face of the disk or other components of the binary and being downgraded in energy.

Written in frequencies, instead of periods, we have $\omega_b = \omega_{rot} - \Omega_{orb}$, and ω_b is referred to as an orbital sideband frequency. In some IPs, extensive photometry discloses more sidebands, displaced from ω_{rot} by other integer multiples of Ω_{orb}. They arise from a number of effects, the most obvious of which is that the signal that is, for example, "reflected" from the secondary is itself modulated in strength at frequencies of Ω_{orb} and $2\Omega_{orb}$ as the cross section of the secondary varies around orbit (the Roche lobe is seen twice end-on and twice side-on). Such amplitude modulation generates the set of frequencies ω_{rot}, $\omega_{rot} \pm \Omega_{orb}$, $\omega_{rot} - 2\Omega_{orb}$, all of which are seen in some IPs (Warner, 1986a); harmonics of these are also occasionally observed. An example is given in Figure 3.23 (Patterson et al., 1998): although the lowest amplitude components may not appear significant, it is their exact numerical coincidence with the foregoing frequency set that gives them conviction. (Note that Fourier transforms usually contain a window function – the pattern of aliases that results from interrupted observations.)

The material and magnetic torques acting on the primary can lead to a zero torque equilibrium balance if $\dot{M}$ is constant, but slow changes in $\dot{M}$ result in P_{rot} varying as the

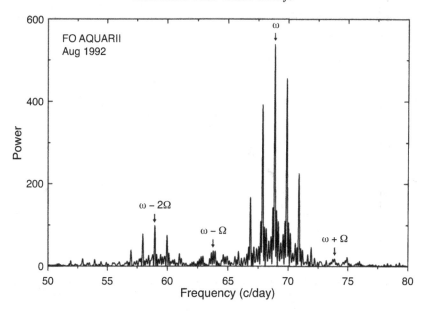

FIGURE 3.23. Power spectrum of FO Aquarii (Patterson *et al.*, 1998). Reproduced by permission of the AAS.

system seeks the equilibrium, spinning up and down by small amounts on time scales of years (Fig. 3.24), with the evolutionary time scale $P_{rot}/(dP_{rot}/dt) \sim 10^7$ yr.

The accretion zones on the primaries of IPs are large and are fed by gas attaching itself from around the inner radius of the disk, with the result that the gas is less "blobby" and X-rays are largely emitted from a stand-off shock; consequently, the majority of IPs are sources of hard X-rays (primarily 2 to 20 keV, but up to 90 keV in some bright sources), in contrast to polars, which are predominantly soft sources ∼40 eV. The X-ray spectra of IPs show a strong energy dependence as a function of spin phase, so the large-amplitude rotational modulation (Fig. 3.25) is due to optical depth effects, not to geometric occultation. This is direct evidence for accretion curtains.

There is, however, a rare class of IPs whose spectra peak in the 40 to 60 eV range. These are interpreted as low accretion rate polars (LARPS), being polars that have not yet synchronized their rotations (Haberl and Motch, 1996; Schwope *et al.*, 2002). On an allied subject, there are four polars in which the primaries are not quite synchronously rotating, and in the case of V1500 Cyg, the desynchronization was the result of angular momentum loss during its eruption. They are not classified as IPs because they are only very temporarily desynchronized – they will return to being polars within 10^2 to 10^3 yr (see Warner, 2002, for magnetic novae). There are a dozen IPs that have known nova eruptions.

The first of the IPs, DQ Herculis, itself the remnant of a nova eruption in 1934, was observed in 1954 as a then uniquely fast variable star with a 71 s sinusoidal modulation of low amplitude (Walker, 1956) and high stability ($Q = \left(\frac{dP}{dt}\right)^{-1} \sim 10^{14}$). Later observations showed that the phase and amplitude of the 71 s oscillations vary systematically through eclipse (Warner *et al.*, 1972), and modeling of these supported the general IP structure (Fig. 3.26; Petterson, 1980 – note that the deduced inclination of DQ Her is 89°, which is why it is not a strong X-ray source). Recent work using eclipse mapping suggests that the 71.1 s observed periodicity is not directly the rotation of the WD but is a reprocessed signal from the disk thickening at the bright spot (Saito and Baptista, 2009).

There are only a few other IPs known with such short periods, namely WZ Sge (27.8 s), AE Aqr (33.1 s), V842 Cen (57 s), V533 Her (63.6 s), and 1RXS J1730-0559 (128 s);

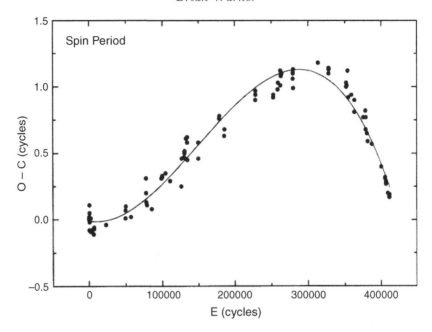

FIGURE 3.24. Observed minus calculated peak arrival times for the 21 min periodicity in FO Aqr (Patterson *et al.*, 1998). Reproduced by permission of the AAS.

the remainder have $P_{rot} > 250$ s, and the majority are $\sim$15 min. Primary spins at these shortest periods indicate that the field strengths must be low (the magnetosphere has been compressed to only a few WD radii): estimates of these fields are $\sim$1 MG. It is natural to ask what would be the observational signatures of CVs with primary fields weaker than this.

3.2.4 Low field accretion

In searching for further examples of short period oscillations among CVs an unexpected discovery was made: lower coherence brightness variations associated with high $\dot{M}$ disks, that is, DNe in outburst and nova-like variables (Warner and Robinson, 1972). They are known as dwarf nova oscillations (DNOs) and have $Q \sim 10^3$ to 10^7. An example, at the highest signal to noise so far achieved, is shown in Figure 3.27. Two further kinds of oscillations were later found, one of very low coherence ($Q \sim 10 - 30$), known as quasi-periodic oscillations (QPOs; Fig. 3.28), and one of similar coherence to the DNOs but of longer period, known as longer period DNOs (lpDNOs). For a general review, see Warner (2004).

From many years of study, we can summarize some of their properties thus:
(i) DNOs have periods typically in the range 3 to 50 s, with PDNO correlated with the $\dot{M}$ that is falling onto the WD: thus, the minimum period in DNe occurs at maximum outburst accretion luminosity, with period changes of up to a factor of 2 during outburst.
(ii) There are two types of QPO, one with $P_{QPO} \sim 2000$ s, and the other maintaining a relationship $P_{QPO} \sim 16 \, P_{DNO}$. At the same time, $P_{lpDNO} \sim 4 \, P_{DNO}$.
(iii) Most DNe show DNOs in outburst, but a few, despite having been very well observed during outbursts, have never shown oscillations. DNOs are occasionally seen briefly in DNe during quiescence.
(iv) Phase shifts through eclipse have been observed that support a similar model to that in DQ Her – that is, a rotating beam of high-energy radiation reprocessed into optical wavelengths.

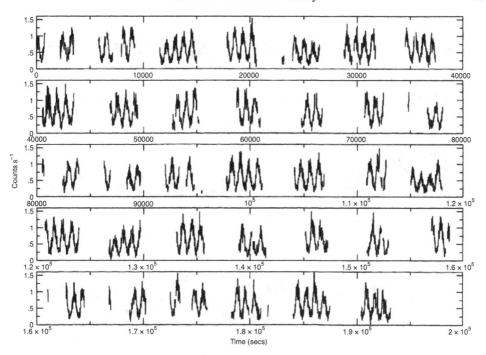

FIGURE 3.25. X-ray light curve of AO Psc, showing 805 s modulation with interruptions caused by the Earth occulting the satellite's view (Hellier *et al.*, 1996).

(v) Sideband oscillations to the DNOs are often seen, with a frequency separation equal to the frequency of the QPO.

(vi) In addition, there is a wide range of other phenomena involving all three kinds of oscillation.

What are we to make of these observations? There are some suggested models that involve oscillations of the accretion disk (which will have periods in the vicinity of the Keplerian period), but I have various objections to these – in particular, if some disks show oscillations, why not all of them (e.g., EK TrA is almost a lookalike for VW Hyi but does not have DNOs). What is the parameter that decides whether a CV has DNOs? My answer is that it is the magnetic field of the primary. The model I have proposed (the Low Inertia Magnetic Accretion model, e.g., Warner and Woudt, 2002) is a simple

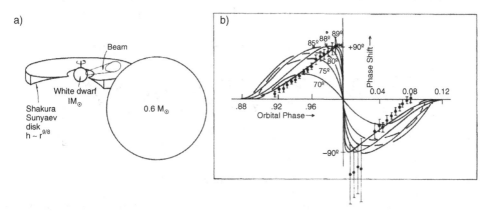

FIGURE 3.26. *a)* Eclipse of DQ Her, *b)* with modeled phase variations (Petterson, 1980). Reproduced by permission of the AAS.

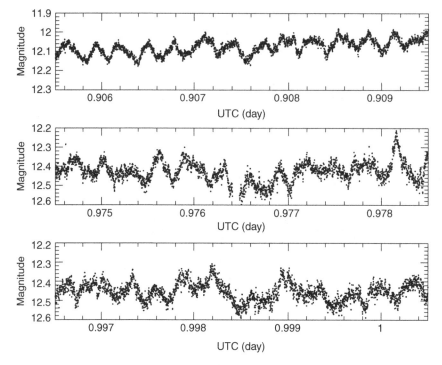

FIGURE 3.27. Light curve of VW Hyi in outburst, observed at 80 ms time resolution with the 11-m SALT telescope, showing a short section of the 28 s oscillations. (From Woudt et al., 2010).

extension of the IP model to lower fields, fields too low to generate DNOs at quiescence in DN, but which become stronger during outburst as the accreting gas runs around the equator, enhancing the field by its shearing action and enabling magnetically controlled accretion very close to the surface of the primary. If the field is low enough ($\lesssim 10^5$ G), the equatorial accreted material is not coupled to the interior (the viscosity of degenerate gas in the interior is almost zero) and so has small moment of inertia and can be spun up and down as $\dot{M}$ varies – in contrast to DQ Her and other IPs, where the accretion torque acts on the inertia of the whole star. Spectra obtained by HST of dwarf novae during and after outburst do indeed show the presence of a long-lived hot equatorial belt on the primary (e.g., Sion et al., 1996), which is where I suggest that the field is generated. The DNO-related QPOs are interpreted as progradely rotating traveling waves near the inner edge of the truncated accretion disk, producing the optical QPOs by simple obscuration and reflection (or reprocessing) of radiation from the region of the primary, and generating

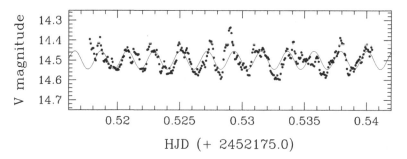

FIGURE 3.28. 185 s QPOs in WX Hyi, with a best fitting sinusoid. From Warner and Woudt (2008).

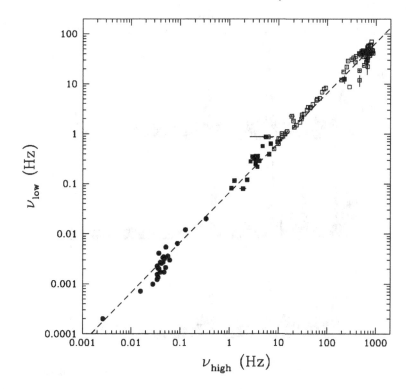

FIGURE 3.29. Correlation between DNO and QPO frequencies in black holes, neutron stars, and white dwarfs. Filled squares: black-hole binaries; open squares: neutron star binaries; filled circles: white dwarf binaries. From Warner and Woudt (2008).

the DNO sidebands by reprocessing of the rotating central radiation beam – as with orbital sidebands in IPs. The lpDNOs are most probably generated by accretion onto high-latitude areas of the primary.

According to the model, the existence of DNOs and QPOs in CVs requires a seed field in the WD primary, which is evidently absent in some systems. This could be the result of a nova explosion in the "recent" past having ejected the magnetic flux (as well as the mass and angular momentum) accreted since the last eruption. The strength of DNOs in a CV would then be an indicator of whether it is nearing another nova event.

Similar sets of luminosity oscillations are seen in X-ray binaries, containing neutron stars and black holes. An identical scaling between DNOs and QPOs is seen in both types of object (Fig. 3.29). In all of these P_{QPO}/P_{DNO} is $\sim$15, which is a ratio not yet understood from any comprehensive model, but would seem to require that all such systems have a common structure. If this is accretion onto a compact magnetized star, it does not conform to the usual belief that black holes cannot possess a magnetic field. However, the recent claim (Barceló et al., 2008) that vacuum polarization can prevent the rapid formation of black holes may allow fields seated in the outer parts of a "black star" to provide the magnetically controlled accretion process.

3.3 Accretion onto pulsating white dwarfs

A small fraction of single DA white dwarfs, those with effective temperatures 10,800 to 12,300 K, are found to be pulsating in nonradial modes, typically with periods in the range 100 to 1000 s; they are known as ZZ Cet stars, and about 150 are currently

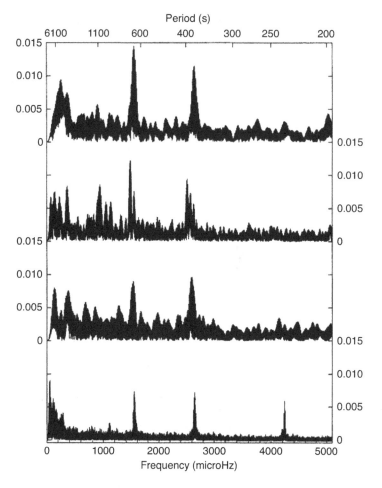

FIGURE 3.30. Fourier transform spectrum of oscillations in GW Lib. From van Zyl *et al.* (2000).

listed. Asteroseismology of these stars enables their masses and internal structures to be determined; for a recent review, see Winget and Kepler (2008).

The first accreting white dwarf with ZZ Cet pulsations – namely the dwarf nova GW Lib – was only found in 1997 (Warner and van Zyl, 1998). Since then, another 10 or so have been found – for a current list, see http://deneb.astro.warwick.ac.uk/phsaap/pulsators.html. In GW Lib, the dominant periods of oscillation are near 230 s, 370 s, and 650 s (Fig. 3.30). The pulsation instability strip for these stars is moved by ∼1000 to 2000 K relative to that for nonaccreting WDs (Arras *et al.*, 2006). For an accreting WD to lie in this temperature range, the accretion rate much be quite low ($\dot{M} \sim 10^{-11}$ $M_\odot y^{-1}$). This produces such a low luminosity for the accretion disk that the absorption spectrum of the primary is easily detectable at visible wavelengths (Fig. 3.31), with the result that optical pulsations can be seen. This is a fortunate circumstance.

The importance of these rare systems is that we may hope eventually, among other things, to determine the masses of their accreted outer hydrogen layers, which are related to the times since the last (or before the next) nova outbursts. Furthermore, a DN outburst heats the outer layers of the WD primary, which moves it out of the instability strip, and we can observe the pulsations starting up again as it cools down. The first of these CV/ZZ hybrids to have an observed outburst was GW Lib itself in April 2007, after which the pulsations disappeared (Copperwheat *et al.*, 2009).

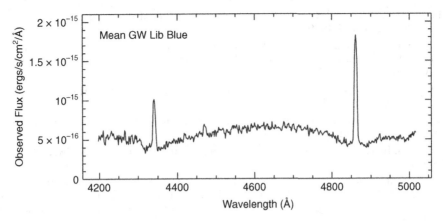

FIGURE 3.31. Spectrum of GW Lib. From Szkody *et al.* (2000). Reproduced by permission of the AAS.

3.4 Acknowledgments

My research is supported by the University of Cape Town, Southampton University, and the National Research Foundation.

REFERENCES

Arras, P., Townsley, D. M., and Bildsten, L. 2006. Pulsational instabilities in accreting white dwarfs. ApJ, **643**(June), L119–L122.

Baptista, R., Horne, K., Wade, R. A., Hubeny, I., Long, K. S., and Rutten, R. G. M. 1998. HST spatially resolved spectra of the accretion disk and gas stream of the nova-like variable UX Ursae Majoris. MNRAS, **298**(Aug.), 1079–1091.

Barceló, C., Liberati, S., Sonego, S., and Visser, M. 2008. Fate of gravitational collapse in semiclassical gravity. Phys. Rev. D, **77**(4), 044032+.

Bondi, H., and Hoyle, F. 1944. On the mechanism of accretion by stars. MNRAS, **104**, 273+.

Copperwheat, C. M., Marsh, T. R., Dhillon, V. S., Littlefair, S. P., Woudt, P. A., Warner, B., Steeghs, D., Gänsicke, B. T., and Southworth, J. 2009. ULTRA-CAM observations of two accreting white dwarf pulsators. MNRAS, **393**(Feb.), 157–170.

Dupuis, J., Fontaine, G., and Wesemael, F. 1993. A study of metal abundance patterns in cool white dwarfs. III – Comparison of the predictions of the two-phase accretion model with the observations. ApJS, **87**(July), 345–365.

Edgar, R. 2004. A review of Bondi-Hoyle-Lyttleton accretion. New A Rev., **48**(Sept.), 843–859.

Epelstain, N., Yaron, O., Kovetz, A., and Prialnik, D. 2007. A thousand and one nova outbursts. MNRAS, **374**(Feb.), 1449–1456.

Fujimoto, M. Y. 1982. A theory of hydrogen shell flashes on accreting white dwarfs. I – Their progress and the expansion of the envelope. II – The stable shell burning and the recurrence period of shell flashes. ApJ, **257**(June), 752–779.

Gänsicke, B. T., Marsh, T. R., Southworth, J., and Rebassa-Mansergas, A. 2008. Metal-rich debris discs around white dwarfs. Pages 149+ of: D. Fischer, F. A. Rasio, S. E. Thorsett, and A. Wolszczan (eds.), *Astronomical Society of the Pacific Conference Series*. Astronomical Society of the Pacific Conference Series, vol. 398.

Ghosh, P., and Lamb, F. K. 1978. Disk accretion by magnetic neutron stars. ApJ, **223**(July), L83–L87.

Haberl, F., and Motch, C. 1996 (Feb.). Two spectrally distinct classes of intermediate polars discovered with ROSAT. Pages 145–146 of: H. U. Zimmermann, J. Trümper, & H. Yorke (eds.), *Roentgenstrahlung from the Universe*. Springer, 2001, Jan.

Harrop-Allin, M. K., Cropper, M., Hakala, P. J., Hellier, C., and Ramseyer, T. 1999. Indirect imaging of the accretion stream in eclipsing polars – II. HU Aquarii. MNRAS, **308**(Sept.), 807–817.

Heerlein, C., Horne, K., and Schwope, A. D. 1999. Modelling of the magnetic accretion flow in HU Aquarii. MNRAS, **304**(Mar.), 145–154.

Heise, J., Brinkman, A. C., Gronenschild, E., Watson, M., King, A. R., Stella, L., and Kieboom, K. 1985. An X-ray study of AM Herculis. I – Discovery of a new mode of soft X-ray emission. A&A, **148**(July), L14–L16.

Hellier, C. 2001. *Cataclysmic Variable Stars – How and Why they Vary* (Springer Praxis Books / Space Exploration), Springer.

Hellier, C., Mukai, K., Ishida, M., and Fujimoto, R. 1996. The X-ray spectrum of the intermediate polar AO Piscium. MNRAS, **280**(June), 877–887.

Horne, K. 1985. Images of accretion discs. I – The eclipse mapping method. MNRAS, **213**(Mar.), 129–141.

Iben, Jr., I. 2003. Lessons from and about symbiotic novae (invited review talks). Pages 177+ of: R. L. M. Corradi, J. Mikolajewska, and T. J. Mahoney (eds.), *Astronomical Society of the Pacific Conference Series*. Astronomical Society of the Pacific Conference Series, vol. 303.

Ireland, M. J., Monnier, J. D., Tuthill, P. G., Cohen, R. W., De Buizer, J. M., Packham, C., Ciardi, D., Hayward, T., and Lloyd, J. P. 2007. Born-again protoplanetary disk around Mira B. ApJ, **662**(June), 651–657.

Karovska, M., Hack, W., Raymond, J., and Guinan, E. 1997. First Hubble Space Telescope observations of Mira AB wind-accreting binary systems. ApJ, **482**(June), L175+.

Kato, M. 2010. Accreting white dwarfs as supersoft X-ray sources. Astronomische Nachrichten, **331**, 140+.

Kenyon, S.J. 1988. Book Review: The symbiotic stars. Cambridge University Press, 1986. Bulletin of the Astronomical Institutes of Czechoslovakia, **39**(Mar.), 128+.

Latham, D. W., Liebert, J., and Steiner, J. E. 1981. The 1980 low state of AM Herculis. ApJ, **246**(June), 919–934.

Marsh, T. R., and Horne, K. 1988. Images of accretion discs. II – Doppler tomography. MNRAS, **235**(Nov.), 269–286.

Mineshige, S., and Wood, J. H. 1989. Viscous evolution of accretion discs in the quiescence of dwarf novae. MNRAS, **241**(Nov.), 259–280.

O'Donoghue, D. 1986. The radius of the accretion disk in Z Cha between outbursts. MNRAS, **220**(May), 23P–26P.

O'Donoghue, D., Buckley, D. A. H., Balona, L. A., Bester, D., Botha, L., Brink, J., Carter, D. B., Charles, P. A., Christians, A., Ebrahim, F., Emmerich, R., Esterhuyse, W., Evans, G. P., Fourie, C., Fourie, P., Gajjar, H., Gordon, M., Gumede, C., de Kock, M., Koeslag, A., Koorts, W. P., Kriel, H., Marang, F., Meiring, J. G., Menzies, J. W., Menzies, P., Metcalfe, D., Meyer, B., Nel, L., O'Connor, J., Osman, F., Du Plessis, C., Rall, H., Riddick, A., Romero-Colmenero, E., Potter, S. B., Sass, C., Schalekamp, H., Sessions, N., Siyengo, S., Sopela, V., Steyn, H., Stoffels, J., Scholtz, J., Swart, G., Swat, A., Swiegers, J., Tiheli, T., Vaisanen, P., Whittaker, W., and van Wyk, F. 2006. First science with the Southern African Large Telescope: peering at the accreting polar caps of the eclipsing polar SDSS J015543.40+002807.2. MNRAS, **372**(Oct.), 151–162.

Patterson, J. 1998. Late evolution of cataclysmic variables. PASP, **110**(Oct.), 1132–1147.

Patterson, J., Kemp, J., Richman, H. R., Skillman, D. R., Vanmunster, T., Jensen, L., Buckley, D. A. H., O'Donoghue, D., and Kramer, R. 1998. Rapid oscillations in cataclysmic variables. XIV. Orbital and spin ephemerides of FO Aquarii. PASP, **110**(Apr.), 415–419.

Petterson, J. A. 1980. Accretion disks in cataclysmic variables. I – The eclipse-related phase shifts in DQ Herculis and UX Ursae Majoris. ApJ, **241**(Oct.), 247–256.

Potter, S. B., Romero-Colmenero, E., Watson, C. A., Buckley, D. A. H., and Phillips, A. 2004. Stokes imaging, Doppler mapping and Roche tomography of the AM Herculis system V834 Cen. MNRAS, **348**(Feb.), 316–324.

Prialnik, D. 1986. The evolution of a classical nova model through a complete cycle. ApJ, **310**(Nov.), 222–237.

Prialnik, D., and Kovetz, A. 1995. An extended grid of multicycle nova evolution models. ApJ, **445**(June), 789–810.

Reimers, D., and Cassatella, A. 1985. The ultraviolet spectrum of the companion of Mira (o Ceti). Observational evidence for a disk formed by wind accretion. ApJ, **297**(Oct.), 275–287.

Robertson, J. W., Honeycutt, R. K., and Turner, G. W. 1995. RZ Leonis Minoris, PG 0943+521, and V1159 Orionis: Three cataclysmic variables with similar and unusual outburst behavior. PASP, **107**(May), 443–+.

Rutten, R. G. M., Dhillon, V. S., Horne, K., and Kuulkers, E. 1994. Spectral eclipse mapping of the accretion disk in the nova-like variable UX Ursae Majoris. A&A, **283**(Mar.), 441–454.

Saito, R. K., and Baptista, R. 2009. Spin-cycle eclipse mapping of the 71 s oscillations in DQ Herculis: reprocessing sites and the true white dwarf spin period. ApJ, **693**(Mar.), L16–L18.

Sanad, M. R., Bobrowsky, M., Hamdy, M. A., and Abo Elazm, M. S. 2009. Density effects on Mg II emission lines of Mira AB. AJ, **137**(Mar.), 3479–3486.

Schwope, A. D., Beuermann, K., Jordan, S., and Thomas, H.-C. 1993. Cyclotron and Zeeman spectroscopy of MR Serpentis in low and high states of accretion. A&A, **278**(Nov.), 487–498.

Schwope, A. D., Brunner, H., Hambaryan, V., and Schwarz, R. 2002 (Jan.). LARPs – low-accretion rate polars. Pages 102+ of: B. T. Gänsicke, K. Beuermann, and K. Reinsch (eds.), *The Physics of Cataclysmic Variables and Related Objects.* Astronomical Society of the Pacific Conference Series, vol. 261.

Schwope, A. D., Mantel, K.-H., and Horne, K. 1997. Phase-resolved high-resolution spectrophotometry of the eclipsing polar HU Aquarii. A&A, **319**(Mar.), 894–908.

Sion, E. M., Cheng, F.-H., Huang, M., Hubeny, I., and Szkody, P. 1996. The cooling white dwarf in VW Hydri after normal outburst and superoutburst: HST evidence of a sustained accretion belt. ApJ, **471**(Nov.), L41+.

Sion, E. M., Greenstein, J. L., Landstreet, J. D., Liebert, J., Shipman, H. L., and Wegner, G. A. 1983. A proposed new white dwarf spectral classification system. ApJ, **269**(June), 253–257.

Smak, J. 1994. Eclipses in cataclysmic variables with stationary accretion disks. IV. On the peculiar T(R) distributions. Acta Astron., **44**(July), 265–276.

Szkody, P., Desai, V., and Hoard, D. W. 2000. Spectroscopy of GW Librae at quiescence. AJ, **119**(Jan.), 365–368.

van Zyl, L., Warner, B., O'Donoghue, D., Sullivan, D., Pritchard, J., and Kemp, J. 2000. GW Librae: an accreting variable white dwarf. Baltic Astronomy, **9**, 231–246.

Vogt, N. 1974. Photometric study of the dwarf nova VW Hydri. A&A, **36**(Dec.), 369–378.

Walker, M. F. 1956. A Photometric investigation of the short-period eclipsing binary, Nova DQ Herculis (1934). ApJ, **123**(Jan.), 68+.

Warner, B. 1972. Observations of rapid blue variables – VIII. The companion to Mira. MNRAS, **159**, 95–100.

Warner, B. 1975. Observations of rapid blue variables – XV. VW Hydri. MNRAS, **170**(Jan.), 219–228.

Warner, B. 1986a. Multiple optical orbital sidebands in intermediate polars. MNRAS, **219**, 347–356.

Warner, B. 1986b. Accretion disk inclinations and absolute magnitudes of classical nova remnants. MNRAS, **222**, 11–18.

Warner, B. 1987. Absolute magnitudes of cataclysmic variables. MNRAS, **227**(July), 23–73.

Warner, B. 1995a. *Cataclysmic Variable Stars*. Cambridge University Press.

Warner, B. 1995b. The AM Canum Venaticorum Stars. Ap&SS, **225**(Mar.), 249–270.

Warner, B. 2002 (Nov.). General properties of quiescent novae. Pages 3–15 of: M. Hernanz and J. José (eds.), Classical Nova Explosions. American Institute of Physics Conference Series, vol. 637.

Warner, B. 2004. Rapid oscillations in cataclysmic variables. PASP, **116**(Feb.), 115–132.

Warner, B., and Cropper, M. 1984. High-speed photometry of the Intermediate Polar V1223 SGR. MNRAS, **206**(Jan.), 261–271.

Warner, B., and Robinson, E. L. 1972. White dwarfs – more rapid variables. Nature, **239**(Sept.), 2–7.

Warner, B., and van Zyl, L. 1998. Discovery of non-radial pulsations in the white dwarf primary of a cataclysmic variable star. Pages 321+ of: F.-L. Deubner, J. Christensen-Dalsgaard, and D. Kurtz (eds.), *New Eyes to See Inside the Sun and Stars*. IAU Symposium, vol. 185.

Warner, B., and Woudt, P. A. 2002. Dwarf nova oscillations and quasi-periodic oscillations in cataclysmic variables – II. A low-inertia magnetic accretor model. MNRAS, **335**(Sept.), 84–98.

Warner, B., and Woudt, P. A. 2008. QPOs in CVs: an executive summary. *AIP Conference Proceedings*, **1054**(1), 101–110.

Warner, B., and Woudt, P. A. 2009. The eclipsing intermediate polar V597 Pup (Nova Puppis 2007). MNRAS 2009(Aug.), 979–984.

Warner, B., Peters, W. L., Hubbard, W. B., and Nather, R. E. 1972. Observations of rapid blue variables – XI. DQ Herculis. MNRAS, **159**, 321–335.

Watson, C. A., Dhillon, V. S., Rutten, R. G. M., and Schwope, A. D. 2003. Roche tomography of cataclysmic variables – II. Images of the secondary stars in AM Her, QQ Vul, IP Peg and HU Aqr. MNRAS, **341**(May), 129–142.

Wesemael, F. 1979. Accretion from interstellar clouds and white dwarf spectral evolution. A&A, **72**(Feb.), 104–110.

Wesemael, F., and Truran, J. W. 1982. Accretion of grains and element abundances in cool, helium-rich white dwarfs. ApJ, **260**(Sept.), 807–814.

Wesemael, F., Greenstein, J. L., Liebert, J., Lamontagne, R., Fontaine, G., Bergeron, P., and Glaspey, J. W. 1993. An atlas of optical spectra of white-dwarf stars. PASP, **105**(July), 761–778.

Wickramasinghe, D. T., and Ferrario, L. 2000. Magnetism in isolated and binary white dwarfs. PASP, **112**(July), 873–924.

Wickramasinghe, D. T., and Ferrario, L. 2005. The origin of the magnetic fields in white dwarfs. MNRAS, **356**(Feb.), 1576–1582.

Winget, D. E., and Kepler, S. O. 2008. Pulsating white dwarf stars and precision asteroseismology. ARA&A, **46**(Sept.), 157–199.

Woudt, P. A., Warner, B., O'Donoghue, D., Buckley, D. A. H., Still, M., Romero-Colemero, E., and Väisänen, P. 2010. Dwarf nova oscillations and quasi-periodic oscillations in cataclysmic variables – VIII. VW Hyi in outburst observed with the Southern African Large Telescope. MNRAS, **401**(Jan.), 500–506.

4. Multiwavelength observations of accretion in low-mass X-ray binary systems

ROBERT I. HYNES

Abstract

This work is intended to provide an introduction to multiwavelength observations of low-mass X-ray binaries and the techniques used to analyze and interpret their data. The focus primarily is on ultraviolet, optical, and infrared observations and their connections to other wavelengths. The topics covered include outbursts of soft X-ray transients, accretion disk spectral energy distributions, orbital light curves in luminous and quiescent states, superorbital and suborbital variability, line spectra, system parameter determinations, and echo mapping and other rapid correlated variability.

4.1 Introduction

The first X-ray binary to be observed and identified as such was Scorpius X-1 (Giacconi et al., 1962), although several other systems were known as optical stars or novae before this. Within a few years, optical and radio counterparts to Sco X-1 were discovered (Sandage et al., 1966; Andrew and Purton, 1968), and the topic has remained multiwavelength in nature since then.

This work is intended to provide an introduction to some of the observational characteristics of X-ray binaries suitable for a graduate student or an advanced undergraduate. My aim was to produce a primer for someone relatively new to the field rather than a comprehensive review. Where appropriate, I also discuss techniques for analysis and interpretation of the data. The focus is almost exclusively on low-mass X-ray binaries, in which the accretion disk is most accessible to multiwavelength observations, and is predominantly biased toward ultraviolet, optical, and infrared observations and their relation to observations at other wavelengths. For a textbook treatment of accretion astrophysics in general, the reader is referred to Frank et al. (2002) and for more comprehensive reviews of X-ray binaries to Lewin et al. (1995) and Lewin and van der Klis (2006).

We begin in Section 4.2 by presenting an overview of the main classes of X-ray binaries and their accretion geometries, and in Section 4.3 describe observations of transient systems. We then address different types of observations in turn. In Section 4.4 we examine expected and observed spectral energy distributions; in Section 4.5 we move on to consider orbital light curves, together with variability on superorbital and suborbital (but still relatively long) time scales. Section 4.6 then examines spectroscopic observations, with a focus on determination of binary system parameters, and finally Section 4.7 looks at shorter time scale variations and in particular correlations between the X-ray and UV/optical/IR behavior. For reference, we include a glossary of notable objects in Section 4.10.

4.2 Classification and geometry of X-ray binaries

The family of X-ray binaries is diverse. A single hierarchical classification system is insufficient, and depending on the question to be asked one may desire to classify objects by the nature of their donor star and mode of accretion, by the nature of the compact object, or by whether their X-ray activity is persistent or transient. Additionally, a subset of these objects produce relativistic jets and are classified as microquasars.

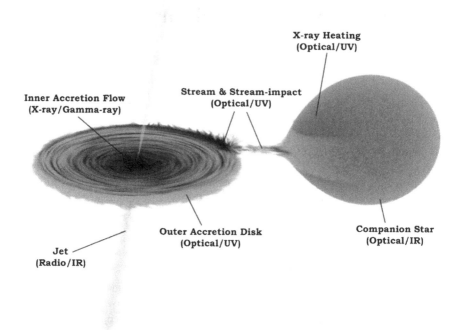

FIGURE 4.1. Geometry of a Roche lobe overflow low-mass X-ray binary. The color scale is inverted, with the brightest areas appearing darkest.

The most common division made is into high-mass and low-mass X-ray binaries: HMXBs and LMXBs. This is based crudely on the mass of the donor star, with HMXBs typically having O or B type donors of mass over $10\,M_\odot$ and LMXBs having late-type G–M donors with masses below $\sim 1.5\,M_\odot$. Classification of those systems with A or F donors has been more vague, but these objects such as Her X-1 and GRO J1655–40 have often been grouped as intermediate-mass X-ray binaries, or IMXBs. The donor mass also provides the primary division in accretion mode, with LMXBs and IMXBs mostly accreting by Roche lobe overflow (Fig. 4.1) and HMXBs by lobe overflow, capture of wind from the donor star, or interaction of the compact object with a circumstellar disk. A small group of LMXBs with low-mass red giant donors do accrete from the donor wind. These are referred to as symbiotic X-ray binaries (Corbet *et al.*, 2008). Among the LMXBs, those with exceptionally short periods and hydrogen-deficient donors are often broken out into the subclass of ultracompact X-ray binaries (UCXRBs).

Another distinction can be made between systems containing black holes and those containing neutron stars. For many purposes, this is more important than the donor nature or accretion mode, and we observe many similarities in X-ray behavior between black holes in LMXBs and HMXBs.

Observationally, an important difference is found between persistent and transient sources. As time has passed, this distinction has blurred, and although there remain simple classical transients such as A 0620–00 that undergo dramatic outbursts every few decades, other systems seem to turn on into a semipersistent state, or execute a whole series of outbursts in succession after emerging from quiescence. As our baseline extends, this kind of quasi-persistent or quasi-recurrent behavior with distinct on and off periods may become a more common characteristic.

Finally, we should mention the category of microquasars. This term has had varying usage, but in the context of X-ray binaries it usually refers to those sources showing resolved, expanding, relativistic jets such as GRS 1915+105 (Mirabel and Rodríguez,

1994). Its usage has varied from this fairly exclusive definition to including all jet sources, then to potentially all black-hole binaries, and possibly also to some neutron star sources.

4.3 Transient X-ray binaries

4.3.1 *Classical soft X-ray transients*

The outburst of the transient LMXB A 0620–00 in 1975 (Kuulkers, 1998, and references therein) opened up what was to become a major area of X-ray binary research. It was not the first transient X-ray source found, but was the first to be studied in great detail and the brightest yet seen.

Among the many reasons for the modern importance of transient systems, their large dynamic range is invaluable. In a single object, on a practical time scale of months to a few years we can watch the evolution of the system through the full range of accretion states and follow causal sequences between them. This is impossible with persistent X-ray binaries and active galactic nuclei (AGN), both of which individually sample a smaller range of parameter space, leaving us to attempt to build a complete picture from snapshots of different objects. Another important characteristic of transient LMXBs is that in quiescence, their light usually becomes dominated by their companion star. Radial velocity studies then allow measurement of system parameters, and particularly the compact object mass. This is how we know of the most compelling examples of stellar-mass black holes and that the majority of known transient LMXBs are black hole systems. We consider system parameter determinations in more detail in Section 4.6.3.

These transient LMXBs are often referred to as soft X-ray transients (SXTs), based on the ultrasoft spectra that are sometimes seen in outburst, or black hole X-ray transients (BHXRTs) due to their high incidence of black holes. Among them is a subset that shows a relatively orderly behavior with common features repeated between several objects. The outbursts of these objects are known as FRED – fast rise exponential decay outbursts. FRED outbursts are often considered to be typical of SXTs, but in fact there are many exceptions. In the sample of outburst light curves compiled by Chen *et al.* (1997), there are both FREDs and irregular outbursts. Since then, we have seen a preponderance of irregular outbursts, and the orderly FRED behavior now seems the exception rather than the rule. Nonetheless, such a repeating pattern is a natural place to begin in trying to understand their behavior and has provided the benchmark for numerical simulations of outburst light curves (e.g., Cannizzo *et al.*, 1995; Dubus *et al.*, 2001; Truss *et al.*, 2002).

We show the hard X-ray (CGRO/BATSE) and optical light curves of GRO J0422+32 in Figure 4.2 as an example of a FRED light curve showing most of the common characteristics. The outburst exhibits a very fast rise followed by a slow exponential decay for about 200 days. During the decay, there are several secondary maxima. At the end of the exponential decay comes a rapid dropoff. After this, the outburst is below the CGRO/BATSE threshold, but continued evolution can be seen in the optical, with a continued decline interrupted by at least two mini-outbursts.

The accepted framework for understanding SXT outbursts is believed to be the disk instability model (DIM; Lasota, 2001), originally developed to explain outbursts of dwarf novae (DNe). The latter usually have a much shorter recurrence time and nonexponential decays. The exponential decay in SXTs has been explained by King and Ritter (1998) as a consequence of irradiation maintaining the whole disk in the hot, high-viscosity state of the DIM. The accretion rate is proportional to the active disk mass, and so if the luminosity traces the accretion rate, then the exponential decay in luminosity reflects the decay of the disk mass. For short-period systems where it is possible to maintain the whole disk in a high state in this way, it is possible to accrete a large fraction of the disk mass during an outburst. Consequently, the recurrence time is longer than in DNe, where only a small fraction of the disk mass is accreted in a single outburst.

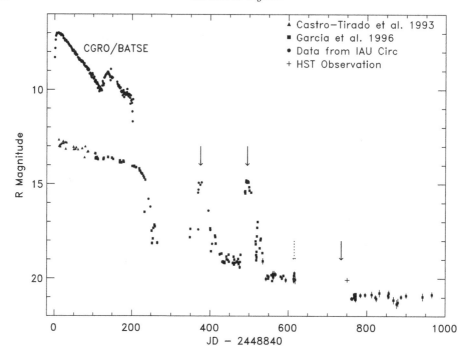

FIGURE 4.2. Outburst light curves of GRO J0422+32 (adapted from Hynes and Haswell, 1999). This shows many of the features of a canonical SXT outburst: a fast rise, exponential decay, and mini-outbursts. Note that the secondary maximum seen prominently in the CGRO/BATSE light curve is not the one discussed in the text. The one discussed occurs earlier and is much less prominent in this object.

Several explanations have been advanced for the secondary maxima, and different features at different points in the decay may have different explanations. In the earliest explanations, the turn-on of X-rays during the outburst irradiated the companion star and stimulated a burst of enhanced mass transfer into the disk. When this material reached the compact object, the X-ray flux was further increased (Chen et al., 1993; Augusteijn et al., 1993). King and Ritter (1998) instead proposed that initially there is an outer region of the disk that begins the outburst in the cool low viscosity state of the DIM. Irradiation at the onset of the outburst then raises this to the high state, and again, its effect on the X-ray light curve is delayed by one viscous time. Finally, Truss et al. (2002) modified this mechanism by instead invoking the growth of tidal instabilities as a mechanism to stimulate accretion from the outermost part of the disk.

The explanation for mini-outbursts remains less conclusive. This phenomenon has been less commonly observed in SXTs, as it happens later when the source has passed below the threshold of all-sky monitors and so is typically mostly covered by sparser optical coverage. These mini-outbursts are also characterized by a higher ratio of optical to X-ray flux. The phenomenon closely resembles the "echo-outbursts" seen in some cataclysmic variables (CVs), where on occasion as many as six roughly uniformly spaced mini-outbursts have been seen (Patterson et al., 1998). In attempting to explain mini-outbursts, Hameury et al. (2000) again invoked irradiation and Osaki et al. (2001) attributed them to failed viscosity decay attempts at the end of the outburst, whereas Hellier (2001) associated them with tidal effects.

Once again, it should be emphasized that only a minority of SXT outbursts follow the classic FRED form (and indeed that not all SXT outbursts even exhibit the defining ultrasoft state; Brocksopp et al. 2004). In many cases the irregular outbursts are associated with longer-period systems where it may not be possible to raise all of the accretion disk into the hot state, resulting in interactions between the hot inner disk and the permanently cold outer region. However, some short-period systems such as XTE J1118+480 also exhibit irregular (and recurrent) outburst activity.

4.3.2 Recurrent transients and semipersistent systems

It is likely that all transient systems are recurrent, of course, but some recur on shorter time scales than others. The best defined of these are systems where outbursts occur once per orbital period, usually near closest approach (periastron). The Be star plus neutron star HMXBs are the most common example of this type, but Cir X-1 is another eccentric transient system.

Some transient LMXB sources emerge from quiescence to undergo multiple recurrent outbursts, like GRO J1655−40, or turn on without subsequently returning to quiescence, like GRS 1915+105. GX 339−4 is sometimes thought of as a rare persistent black hole but is actually more like a very active recurrent transient and, given a longer baseline, may turn out to have been discovered in the midst of a short-lived series of outbursts.

Among neutron stars, we see several that exhibit a similar kind of semipersistent behavior. 4U 2129+47 was identified as a 5.2 hr LMXB by Thorstensen et al. (1979) only to fade to quiescence in 1983 (Pietsch et al., 1983). EXO 0748−676 went into outburst in 1985 (Parmar et al., 1985) and remained as an apparently persistent source for 24 years, before finally returning to quiescence in 2008.

This behavior among neutron stars provides a unique tool to probe the properties of the neutron star. During an outburst, the crust is heated by accretion and is no longer in thermal equilibrium with the core. The cooling curve of the neutron star after outburst then depends sensitively on the thermal conductivity of the crust, so these observations can be used to test models for the structure and composition of neutron star crust and the crust-core interface (Brown and Cumming, 2009).

4.4 Spectral energy distributions

4.4.1 Overview

X-ray binaries are multiwavelength objects with detectable emission from radio to gamma-rays and at all wavelengths in between. The spectral energy distribution (SED) measures the relative energy contributed in different bands and crudely characterizes the shape of the spectra. The SED can inform us of the different components emitting, in particular the disk, and the structure of those components. Without additional spatial constraints, such as might be provided by eclipse mapping, we do not obtain any information about the spatial arrangement of emitting regions, so there may be degenerate models. In the simplest case of blackbody emission from an accretion disk, the information gained is really a measure of how much emitting area is present at each temperature. With the common assumptions that the disk is axisymmetric and that the temperature increases monotonically inward, however, we can translate this information into the temperature as a function of radius. In practice, it is more common to make these assumptions and fit simple derived models to the SED, and so we focus on the basis of these models.

At this point there are many different representations of the SED in use, and it is unlikely that a common standard will be adopted. It is worthwhile to be familiar with the different conventions and to be able to mentally transform between them. In order to show the order-of-magnitude range of values typically present, almost all authors use logarithmic axes for both the energy/wavelength axis and for flux. In the X-ray binary community it is common to use an energy-like x-axis, based on either photon frequency (ν) or energy (E). The y-axis is either the flux per unit frequency, F_ν (or equivalently per unit energy, F_E), or this multiplied by the photon energy. Examples of SEDs in the ν–F_ν representation are shown in Figures 4.3 and 4.4. A straight F_ν representation makes it easier to estimate the spectral index, α, of a power law expressed as $F_\nu \propto \nu^\alpha$. In particular, this representation is well suited to the typically flat ($F_\nu \simeq$ constant) radio spectra often seen and associated with jets. On the other hand, in a plot of $\log \nu F_\nu$ versus $\log \nu$, the height in νF_ν is a direct measure of the amount of energy emerging per logarithmic frequency interval at that frequency; hence, the peak of an SED in νF_ν is the energetically most important component. A ν–νF_ν SED is also shown in Figure 4.3. The most common SED forms encountered in X-ray binary work are then ν–F_ν, ν–νF_ν, E–F_E, and E–EF_E, with the latter two most common in dealing with X-ray and gamma-ray data. Wavelength-based forms are sometimes used, λ–F_λ or λ–λF_λ. To mentally transform between them, it is helpful to remember that a power-law of $F_\nu \propto \nu^\alpha$ becomes $F_\lambda \propto \lambda^{2-\alpha}$ in this representation. The final variation on these representations is unique to X-ray and gamma-ray astronomy where the spectrum is sometimes described in an N_E–E representation where N_E is the number of photons per unit energy interval. Power-law spectra are then specified by the photon index, Γ, where $N_E \propto E^{-\Gamma}$. Note the negative sign in the definition. The photon index is related to the F_ν–ν spectral index by $\Gamma = 1 - \alpha$.

4.4.2 The blackbody disk model

The spectral model that has most commonly been fitted to optical and UV SEDs of X-ray binaries is a blackbody-based model (Lynden-Bell, 1969; Shakura and Sunyaev, 1973; Frank et al., 2002). The disk is assumed to be axisymmetric, with temperature increasing inward. Having defined the functional form of $T(R)$, the SED is evaluated by summing emission from each concentric annulus. Simple limiting cases can be derived analytically, but with realistic assumptions (for example, that both viscous and irradiative heating are present), a numerical integration is more useful.

The problem can be defined for a power-law temperature distribution:

$$T(R) = T_0 \left(\frac{R}{R_0}\right)^{-n}. \tag{4.1}$$

In the commonly discussed steady-state disk, $n = 3/4$ (Shakura and Sunyaev, 1973). We assume a local blackbody spectrum, B_ν,

$$B_\nu(R) = \frac{2h\nu^3}{c^2 \left(e^{h\nu/kT(R)} - 1\right)}, \tag{4.2}$$

and write the integrated SED of the disk as

$$F_\nu = \frac{2\pi \cos i}{D^2} \int_{R_{\rm in}}^{R_{\rm out}} B_\nu(R) R dR, \tag{4.3}$$

where D is the source distance and i is the binary inclination. This leads to the general solution:

$$F_\nu = \nu^{3-2/n} \frac{4\pi h^{1-2/n} \cos i \, k^{2/n} T_0^{2/n} R_0^2}{nc^2 D^2} \int_{x_{\rm in}}^{x_{\rm out}} \frac{x^{2/n-1} dx}{e^x - 1}, \tag{4.4}$$

where the substitution $x = h\nu/kT$ has been made. In the limiting case of an unbounded disk extending from zero radius to infinity, the integral is just a numerical value and the frequency dependence extracted is $F_\nu \propto \nu^{3-2/n}$. For the Shakura-Sunyaev disk, $n = 3/4$, we obtain the well-known result $F_\nu \propto \nu^{1/3}$. If the disk is irradiatively heated, then the simplest assumption is that the irradiating flux drops off as the inverse square of the radius, leading to $T^4 \propto R^{-2}$ and $T \propto R^{-0.5}$. The resulting SED should then exhibit $F_\nu \propto \nu^{-1}$, a red spectrum actually decreasing into the UV.

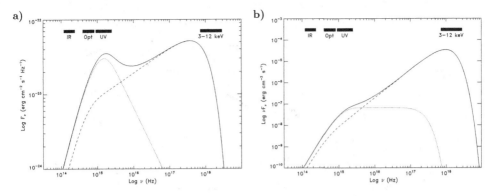

FIGURE 4.3. Blackbody SED for cases with viscous heating (dotted), irradiative heating (dashed), and a combination of the two (solid). a) shows the ν–F_ν representation. b) shows the same models in the ν–νF_{nu} representation.

In practice, since the irradiation temperature $T_{\rm irr} \propto R^{-0.5}$ drops off more slowly as a function of radius than the temperature due to viscous heating, $T_{\rm visc} \propto R^{-3/4}$, we expect the inner disk to be dominated by viscous heating and the outer by irradiation, leading to a spectrum that turns over between the two limits at some intermediate frequency (see Fig. 4.3). Furthermore, the disk does not extend from zero radius to infinity, and both inner and outer bounds are important in defining the spectrum. The inner radius influences the X-ray spectrum and the outer radius the UV and optical. Modeling of real X-ray data is more complex than described here, first because the temperature distribution of the inner disk is modified by the boundary condition applied at the inner edge of the disk, and second because Comptonization changes the X-ray spectrum substantially (see Chapter 6).

4.4.3 Multiwavelength observations of black holes in outburst

The first really high-quality dataset with which to test the picture just described came from IUE and HST observations of the black-hole system X-ray Nova Muscae 1991 (Cheng et al., 1992). Multiple epochs of data as the source faded revealed an approximately $F_\nu \propto \nu^{1/3}$ spectrum consistent with a viscously heated accretion disk. Taken at face value, however, this interpretation requires extremely high accretion rates comparable to or in excess of the Eddington limit (the accretion rate at which radiation pressure balances gravity and can suppress further accretion).

Improvements in HST capabilities and target of opportunity strategies led to a much richer dataset on XTE J1859+226 (see Fig. 4.4; Hynes et al., 2002), both in wavelength and in temporal coverage. Early observations showed a smooth spectrum with a broad hump in the UV well fitted by blackbody disks with a relatively flat temperature distribution consistent with an irradiated disk. Later in the outburst, the far-UV spectrum transformed from a UV-soft form, where f_ν declined with ν, to a rising UV-hard form, consistent with expectations of a purely viscously heated disk and similar to that seen in X-ray Nova Mus 1991.

Other black hole transients have had less complete coverage but show similar characteristics. A 0620–00 showed a UV-hard SED resembling X-ray Nova Mus 1991 or XTE J1859+226 at late times, whereas GRO J0422+32 was UV soft like XTE J1859+226 at early times (Hynes, 2005, and references therein).

It should be emphasized that the primary diagnostic between the UV-hard and UV-soft states (and by inference R^{-3} and R^{-2} heating) is the turnover or its absence in the UV, and the slope of the far-UV spectrum. Hynes (2005) showed SEDs of these systems and others based only on the optical portion of the SED, and no such discrimination was possible. To obtain a reliable SED requires not only good coverage in the satellite

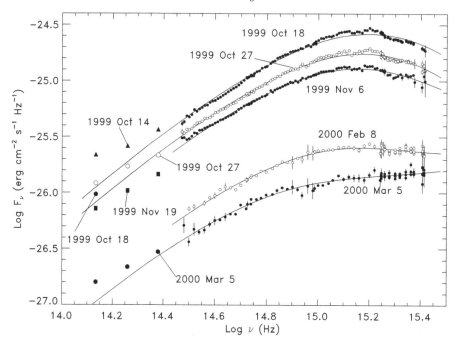

FIGURE 4.4. Observed SEDs of XTE J1859+226 from Hynes *et al.* (2002). Different symbols are used to differentiate different observation dates.

UV (for which HST is ideal, but Swift/UVOT lacks far-UV capability), but also a good understanding of the effect of interstellar reddening on the SED. For a discussion of the latter, see Fitzpatrick (1999) and references therein.

4.4.4 Irradiation of disks

The amount of irradiation of the accretion disk that is actually expected has been examined by Dubus *et al.* (1999) and expressed somewhat differently by Dubus *et al.* (2001). They express the local irradiation temperature for a disk element at radius R as

$$\sigma T_{\text{Irr}}^4 = \mathcal{C} \frac{L_{\text{X}}}{4\pi R^2}, \qquad (4.5)$$

where σ is the Stefan-Boltzmann constant, L_{X} is the irradiating (X-ray) luminosity, and $\mathcal{C}$ is a dimensionless measure of the efficiency of irradiation. $\mathcal{C}$ parametrizes our ignorance of the illumination geometry and the (energy-dependent) albedo of the disk element. Empirically, Dubus *et al.* (2001) find that a value of $\mathcal{C} \simeq 5 \times 10^{-3}$ is consistent with observations.

One difficulty with this simple picture that has been appreciated for some time is that models of disk structure usually predict a disk profile that is convex as a function of radius (Dubus *et al.*, 1999, and references therein). Such a disk should be self-shielding and the inner portion should be unable to irradiate the outer regions, in spite of considerable evidence (such as echo mapping; Section 4.7) that accretion disks do experience irradiation. Dubus *et al.* (1999) suggest that either X-rays are scattered by material out of the plane, or warping of the accretion disk (see Section 4.5.5) can expose portions of the outer disk to irradiation. In this case, $\mathcal{C}$ also parametrizes our ignorance about how irradiation of the disk is mediated. The caveat here is that for many accretion geometries, $\mathcal{C}$ itself becomes a function of radius.

One example where the radial dependence of irradiation can be significantly modified is important to discuss, and to highlight how an inferred temperature distribution may

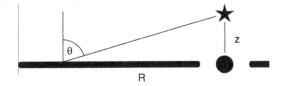

FIGURE 4.5. Irradiation geometry appropriate for a "lamp post" above the accretion disk.

not be sufficient to constrain the underlying astrophysics. This is the case of irradiation by a "lamp post" above the disk: for example, flares in a vertically extended corona, or back-irradiation from a jet. This case is well known in the AGN and young star communities, where it is sometimes considered the normal irradiation geometry. It is illustrated in Figure 4.5. In this case, the intensity of irradiating flux decreases according to the inverse square law as considered earlier, but in addition the angle of incidence becomes steeper at larger radii, resulting in spreading the irradiation over a larger area. This introduces another factor proportional to $\cos\theta \simeq z/R$ for $z \ll R$ (where z is the height of the irradiating source above the disk). The additional $1/R$ dependence results in heating varying as R^{-3} just as for the viscously heated steady-state accretion disk. It is therefore possible that the UV-hard spectra discussed earlier could also result from irradiation, and this may avoid the difficulty noted, for example, in the case of X-ray Nova Muscae 1991, where purely viscous heating would require accretion at or above the Eddington limit.

4.4.5 Evidence for jets in spectral energy distributions

Not all emission in the spectral energy distributions of LMXBs originates in the disk. There is now very persuasive evidence in a number of black hole LMXBs for IR synchrotron emission from a jet. This is seen as a flat IR spectrum extending to lower frequencies than possible for the disk (XTE J1118+48; Hynes et al., 2000) or even a two-component spectrum with a red IR component from the jet and a blue disk component (GX 339–4; Corbel and Fender, 2002). In both cases the IR flux is comparable to that seen in the radio, as expected from extrapolation of the characteristic flat spectrum of a compact radio jet (Blandford and Konigl, 1979). More recently a similar SED has also been seen in the neutron star LMXB 4U 0614+091 (Fig. 4.6; Migliari et al. 2006, 2010). The second observation was based on quasi-simultaneous observations of a persistent source and so is a robust measure of the SED.

Another signature expected for synchrotron emission from a jet would be polarization. To date no large IR polarizations have been found, although Shahbaz et al. (2008a) saw polarizations in Sco X-1 and Cyg X-2 of a few percent that increased with wavelength, as would be expected from the increasing fractional contribution of a jet at longer wavelengths, and which were not consistent with expected interstellar polarization.

4.4.6 Quiescent SEDs

Observations of quiescent LMXBs also reveal emission across most of the observable range, although radio detections are extremely challenging, and those at gamma-ray energies have been impossible so far. A major complication in obtaining SEDs of the accretion emission in quiescent LMXBs is that the companion star can completely dominate the optical and IR range of the spectrum. Unfortunately, the brightest quiescent system, V404 Cyg, also exhibits almost complete dominance of the UV to near-IR SED by the companion (Hynes et al., 2009b).

The accretion flow is clearly detected at X-ray energies in many systems and appears to be characterized by a soft power-law of photon index $\Gamma \sim 2$, corresponding to $F_\nu \propto \nu^{-1}$ (e.g., Kong et al., 2002). This almost certainly originates from the inner region of the accretion flow, which is believed to form a hot, evaporated, advective flow. Spectral models of such flows variously attribute the X-ray emission to either bremsstrahlung or

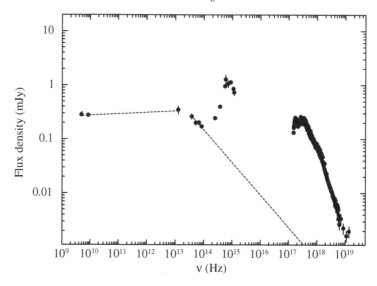

FIGURE 4.6. SED of 4U 0614+091 (adapted from Migliari et al., 2010). Reproduced by permission of the AAS.

Comptonization, but there remain many uncertainties in these models, and there is no uniquely accepted solution (Narayan et al., 1996, 1997; Esin et al., 1997; Quataert and Narayan, 1999). X-ray emission lines could provide a valuable discriminant (Narayan and Raymond, 1999), but the most sensitive observations of a quiescent SXT, XMM-Newton observations of V404 Cyg, yielded only upper limits on the iron Kα emission line that is expected to be strongest (Bradley et al., 2007).

A UV excess is sometimes detectable (McClintock et al., 1995; McClintock and Remillard, 2000; McClintock et al., 2003; Hynes et al., in preparation). The spectral shape can be characterized as a hot quasi-blackbody of temperature $\sim 10,000$ K. Our best explanation for this component currently is that it originates either from the accretion stream-impact point or from a relatively hot region of the disk (hot enough that it should be locally in the high state of the DIM).

A mid-IR excess is also seen (Muno and Mauerhan, 2006), but the origin remains debated. In some sources such as V404 Cyg, the excess is rather subtle, but in others like A 0620−00 and XTE J1118+480, it is very pronounced and clearly real. The favored explanation of Muno and Mauerhan (2006) for most sources was that it originates in a *circumbinary* accretion disk comprising material lost during outbursts or even the original supernova that formed the compact object. At longer wavelengths, radio emission appears to exhibit a flat spectrum, as is often seen in more luminous states (Gallo et al., 2005). The most natural explanation for this is that a weak jet continues in quiescence. It is even possible that this dominates the quiescent energy budget, with much of the accretion power carried away in the bulk kinetic energy of the jet (Fender et al., 2003). Interestingly, an extrapolation of the flat radio spectrum into the mid-IR region comes reasonably close to the IR excess flux, leading Gallo et al. (2007) to interpret the mid-IR excess as due to the jet instead of a circumbinary disk, resulting in a situation similar to the hard state systems discussed in Section 4.4.5.

4.5 Light curves

4.5.1 *Ellipsoidal variations*

The simplest form of optical light curve of an LMXB is that from ellipsoidal variations in quiescence where the light is often dominated by the companion star. This star is tidally

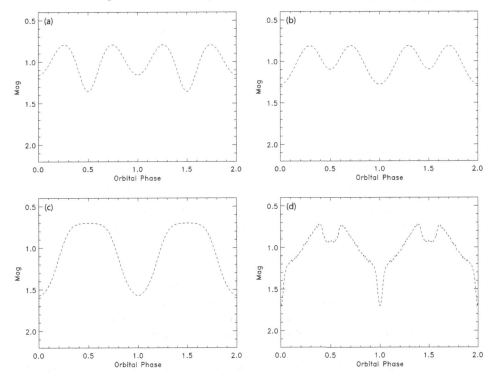

FIGURE 4.7. Orbital light curves of LMXBs. *a)* Pure ellipsoidal variations. *b)* Ellipsoidal variations plus weak irradiation of the donor. *c)* Strong irradiation of the donor. *d)* Irradiation of disk and donor with mutual eclipses. These figures were generated by the XRBinary code written by E. L. Robinson.

distorted into a teardrop shape. When viewed side-on at orbital phases 0.25 and 0.75, we see a large cross-sectional area and hence maximum light. When viewed end-on at phases 0.0 and 0.5, we see a smaller cross-sectional area and hence minimum light. In essence, ellipsoidal variations then take a near-sinusoidal form with two cycles per binary orbit. An additional complication is that the surface of the companion is not uniformly bright because of gravity darkening (von Zeipel, 1924). Less flux emerges from regions of the companion with a lower surface gravity, and so these regions have a lower effective temperature. In the absence of significant X-ray heating of the donor star, this results in maximum surface temperature at the poles and minimum at the inner Lagrangian point facing the compact object. This breaks the symmetry between phase 0.0 and 0.5 (but not between 0.25 and 0.75). Consequently, ellipsoidal variations should exhibit two equal maxima but unequal minima, with the phase 0.5 minimum being deeper. A model ellipsoidal light curve is shown in Figure 4.7a.

The amplitude of ellipsoidal modulations is mostly determined by the binary inclination, with a lesser dependence on mass ratio and an even weaker dependence on stellar temperature and surface gravity (via limb darkening changes). This has made measurement of ellipsoidal modulations the technique of choice for determining inclinations of non-eclipsing quiescent LMXBs. An example is shown in Fig. 4.8.

There are several caveats to this approach. The first and most obvious is that the observed variations must actually be produced by ellipsoidal modulations. Other mechanisms for periodic variability include superhumps (see Section 4.5.4), variations in the visibility of the accretion stream impact point, and starspots on the companion star. There are certainly light curves that cannot fully be explained by ellipsoidal variations. For example, many studies have now been made of A 0620–00, finding that the light curves clearly change in morphology from epoch to epoch, often showing asymmetric maxima

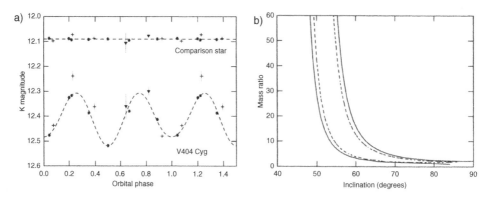

FIGURE 4.8. *a)* IR ellipsoidal modulations in the quiescent SXT V404 Cyg with a fitted model. *b)* Derived joint constraints on inclination and mass ratio. Both figures are from Shahbaz *et al.* (1994).

(McClintock and Remillard, 1986; Haswell *et al.*, 1993; Gelino *et al.*, 2001; Cantrell *et al.*, 2008). The authors have variously invoked state changes, disk asymmetries, starspots, and eclipses to explain the behavior seen. Although the system is no longer believed to be at a high enough inclination to eclipse, it is likely that several of the other effects may all play a role.

Even in the absence of periodic contamination of the ellipsoidal light curves, the presence of nonvariable disk emission will dilute the ellipsoidal variations. This will make the amplitude appear smaller, leading to an underestimate of the binary inclination and ultimately an overestimate of the compact object mass. It is often argued that this contamination can be minimized by measuring IR ellipsoidal variations (e.g., Gelino *et al.*, 2001), but an IR contamination from a cool disk might well be *expected* in quiescence (Hynes *et al.*, 2005), and the presence of IR excesses in quiescent SXTs discussed in Section 4.4.6 reinforces this expectation. To attempt to compensate for this, one can try to measure the amount of disk light by measuring dilution of the photospheric absorption lines of the companion star; this is termed the veiling. Although some measurements suggest small or undetectable IR veiling, such as V404 Cyg (Shahbaz *et al.*, 1996), in other objects the IR veiling can be significant (Froning *et al.*, 2007). In any case, we know that flickering light is present in quiescence which can have large amplitudes (e.g., Hynes *et al.*, 2003b), that IR flickering can be detectable (Reynolds *et al.*, 2007), and that longer time scale changes in the overall flux level also occur (Cantrell *et al.*, 2008). These facts suggest that even if we believe the veiling can be measured reliably, it is strictly only valid if measured simultaneously with the ellipsoidal variations, and ideally during the passive state identified by Cantrell *et al.* (2008).

4.5.2 X-ray heating effects

As we move from quiescence into luminous or outburst states, the accretion rate onto the compact object, and hence the X-ray luminosity, increases. The first effect on the light curve is that the temperature of the inner face of the companion star is raised by irradiation from the increased X-ray luminosity. This offsets the gravity darkening effect and will first weaken the phase 0.5 minimum, then reverse the depths of the minima (so that phase 0.0 is the deeper minimum). With strong enough heating, such as for typical short-period luminous LMXBs, the ellipsoidal effects are no longer discernible, and we see a single-humped modulation with a maximum at phase 0.5 and a minimum at phase 0.0. This progression from pure ellipsoidal modulations to strong heating of the companion star is illustrated in Figure 4.7a–c.

This effect is commonly observable in neutron star LMXBs, allowing the orbital period to be determined from optical photometry. In many cases, in the absence of X-ray eclipses or dips, this may be the first or only evidence for the orbital period. For example, in Sco X-1 the orbital period of 0.78 days was first identified in this way (Gottlieb et al., 1975).

4.5.3 Eclipses

If an LMXB has a high enough inclination, then eclipses of the accretion disk by the companion star can occur, sharpening the minimum at phase 0.0, whereas eclipses of the irradiated face of the companion star by the disk may introduce a secondary eclipse at phase 0.5 (Fig. 4.7d). Real light curves can look very much like this model – for example, XTE J2123–058 (Zurita et al., 2000). At higher inclinations, self-obscuration of the disk becomes an important factor, and additional structure appears in the light curves reflecting deviations from axisymmetry of the disk. The most pronounced manifestation of this is X-ray dipping, discovered in 4U 1915–05 (White and Swank, 1982) and subsequently seen in about 10 other neutron star LMXBs. X-ray dips are characterized by rather irregular dipping, usually strongest around phases 0.7 to 1.0, likely indicating absorption by inhomogeneous material associated with the accretion stream impact point and/or material from it that overflows the disk. At even higher inclinations, the central X-ray source is *permanently* obscured by the disk structure, resulting in an accretion disk corona (ADC) source where only scattered X-rays are seen. The prototypical ADC source is 2A 1822–371 (Mason et al., 1980); see Bayless et al. (2010) for a recent multiwavelength study. Such sources are characterized by a much lower ratio of X-ray to optical brightness than in lower-inclination objects, and by broad partial X-ray eclipses rather than narrow total ones.

Eclipsing systems provide much more precise constraints on the binary inclination, and so have the potential to yield more accurate system parameters than are possible with ellipsoidal variations alone. Unfortunately, no eclipsing black hole LMXBs are known in our galaxy, so this potential benefit for black hole mass determination has yet to be fully realized. The only eclipsing black hole candidate known with confidence is the HMXB M33 X-8 (Orosz et al., 2007). The lack of eclipsing black hole systems is unlikely to be coincidence and reflects a selection effect leading to an absence of black hole LMXBs with inclination $i > 75°$. It is proposed that higher inclination black hole systems are ADC sources where much of the X-ray luminosity is hidden from view, and so we are less likely to detect them (Narayan and McClintock, 2005). Neutron star ADC sources have been found in spite of this selection effect because they are persistently active, whereas black holes are usually transient.

In CVs, eclipsing systems facilitate eclipse mapping of the accretion disk emission structure and radial temperature dependence (see Warner's review in Chapter 3). This tomographic eclipse-mapping methodology developed for CVs depends on the known geometry of the companion Roche lobe occulting the disk. It has not proved directly applicable in LMXBs because of the disk self-obscuration effects together with the additional light from heating of the companion star. Instead, LMXB light curves have been interpreted by developing a model of the binary geometry, including irradiation of the disk and companion and possibly asymmetric structure in the accretion disk, and then predicting multiwavelength light curves. The nature of the self-obscuration remains a matter of debate. Early models interpreted it as an elevated rim to the accretion disk. The major flaw of this explanation is that to maintain outer disk material at this elevation would require temperatures much higher than expected, so such material could not be in hydrostatic equilibrium. Bayless et al. (2010) instead suggest that the obscuring material is a wind from the inner disk and so is not in hydrostatic equilibrium.

Another benefit to eclipsing LMXBs is that eclipses can provide very precise timing of the orbital period, with ingress and egress times of the total eclipse of the neutron star lasting just a few seconds. The duration here reflects the scale height of the companion star atmosphere, which is larger than the neutron star size. The best studied case by far is EXO 0748–676 (Wolff et al., 2009). In this system, an ongoing eclipse monitoring program through the history of the Rossi X-ray Timing Explorer (RXTE), coupled with less well sampled observations from the preceding decade, has resulted in an orbital period history with exquisite precision. Surprisingly, rather than showing a gradual period evolution, this monitoring has revealed distinct epochs of duration 5 to 10 years with no obvious changes in the X-ray behavior corresponding to the abrupt transitions between epochs. Wolff et al. (2009) attributed these changes to magnetic cycles in the companion star that modify its internal structure, and hence redistribute a small amount of angular momentum. In support of this interpretation, similar behavior is seen in magnetically active RS CVn systems.

4.5.4 Superhumps

It should not be assumed that all periodicities are actually orbital modulations. The tidal interaction of the companion star with the outer disk can excite eccentric modes within the disk. The most commonly encountered mode is excited when the disk grows to a radius where there is a 3:1 resonance between the local Keplerian frequency and the orbital frequency (Whitehurst and King, 1991). This is possible only in systems with relatively small mass ratios ($q = M_2/M_1 < 0.25$), as larger mass ratios result in the disk being truncated inside the 3:1 resonance radius by tidal stresses from the companion star. In even more extreme mass ratio systems, it may be possible to also excite the 2:1 resonance. Once eccentricity develops, the eccentric disk will slowly precess. When the eccentric disk reaches closest to the companion star, it experiences increased tidal stresses and hence more heating. This leads to variations in the optical and UV light from the outer disk, termed superhumps, that occur at the beat period between the orbital period and the precession period. Superhumps are commonly seen in CVs and are discussed in more detail by Warner in Chapter 3, but they can occur in LMXBs as well.

The most likely LMXBs to exhibit superhumps are black hole systems, as these tend to have more extreme mass ratios than most neutron star LMXBs. O'Donoghue and Charles (1996) examined the cases for several black hole transients and concluded that superhumps had definitely been seen in two systems, GRO J0422+32 and X-ray Nova Muscae 1991, and had possibly also been seen in GS 2000+25. Unlike the CV case, however, where superhumps are uniquely present during outburst with the orbital period absent, in LMXBs both orbital and superhump modulations may be present, at different times or even simultaneously. The superhump period is very similar to the orbital period (to within a few percent), and discrimination between the two can be difficult. More recent and very interesting examples of superhumping LMXBs have included XTE J1118+480 (Uemura et al., 2000, Zurita et al., 2002) and GRS 1915+105 (Neil et al., 2007).

Another difference between CVs and LMXBs is that the mechanism for producing superhump light in CVs, enhanced heating of the disk by tidal stresses induced by the companion star, should not be important in LMXBs (Haswell et al., 2001). This is because the optical light should not be dominated by viscous dissipation, but instead by irradiation by X-rays. This suggests that instead the superhump arises from coupling of irradiation to tidal distortion of the disk. In support of this, Haswell et al. (2001) show that disk simulations do produce a ~10 % increase in disk area at the time of superhump maximum. Smith et al. (2007) note that the period excess in LMXBs is also smaller than that in CVs and suggest that the precession period is longer because LMXB disks should be hotter and thicker due to irradiation.

The case of XTE J1118+480 is especially interesting. Superhumps were seen both during outburst, with varying morphology and period (Uemura et al., 2000), and late in the decay to quiescence (Zurita et al., 2002). The superhump periods inferred were just 0.1% to 0.6% of the orbital period. Coupled with the extremely small mass ratio inferred spectroscopically (0.037 ± 0.007; Orosz, 2001), this makes XTE J1118+480 an important object for studying the extreme limits of superhump behavior in which the 2:1 resonance may become the dominant mechanism for exciting disk eccentricity.

4.5.5 Superorbital periods

Other periods longer than the orbital period can arise. The most famous is the 35-day period seen in the neutron star LMXB Her X-1. This manifests in both a long modulation in optical light curve morphologies (Gerend and Boynton, 1976) and changes in X-ray pulse profiles. The optical behavior was modeled successfully with a precessing tilted accretion disk that modulates shadowing of the companion star on a 35-day period. The mechanism for inducing this tilt is now believed to be radiation-driven warping of the accretion disk (Wijers and Pringle, 1999).

Ogilvie and Dubus (2001) investigated the stability of accretion disks to radiation-driven warping in systems of a wide range of mass ratios and binary separations. They focused on the two lowest-order warping modes, 0 and 1, and identified several regimes where disks were either stable or unstable to mode-0 perturbations, to mode-1 perturbations, or to both. Reassuringly, Her X-1 lies in the region unstable to mode-0 only, and most other systems in the unstable regime also exhibit superorbital periodicities of some form. Clarkson et al. (2003) tested this description more thoroughly with a dynamic power-spectral analysis of RXTE light curves. Systems such as Her X-1 and LMC X-4 that are in the pure mode-0 regime exhibit relatively stable periodicities. SMC X-1, which lies close to the region of mode-1 instability, exhibits a less stable long period, whereas Cyg X-2, which is unstable to both mode-0 and mode-1 oscillations, exhibits complex and multiperiodic behavior.

4.5.6 Quasi-periodic oscillations

Quasi-periodic oscillations, or QPOs, are common on many time scales in LMXBs. Some of the superorbital periods discussed previously would qualify as long-period QPOs rather than strict periodicities. At the other extreme of time scale, there is a rich phenomenology of high-frequency (milliseconds to seconds) QPO behavior seen in X-ray observations of LMXBs arising from the inner accretion flow. Rather than discuss that here, we focus on those QPOs visible in the optical and/or UV; for a review of X-ray QPOs, see van der Klis (2006).

A QPO is a repeating signal that is not strictly periodic. It may wander in frequency or exhibit changes in phase. A transient signal, for example a decaying series of pulses, will also manifest as a QPO. In a Fourier transform, a QPO appears as a peak with finite width, whereas a true coherent periodicity has width limited only by the time period sampled by the data. QPOs are often characterized by their coherence, $Q = \nu/\Delta\nu$, which is a measure of how broad the QPO is in frequency space. A low-coherence QPO is essentially an excess of noise around a peak frequency, defining a preferred but not unique time scale.

One class of optical QPOs has now been seen in two UCXRBs. Chakrabarty et al. (2001) identified a strong optical and UV QPO around 1 mHz in the 42 min orbital period system 4U 1626–67. This was seen to be moderately coherent ($Q \sim 8$), stronger in the UV than optical, and completely absent in X-rays. The authors attributed the effect to a precessing warp in the inner accretion disk. Subsequently a similar feature also around 1 mHz has been seen in 4U 0614+091 (Zhang et al., in preparation). This

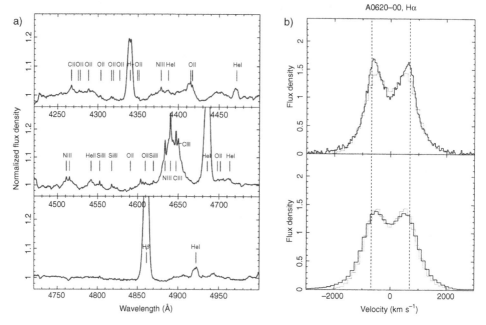

FIGURE 4.9. *a)* Blue spectrum of Sco X-1 (from Steeghs and Casares, 2002). This shows the typical range of atomic emission lines present in the spectrum of a bright LMXB. *b)* Hα profiles of A 0620–00 (from Marsh *et al.*, 1994). This illustrates the typical double-peaked disk profile especially seen in quiescent LMXBs. Solid data are that of Marsh, dotted are from Johnston *et al.* (1989). In the lower panel, Marsh's data have been binned at blurred to match that of Johnston. Dashed lines indicate the velocity of the outer edge of the disk inferred by Marsh. Reproduced by permission of the AAS.

system is believed to have an orbital period of 51 min (Shahbaz *et al.*, 2008b), so the similarity of time scales to 4U 1626–67 is striking.

A time scale around 1 mHz also appears to be significant in quiescent SXTs, although this is probably coincidence, given the very different orbital periods and physical conditions. Hynes *et al.* (2003b) identified a break in the power-density spectrum of the 8 hr period system A 0620–00 at around 1 mHz. An actual QPO at 0.78 mHz was seen in the 6.5 day period V404 Cyg (Shahbaz *et al.*, 2003) and at 2 mHz in the 4.1 hr system XTE J1118+480 (Shahbaz *et al.*, 2005). The favored explanation for QPOs in quiescent SXTs has been that they are associated with the transition from a thin accretion disk to an evaporated advective flow, although the interpretation is probably not as simple as associating the QPO frequency with the Keplerian rotation period at the transition radius.

4.6 Spectroscopy

4.6.1 *Emission and absorption line spectra*

Luminous X-ray binaries can show a range of emission lines, although their spectra are nowhere near as rich as those of planetary nebulae or active galactic nuclei can be. We show an example, the blue spectrum of Sco X-1, in Figure 4.9. H I lines of the Balmer series in the optical and Brackett series in the infrared are common, but by no means ubiquitous. They can be absent, and sometimes show absorption structure. He I lines often accompany H I. More reliably present are He II lines in the optical (dominated by 4686 Å) and ultraviolet (1640 Å).

Besides hydrogen and helium, the main other lines seen are of carbon, nitrogen, oxygen, and silicon. In the optical, the strongest of these features is a blend of N III and C III lines around 4640 Å commonly referred to as the Bowen blend (McClintock et al., 1975), but many weaker lines are present and can be seen in high-quality spectra of Sco X-1 (Steeghs and Casares, 2002). In the UV, several strong resonance lines dominate: C IV 1548,1551 Å, N V 1239,1243 Å, Si IV 1394,1403 Å, and O V 1371 Å are the most prominent. Dramatic differences in the relative intensities of these have been seen between objects with carbon and oxygen lines sometimes completely absent. This has been interpreted as evidence for CNO processing in the donor star (Haswell et al., 2002). Among ultracompact systems, hydrogen and sometimes even helium may be absent, leaving spectra dominated by carbon and oxygen (Nelemans et al., 2004, 2006).

In quiescence, higher excitation lines of He II together with the CNO lines are absent, and only H I and very weak He I are present in emission. Provided the disk is sufficiently dim, the photospheric absorption spectrum of the companion emerges, facilitating measurement of its radial velocity curve and rotational broadening and derivation of the system parameters. This is discussed in Section 4.6.3.

4.6.2 Emission line profiles and Doppler tomography

Quiescent LMXBs present the simplest emission line profiles to understand. Optical hydrogen and helium lines are seen with a double-peaked profile (e.g., Fig. 4.9) consistent with expectations from an approximately axisymmetric Keplerian accretion disk as also seen in CVs (Horne and Marsh, 1986). The blue wing of the profile comes from the side of the disk approaching us, and the red wing from that which is receding. Most sensitively, the peak separation can measure the velocity at the outer edge of the disk. More close examination, however, reveals changes in the line profiles, with the relative strengths of the two peaks varying. To examine how this relates to the binary orbital phase, it is usual to plot a trailed spectrogram, with wavelength (equivalent to velocity for a single line) on the x-axis and orbital phase on the y-axis. Good-quality data then reveal that the asymmetries arise from a third peak moving back and forth sinusoidally, tracing out an S-wave in the trailed spectrum (Fig. 4.10; Marsh et al., 1994).

To better understand how such components can be associated with locations in the binary, the technique of Doppler tomography has been developed. A number of documents have reviewed this well; for example, Marsh (2001). The key idea is that any given component of the binary should have a defined velocity in the orbital plane that can be resolved into x and y components, V_x and V_y. Each such component will also lead to a single S-wave in a trailed spectrogram. Doppler tomography is essentially a transformation from visualizing the data in terms of radial velocity and phase to an alternative visualization in terms of V_x and V_y. Thinking of the V_x–V_y plane in terms of polar coordinates, the phasing of an S-wave component dictates the azimuth of a spot in the $V_x x$–V_y plane, and its amplitude determines the distance of the point from the origin ($V_x = V_y = 0$). Doppler tomography can also be thought of as representing the trailed spectrogram as a sum of an arbitrary number of S-waves.

The dominant component that is expected in a Doppler tomogram is a ring of emission in velocity space corresponding to the accretion disk (see Fig. 4.10; Marsh et al., 1994). Since this is represented in velocity space, the inner disk has high velocities and so is on the outside of the ring in the tomogram; in this sense, tomograms are inside-out compared to visualizing things in real space. The other component most often seen is a structure on the left-hand side corresponding to the accretion stream, the stream-impact point, or a bulge in the disk downstream of it. In quiescent LMXBs such as A 0620–00, this is relatively well behaved and lies at velocities expected for the accretion stream (Marsh et al., 1994). In persistent LMXBs, however, the maximum emission often lies below the expected stream velocities in the tomogram (Casares et al., 2003; Pearson et al., 2006)

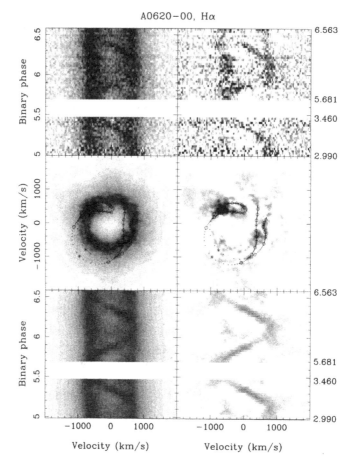

FIGURE 4.10. Doppler tomogram of Hα emission in A 0620–00 (from Marsh et al., 1994). The right-hand panels show the effect of subtracting the nonvarying part of the trailed spectrum (equivalent to the symmetric part of the tomogram). The upper panels show the data; the middle panels are the tomograms; and the lower are reconstructions of the data from the tomogram.

and may even appear in the lower-left quadrant (Hynes et al., 2001). This is usually attributed to a bulge introduced by the stream impact or material overflowing the disk, but its origin is not yet fully understood. The final component sometimes revealed in some emission lines is the companion star. This is discussed more in Section 4.6.4.

There are several limitations of the technique to keep in mind, especially as X-ray binaries can violate a number of these. Doppler tomography assumes that emission lines are optically thin and that all of the flux is visible all of the time. This requires lines that do not modulate in integrated intensity over the binary orbit; yet observationally, such modulations usually are present, with eclipses presenting the most dramatic example. The new technique of modulation tomography (Steeghs, 2003) partially addresses this by allowing the line intensity to modulate sinusoidally. This technique has been successfully applied to the quiescent LMXB XTE J1118+480 by Calvelo et al. (2009), where it is found that the disk component, as expected, does not modulate in intensity but the accretion stream impact point does. Another limitation that can more justifiably be assumed satisfied is that all motions are in the plane (i.e., there is no V_z component). Even so, a disk wind, for example, violates this assumption, and wind emission would confuse the reconstruction. A final thing to keep in mind, but not a limitation of the technique as such, is that a Doppler tomogram is reconstructed in velocity space, not real space. The velocities of emitting material can be constrained, but additional assumptions, such as system parameters, are required to associate these velocity components with physical components of the binary. Conversely, of course, this means that Doppler tomograms can be used to constrain those system parameters, and we return to this idea in Section 4.6.4. Even knowing system parameters, the mapping between real and velocity space is not one-one, and so there may be ambiguity about the physical location of emitting material.

4.6.3 Mass determinations in quiescent systems

Binary system parameters, and especially masses, are essential for many purposes. Specific parameters provide information about the evolutionary state of a system, whereas distributions in period and compact object mass, for example, test models for the formation and evolution of binary populations (Fryer and Kalogera, 2001; Pfahl et al., 2003). A compact object mass in excess of $3\,M_\odot$ is considered the most convincing evidence for a black hole, while precise measurements of neutron star masses could constrain the equation of state of neutron star matter (e.g. Özel and Psaltis, 2009). Finally, most of the observational techniques described here depend, to a greater or lesser degree, on knowing the system parameters for a quantitative interpretation.

Dynamical parameter determinations are underpinned by Kepler's third law, which can be expressed as:

$$f(M) = \frac{K_2^3 P_{\rm orb}}{2\pi G} = M_1 \frac{\sin^3 i}{(1+q)^2} \tag{4.6}$$

where K_2 is the radial velocity semiamplitude of the donor star, $P_{\rm orb}$ is the binary orbital period, M_1 is the compact object mass, i is the binary inclination, and $q = M_2/M_1$ is the mass ratio. $f(M)$ is termed the mass function. It is of crucial importance because (i) it can be measured from the radial velocity curve alone (K_2 and $P_{\rm orb}$) and (ii) it provides a strict lower limit on M_1 (since $\sin i \leq 1$ and $1+q > 1$). Many objects have been classified as a black hole purely on the basis of mass functions greater than the assumed maximum mass of a neutron star, around $3\,M_\odot$. The most convincing example, and an excellent case study for system parameter determination in general, is V404 Cyg (Casares et al., 1992; Casares and Charles, 1994). It has an orbital period of 6.47 days and $K_2 = 208.5 \pm 0.7\,{\rm km\,s^{-1}}$. This leads to a mass function $f(M) = 6.08 \pm 0.06\,M_\odot$.

Beyond this, it can be seen that determination of the actual compact object mass requires measurement of the binary mass ratio and inclination. The preferred method to estimate the mass ratio is to observe rotational broadening in photospheric absorption lines from the donor star. Since the projected radial velocity semiamplitude, K_2, and the rotational broadening, $v\sin i$, are both affected by orbital inclination in the same way, the ratio of $v\sin i/K_2$ is purely a function of the ratio of the radius of the donor star to the binary separation and hence is a function of q only (although details such as limb darkening are important for precise measurement of $v\sin i$ from observed rotationally broadened lines). A good approximation (see, e.g., Wade and Horne, 1988) is

$$V_{\rm rot} \sin i = 0.462 K_2 q^{1/3} (1+q)^{2/3}. \tag{4.7}$$

In practice, rotational broadening measurements are challenging because the expected values, typically below $100\,{\rm km\,s^{-1}}$, are often close to the spectral resolution of the data used. The usual method to extract this information is to observe both the target and a template star of matched spectral type and then convolve the template spectrum with rotational broadening profiles until an optimal match is found. In the case of V404 Cyg, Casares and Charles (1994) measured $V_{\rm rot} \sin i = 39.1 \pm 1.2\,{\rm km\,s^{-1}}$, which, together with K_2 cited earlier, yields $q = 0.060^{+0.004}_{-0.005}$.

An alternative approach to the mass ratio is to attempt to measure the radial velocity semiamplitude of the compact object, K_1, from disk emission lines, since $q = K_1/K_2$.

In practice this has been fraught with errors, with emission line radial velocity curves exhibiting incorrect phasings and different systemic velocities compared with the companion star. It is argued that these effects arise from asymmetries far out in the disk, and that they can be minimized by measuring just the high-velocity wings of a line, but even then, mass ratios from rotational broadening are preferable.

The preferred method to determine the inclination would be to model eclipses (see Section 4.5.3). Most systems (including all galactic black hole candidates) do not eclipse, however, so instead the usual method has been to measure ellipsoidal variations (see Section 4.5.1). Continuing our case study of V404 Cyg, the ellipsoidal variation study of Shahbaz et al. (1994) was illustrated earlier in Figure 4.8. These authors model IR ellipsoidal modulations to deduce $i = 56 \pm 4°$. Combined with the mass function and mass ratio measurements discussed previously, this implies $10\,M_\odot < M_1 < 15\,M_\odot$ with a preferred value of $M_1 = 12\,M_\odot$. As discussed in Section 4.5.1, an important concern with ellipsoidal studies is that the modulations may be diluted by disk contamination. Shahbaz et al. (1996) measured this in the IR and placed an upper limit on the disk contribution of 14%. This at most reduces the derived mass by $2\,M_\odot$ from $12\,M_\odot$ to $10\,M_\odot$.

It should be emphasized that we have focused on V404 Cyg because it is both one of the best studied systems and one with the cleanest results. In most quiescent SXTs, measurements are less precise and concerns about disk contamination are more serious; consequently, most mass determinations in these systems are much less secure. This method has now been applied to about 15 quiescent SXTs. For a recent compilation, see Casares (2007).

4.6.4 Mass determinations in luminous systems

As the accretion rate increases from quiescence, the optical light quickly becomes dominated by the heated accretion disk and companion star. This means that in most persistently luminous LMXBs, we never have the opportunity to perform a radial velocity study of photospheric absorption lines from the unheated portions of the companion. There are a few exceptional systems in which the donor star is a giant and is still visible even in luminous states. Cyg X-2 is the most famous example (Casares et al., 1998), and 2S 0921–636 is another (Shahbaz et al., 2004; Jonker et al., 2005).

In other systems we must seek a different source of radial velocity information. A new approach was suggested by observations of Sco X-1 that revealed narrow emission components in the Bowen blend (see Section 4.6.1) that moved in anti-phase with broader components attributed to the disk (Steeghs and Casares, 2002). The radial velocity curve of just the sharp components yielded $K_2 > 77\,\mathrm{km\,s^{-1}}$, and the authors derived a neutron star mass $\sim 1.4\,M_\odot$, consistent with that usually expected. The major caveat here is that the emission lines would originate from the heated inner face of the donor star, and so their center-of-light would not coincide with the donor's center of mass. This is why the K_2 value derived by this method is only a lower limit on the true value. To improve on this requires modeling of the binary geometry to attempt to estimate by how much K_2 is underestimated (Muñoz-Darias et al., 2005).

Sco X-1 was an ideal case. Only one other system, the long-standing black hole candidate GX 339–4, has shown clear, sharp, moving N III and C III components in individual spectra (Hynes et al., 2003a). These observations provided both a convincing orbital period of 1.76 days, and a mass function of $f(M) = 5.8 \pm 0.5\,M_\odot$, finally confirming the black hole nature of this source. In other objects, such as 2A 1822–371, sharp components cannot be identified in the line profiles directly, but the companion star can be picked out very effectively using Doppler tomography (Casares et al., 2003). For a more complete review of what has been achieved by this method, see Cornelisse et al. (2008).

4.7 Rapid variability

4.7.1 Echo mapping

Multiwavelength variability is a near-universal characteristic of X-ray binaries. X-rays vary due to rapid changes in the inner accretion flow on time scales of milliseconds and longer. These X-rays then irradiate the outer accretion disk and companion star, resulting in reprocessed optical and UV radiation that is expected to be imprinted with the same variability as the X-ray signal. An important difference, however, is that the optical emission and X-rays originate from a volume of significant spatial extent, resulting in light travel time delays between the X-rays and the reprocessed emission. It is then possible to infer information about the geometry and scale of the reprocessing region from the lags measured between X-ray and optical/UV variability; this technique is known as reverberation or echo mapping, as the reprocessed light behaves as an echo. Echo or reverberation mapping is not uniquely applied to X-ray binaries. Much of the development and application of the technique has been for active galactic nuclei (AGN); see, for example, Peterson and Horne (2006) for a recent review of the AGN problem, and O'Brien et al. (2002) for the application to X-ray binaries.

The key idea is that local optical (or UV) variability, in either lines or continuum, is induced by reprocessing of X-ray variability, but with a lag varying with reprocessing location. Each such location can be thought of as responding to X-rays with a delta-function response at a delay time determined by the path difference between direct and reprocessed emission. The total optical response is then the sum of lagged responses from all the reprocessing elements. For a delta-function variation in the X-rays, the optical response is then termed the transfer function and measures how strong the response is as a function of the delay, effectively encoding information about the reprocessing geometry. For continuous variability, the optical light curve can be modeled as a convolution of the X-ray light curve with the transfer function:

$$L_{\text{opt}}(t) = L_X * \Psi = \int L_X(t - \tau)\Psi(\tau)d\tau, \tag{4.8}$$

where L_X and L_{opt} are the X-ray and optical luminosities, Ψ is the transfer function, and τ is the X-ray to optical lag.

Several assumptions are inherent in this description. It is assumed that the optical responds linearly to the X-rays, or at least that a nonlinear response can be linearized for small perturbations. It is also implicit that the X-rays originate from a point source, or at least a region much smaller in spatial extent than the reprocessing region. Finally, to determine geometric information, it is necessary that the lags be geometric in origin; significant reprocessing times would compromise this.

Geometrical modeling of the response

In the case of X-ray binaries, we have a clearer expectation of the reprocessing geometry than in AGN. We anticipate reprocessing from the accretion disk around the compact object, possibly enhanced at a bulge where material feeds into the disk from the companion star. We also might expect some reprocessing from the heated inner face of the companion star. O'Brien et al. (2002) and Muñoz-Darias et al. (2005) modeled the reprocessing geometry to predict transfer functions for a variety of binary parameters. An example as a function of orbital phase is shown in Figure 4.11. Simplistically, one expects two components. The disk will extend from zero lag to $r_{\text{disk}}(1 + \sin i)$, where r_{disk} is the disk radius in light seconds and i the binary inclination. Within this range, the shape of the response is strongly sensitive to the inclination and somewhat less so to the degree of disk flaring. The response from the companion star approximately oscillates within the range $a(1 \pm \sin i)$ over the course of the binary orbit, where a is the binary separation in light seconds. The strength and width of the companion response

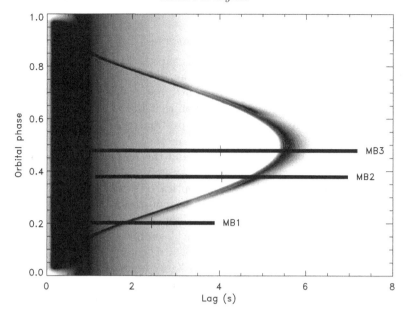

FIGURE 4.11. Strength of response (indicated by grayscale) as a function of orbital phase and lag, as calculated by the binary modeling code of O'Brien et al. (2002). Solid bands indicate the ranges of lags inferred from X-ray bursts by Hynes et al. (2006b) for EXO 0748–676. Reproduced by permission of the AAS.

is a strong function of the mass ratio and disk thickness (which determines how much of the companion is shielded). One of the great appeals of applying echo mapping to X-ray binaries is that with phase-resolved observations of the companion echo over the orbit, one could measure both a and i independently of other techniques and assumptions.

Empirical descriptions of the response

Many echo-mapping studies use more pragmatic and less model-dependent approaches. Ideally, one would take high-quality X-ray and optical light curves and deconvolve them to directly determine the shape of the transfer function. This is the basis of the maximum entropy echo-mapping technique (Horne, 1994), in which a maximum entropy regularization method is used to suppress the problem of fitting the noise. Unfortunately, typical X-ray binary datasets do not have the signal-to-noise ratio for this to work effectively.

A simpler approach has been developed in which a very simple functional form is adopted for the transfer function, either a rectangular response (Pedersen et al., 1982) or a Gaussian (Hynes et al., 1998). Both are introduced as approximations to the response, rather than for a physically motivated reason, and amount to the assumption that the data only constrain the mean lag and the amount of smearing out in time of the variability. The Gaussian formulation essentially yields the first two moments of the delay distribution. There is a potential difficulty with this approach as noted by Muñoz-Darias et al. (2007). If the X-ray light curve is convolved with a Gaussian, this smoothes the data and has the effect of suppressing the noise. If the noise in the X-ray light curve is significant compared to the real variability (or if there is additional variability that does not correlate with that in the optical), then the χ^2 of a fit may be reduced by adopting an artificially high Gaussian width, that is, over-smoothing the data. Thus there is a possibility that the widths of transfer functions derived in this way may be overestimated.

An even simpler but widely used approach is to measure the cross-correlation function (CCF) of the two datasets, defined for continuously sampled data as

$$\mathrm{CCF}(\tau) = \frac{\int [f(t) - \bar{f}][g(t+\tau) - \bar{g}]}{\sigma_f \sigma_g}, \qquad (4.9)$$

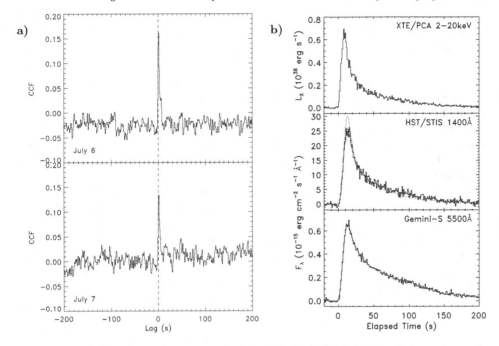

FIGURE 4.12. *a)* X-ray/optical cross-correlations of Swift J1753.5–0127 early in outburst, from Hynes *et al.* (2009a). *b)* Simultaneous burst profiles in EXO 0748–676 from Hynes *et al.* (2006b). Dotted lines assume fixed normalization of the UV flux relative to the optical; solid line fits allow this normalization to float.

where f and g are the driver and echo time-series, respectively; $\bar{f}$ and $\bar{g}$ are their means; σ_f and σ_g are their standard deviations; and τ is the lag at which the CCF is being evaluated. Gaskell and Peterson (1987) and Edelson and Krolik (1988) discuss two approaches to implementing this calculation for discretely and unevenly sampled data. This yields an estimate of the mean lag, but information about the smearing is hard to extract, and care should be taken in interpreting the results of a cross-correlation analysis (Koen, 2003). If the driver-echo relationship is accurately described by the convolution in equation 4.9, then the CCF is equivalent to the driver autocorrelation function (ACF) convolved with the transfer function. In cases where the driver ACF is relatively narrow, the CCF may provide a reasonable approximation to the transfer function. We show an example in Figure 4.12, based on X-ray/optical data from the black hole transient Swift J1753.5–0127 (Hynes *et al.*, 2009a).

4.7.2 *Type I X-ray bursts*

Type I X-ray bursts are thermonuclear explosions on the surface of a neutron star in an LMXB (Strohmayer and Bildsten, 2006). They represent an enormous increase in the X-ray flux, a factor of 20 or more in lower luminosity systems, rising on a time scale of a few seconds. X-ray bursts then have the potential to be an ideal echo-mapping probe. Hence it is using these events that some of the first echo-mapping experiments were performed. Reprocessed optical bursts were discovered in the late 1970s in the LMXBs 4U 1735–444 and Ser X-1 (Grindlay *et al.*, 1978; McClintock *et al.*, 1979; Hackwell *et al.*, 1979). The optical flux was found to rise by nearly a factor of 2 and lag a few seconds behind the X-rays. It was immediately appreciated that the optical flux was several orders of magnitude too high to be due to direct emission from the neutron star surface, and hence that the brightening must be due to reprocessing of X-rays by the much larger projected area of the accretion disk and/or companion star. The 2.8-s lag in 4U 1735–444 (McClintock *et al.*, 1979) supported this interpretation, being consistent with the expected light travel time delays in this short-period binary.

The large amplitudes of X-ray bursts drive a stronger optical response than X-ray flickering does, and so provide additional information not available when variability is a small perturbation. Observations never record the total bolometric luminosity, but always that within a specific bandpass. Consequently, the shape of the observed reprocessed light curve depends on the spectral evolution as the reprocessor cools and the peak of the spectrum moves into or out of the bandpass used. Shorter wavelengths are sensitive to hotter material, and they are expected to decay more rapidly; hence, multicolor observations of X-ray bursts provide some temperature sensitivity, in addition to that lag information which is available even for smaller perturbations. This is not a subtle effect, and the early observations indicate, that the reprocessor temperature typically doubles during a burst (Lawrence et al., 1983).

The best dataset yet obtained for the method corresponds to EXO 0748–676 (Hynes et al., 2006b). Four bursts were observed over two successive nights using RXTE and Gemini-S, providing some phase information, and one of these was also observed at high time-resolution in the far-UV by *HST*/STIS (see Fig. 4.12). The last was a unique observation to date, providing far more sensitivity to high-temperature responses than is possible with optical data alone, and also yielding a time-resolved UV spectrum, facilitating a direct test of the expectation that the reprocessed light should be close to a blackbody. Perhaps most interestingly, the three bursts observed sampled enough of a phase range to see apparent changes in the lag and smearing as a function of orbital phase (see Fig. 4.11). The phasing and amplitude of the changes seen are both consistent with expectations of models in which both the disk and the companion star contribute to reprocessing. This is one of the few observations to date that can claim to be true echo-tomography exploiting different viewing angles of the binary.

4.7.3 *Flickering in persistent neutron star systems*

Although bursts provide an ideal signal for echo mapping in some neutron star binaries, we must seek another technique in black hole systems and nonbursting neutron stars. An alternative source of variability is provided by the flickering that seems a ubiquitous signature of accretion. Since this flickering is always present at some level, unlike bursts that recur only every few hours, flickering variability can potentially provide phase-resolved information in any system. This potential has yet to be fully realized, however. It has been found that the optical response is rather weak, with standard deviations of only a few percent in the optical light curves. Consequently, high signal-to-noise observations are needed to pick out a measurable correlation. Even then, success is typically achieved only when high levels of variability are present, with other datasets yielding a nondetection.

There are several bright and persistently active neutron star X-ray binaries that potentially provide ideal targets for these studies. It is in fact somewhat surprising that more has not yet been achieved with these. The bright neutron star system Sco X-1 was observed by Ilovaisky et al. (1980) and Petro et al. (1981). Both found correlations, with evidence for lags and substantial smearing of the response; Petro et al. (1981) described the optical response as a low-pass filtered version of the X-rays, with variability on time scales < 20 s smoothed out. McGowan et al. (2003) reanalyzed these datasets with the Gaussian transfer function method. In some cases, no good fit could be obtained. The pair of light curves where the method did appear to succeed yielded a lag of 8.0 ± 0.8 s and Gaussian dispersion of 8.6 ± 1.3 s. For comparison, lags of up to 4 to 5 s are expected from the disk and 10 s from the companion star.

We recently obtained some superb quality observations of Cyg X-2. Very clear correlations were seen when the source was on the flaring branch. A Gaussian transfer function analysis of the whole light curve proved unsatisfactory, as we see not only a modulation of the optical light that is absent in X-rays, but also an apparent variation in the efficiency of reprocessing from one event to the next. Much more success was achieved by analyzing

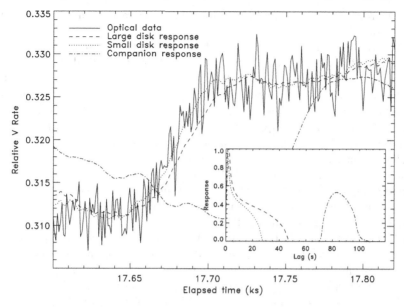

FIGURE 4.13. Fit to optical data for Cyg X-2 in the flaring branch. The dashed line shows the response from the whole disk, the dotted line is that from just the inner disk, and the dash-dot line is from the companion star.

individual events in the light curve independently. A Gaussian fit to the transfer function then suggests lags around 10 s, as might be expected from the accretion disk in this long-period (9.8 day) binary. The data quality is sufficient to permit direct comparison with model transfer functions as well. We show in Fig. 4.13 a small segment of the optical light curve with fits generated by convolving the X-ray light curve with model transfer functions generated using the code of O'Brien et al. (2002) and parameters from Casares et al. (1998) and Orosz and Kuulkers (1999). The relative contribution of reprocessing from disk and companion star is highly sensitive to the amount of shielding of the companion by the disk, hence we have considered the two components separately. The data appear much more consistent with the pure disk model than the pure companion star model suggesting that the disk (and indeed just the inner disk) dominates the response, at least in the optical continuum.

Muñoz-Darias et al. (2007) tried a different approach on Sco X-1. As discussed in Section 4.6.4, we have realized in recent years that the Bowen blend of N III and C III lines around 4640 Å often contain a strong component from the companion star. They used a narrow-band filter encompassing the Bowen blend and He II 4686 Å to attempt emission-line echo mapping. They report results from several flaring branch data segments showing strong correlated variability. As expected, they find longer lags from the narrow-band filter containing the Bowen blend than from a continuum bandpass, suggesting that the Bowen response contains a larger contribution from the companion star. After attempting to remove the continuum contribution from the Bowen light curves, they estimate a line lag of $\tau_0 = (13.5 \pm 3.0)$ s and a continuum lag of $\tau_0 = (8.5 \pm 1.0)$ s. The difficulty of continuum subtraction with narrow-band photometry is a limitation of this approach, and we can expect a significant improvement with rapid spectroscopy with new instrumentation such as ULTRASPEC (Dhillon et al., 2008).

4.7.4 Another mechanism for correlations?

Fundamental to echo mapping is the assumption that correlated X-ray and optical/UV variability indicate reprocessing of the X-rays by relatively cool material. We should not take this for granted, however, and there are some observations that seriously challenge this assumption.

The first indication of difficulties came from fast optical observations of the black-hole binary GX 339–4 (Motch et al., 1982, 1983). Dramatic optical variability was seen extending to extremely short time scales (10 to 20 ms), much shorter than the light travel time scales expected, or the smearing typically observed in other systems described earlier. Fabian et al. (1982) argued that the flares most likely originated in cyclotron radiation, with a brightness temperature $> 9 \times 10^8$ K. Correlations were seen in a short (96 s) simultaneous observation, but of a puzzling nature. The X-ray and optical were anti-correlated, with optical dips apparently preceding the X-rays by 2.8 ± 1.6 s. The connection between X-ray and optical behavior was further reinforced by the presence of quasi-periodic oscillations at the same frequency in both energy bands. The brevity of the simultaneous observation and the ambiguity in the lags introduced by quasi-periodic variability left this result tantalizing, however.

New light was shed on this behavior by the 2000 outburst of the black hole system XTE J1118+480. A much larger time-resolved database was accumulated on this object, including both simultaneous X-ray/optical data (Kanbach et al., 2001) and independent multi-epoch simultaneous X-ray/UV observations (Hynes et al., 2003c). Large-amplitude X-ray variability was present and accompanied by correlated optical and UV variations. In this case, a positive correlation with the optical/UV lagging the X-rays was clearly present, leading to hopes that this would be an ideal echo-mapping dataset. There were serious problems with this interpretation, however. These were most pronounced in the optical data (Kanbach et al., 2001) and included an optical (ACF) narrower than that seen in X-rays, and a cross-correlation function containing a marked "precognition dip" before the main peak. The latter could be interpreted in terms of optical dips leading X-ray flares by a few seconds, as suggested in GX 339–4, suggesting a common origin. Neither of these effects are expected in a reprocessing model. Light travel times should only act to smooth out optical responses, and hence broaden the optical ACF. Also, the continuum responses (as considered here) should generally be positively correlated with the X-rays, not anti-correlated. As in GX 339–4, the variability extended to very short time scales (<100 ms), and hence Kanbach et al. (2001) estimated a minimum brightness temperature of 2×10^6 K. They also suggested that the strange variability properties were the result of optical cyclosynchrotron emission. These properties become weaker at shorter wavelengths (Hynes et al., 2003c), as does the variability, as might be expected if the behavior originates from a very red source of emission such as synchrotron.

Dominant synchrotron emission in this system was not uniquely suggested by the variability properties. The very flat UV to near-IR spectrum had previously been attributed to synchrotron emission (Hynes et al., 2000) and the broadband spectral energy distribution has been successfully accounted for using a simple jet model (Markoff et al., 2001). Striking support for a synchrotron origin for the variability has been provided by time-resolved IR observations during the 2005 outburst of XTE J1118+480 (Hynes et al., 2006a). Simultaneous 2 s images in the IR J, H, and K filters could be used to isolate the color of the IR variability. It was found to be very red ($F_\nu \propto \nu^{-0.78}$), consistent with optically thin synchrotron emission.

These arguments together provide strong evidence that a jet, or at least some kind of outflow, is responsible for much of the IR and even optical emission in XTE J1118+480 and for the correlated variability. By extension, the same interpretation may apply to other objects showing similar properties. GX 339–4 is of course a prime candidate, and recent more extensive observations of this object show similar behavior (Gandhi et al., 2008). A third object has been added to the sample in Swift J1753.5–0127 (Durant et al., 2008; Hynes et al., 2009a). Here early observations near the outburst peak supported a simple reprocessing interpretation (see Fig. 4.12), with a transfer function consistent with the inferred system parameters, but later observations showed both negative and positive correlations rather similar to those in XTE J1118+480. The full explanation for the observed correlations and anti-correlations remains to be established, but this case illustrates that new astrophysics can be uncovered in unexpected places.

4.8 Conclusion

X-ray binaries emit across the electromagnetic spectrum, and a full understanding of accretion processes in this environment depends on multiwavelength observations. Not only do different wavelengths illuminate different aspects of the behavior (inner disk, outer disk, companion star, jet, etc.), but also connections between the different wavelengths provide essential information on causal connections between them. Understanding X-ray binaries to the fullest extent possible requires knowledge not only of X-ray and gamma-ray astronomy, but also of optical, ultraviolet, infrared, and radio wavelengths.

4.9 Acknowledgments

I am grateful to the Instituto de Astrofísica de Canarias (IAC) for the invitation and funding to present a series of lectures at the XXI Canary Island Winter School, on which this work is based. I would also like to gratefully acknowledge Valerie Mikles, Chris Britt, Lauren Gossen, and Chris Dupuis for providing an abundance of helpful comments on this manuscript and catching many mistakes in earlier versions. This is the document I would like to have given them when they began working in this field. Preparation of this work has made extensive use of NASA's Astrophysics Data System.

4.10 Appendix: glossary of objects cited

2A 1822–371: Persistent eclipsing neutron star LMXB. ADC source.
2S 0921–630: Persistent eclipsing neutron star LMXB. ADC source. Long period.
4U 0614+091: Persistent neutron star UCXB.
4U 1626–67: Persistent neutron star UCXB.
4U 1735-44: Persistent neutron star LMXB.
4U 1915–05: Persistent neutron star UCXB. Prototypical dipping source.
4U 2129+47: Quasi-persistent eclipsing neutron star LMXB. ADC source.
A 0620–00: Prototypical transient black hole LMXB.
Cir X-1: Persistent neutron star IMXB or HMXB. Eccentric orbit. Microquasar.
Cyg X-2: Persistent neutron star LMXB. Long period
EXO 0748–676: Quasi-persistent eclipsing neutron star LMXB.
GRO J0422+32: Transient black hole LMXB.
GRO J1655–40: Recurrent transient black hole IMXB. Microquasar.
GRS 1915+105: Quasi-persistent black hole LMXB. Microquasar. Very long period.
GS 2000+250: Transient black hole LMXB.
GX 339–4: Recurrent transient black hole LMXB.
Her X-1: Persistent eclipsing neutron star IMXB. Prototypical warped disk source.
LMC X-4: Persistent neutron star HMXB.
M33 X-7: Extragalactic eclipsing black hole HMXB.
Sco X-1: Persistent prototypical neutron star LMXB. First extrasolar X-ray source discovered. Brightest persistent extrasolar X-ray source in the sky.
Ser X-1: Persistent neutron star LMXB.
SMC X-1: Persistent neutron star HMXB.
Swift J1753.5–0127: Transient or quasi-persistent black hole LMXB.
X-ray Nova Muscae 1991: Transient black hole LMXB.
V404 Cygni: Transient black hole LMXB. Long period.
XTE J1118+480: Recurrent transient black hole LMXB.

XTE J1859+226: Transient black hole candidate LMXB.
XTE J2123−058: Transient eclipsing neutron star LMXB.

REFERENCES

Andrew, B. H., and Purton, C. R. 1968. Detection of radio emission from Scorpio X-1. Nature, **218**, 855–856.

Augusteijn, T., Kuulkers, E., and Shaham, J. 1993. "Glitches" in soft X-ray transients: Echoes of the main burst? A&A, **279**, L13–L16.

Bayless, A. J., Robinson, E. L., Hynes, R. I., Ashcraft, T. A., and Cornell, M. E. 2010. The structure of the accretion disk in the accretion disk corona X-ray binary 4U 1822-371 at optical and ultraviolet wavelengths. ApJ, **709**, 251–262.

Blandford, R. D., and Konigl, A. 1979. Relativistic jets as compact radio sources. ApJ, **232**, 34–48.

Bradley, C. K., Hynes, R. I., Kong, A. K. H., Haswell, C. A., Casares, J., and Gallo, E. 2007. The spectrum of the black hole X-Ray Nova V404 Cygni in quiescence as measured by XMM-Newton. ApJ, **667**, 427–432.

Brocksopp, C., Bandyopadhyay, R. M., and Fender, R. P. 2004. "Soft X-ray transient" outbursts which are not soft. New Astronomy, **9**, 249–264.

Brown, E. F., and Cumming, A. 2009. Mapping crustal Heating with the cooling light curves of quasi-persistent transients. ApJ, **698**, 1020–1032.

Calvelo, D. E., Vrtilek, S. D., Steeghs, D., Torres, M. A. P., Neilsen, J., Filippenko, A. V., and González Hernández, J. I. 2009. Doppler and modulation tomography of XTEJ 1118+480 in quiescence. MNRAS, **399**, 539–549.

Cannizzo, J. K., Chen, W., and Livio, M. 1995. The accretion disk limit cycle instability in black hole x-ray binaries. ApJ, **454**, 880–894.

Cantrell, A. G., Bailyn, C. D., McClintock, J. E., and Orosz, J. A. 2008. Optical State Changes in the X-Ray-quiescent Black Hole A0620-00. ApJL, **673**, L159–L162.

Casares, J. 2007. Observational evidence for stellar-mass black holes. Pages 3–12 of: V. Karas and G. Matt (eds.), IAU Symposium. IAU Symposium, vol. 238.

Casares, J., and Charles, P. A. 1994. Optical studies of V404 Cyg, the X-ray transient GS 2023+338. IV. The rotation speed of the companion star. MNRAS, **271**, L5–L9.

Casares, J., Charles, P. A., and Kuulkers, E. 1998. The mass of the neutron star in Cygnus X-2 (V1341 Cygni). ApJL, **493**, L39–L42.

Casares, J., Charles, P. A., and Naylor, T. 1992. A 6.5-day periodicity in the recurrent nova V404 Cygni implying the presence of a black hole. Nature, **355**, 614–617.

Casares, J., Steeghs, D., Hynes, R. I., Charles, P. A., and O'Brien, K. 2003. Bowen fluorescence from the companion star in X1822-371. ApJ, **590**, 1041–1048.

Chakrabarty, D., Homer, L., Charles, P. A., and O'Donoghue, D. 2001. Millihertz optical/ultraviolet oscillations in 4U 1626-67: evidence for a warped accretion disk. ApJ, **562**, 985–991.

Chen, W., Livio, M., and Gehrels, N. 1993. The secondary maxima in black hole X-ray nova light curves – clues toward a complete picture. ApJL, **408**, L5–L8.

Chen, W., Shrader, C. R., and Livio, M. 1997. The properties of X ray and optical light curves of X-ray novae. ApJ, **491**, 312–338.

Cheng, F. H., Horne, K., Panagia, N., Shrader, C. R., Gilmozzi, R., Paresce, F., and Lund, N. 1992. The Hubble Space Telescope observations of X-ray Nova Muscae 1991 and its spectral evolution. ApJ, **397**, 664–673.

Clarkson, W. I., Charles, P. A., Coe, M. J., and Laycock, S. 2003. Long-term properties of accretion discs in X-ray binaries – II. Stability of radiation-driven warping. MNRAS, **343**, 1213–1223.

Corbel, S., and Fender, R. P. 2002. Near-infrared synchrotron emission from the compact jet of GX 339-4. ApJL, **573**, L35–L39.

Corbet, R. H. D., Sokoloski, J. L., Mukai, K., Markwardt, C. B., and Tueller, J. 2008. A comparison of the variability of the symbiotic X-ray binaries GX 1+4, 4U 1954+31, and 4U 1700+24 from Swift BAT and RXTE ASM Observations. ApJ, **675**, 1424–1435.

Cornelisse, R., Casares, J., Muñoz-Darias, T., Steeghs, D., Charles, P., Hynes, R., O'Brien, K., and Barnes, A. 2008 (May). An overview of the Bowen Survey: Detecting donor star signatures in low mass X-ray binaries. Pages 148–152 of: R. M. Bandyopadhyay, S. Wachter, D. Gelino, and C. R. Gelino (eds.), *A Population Explosion: The Nature & Evolution of X-ray Binaries in Diverse Environments*. American Institute of Physics Conference Series, vol. 1010.

Dhillon, V. S., Marsh, T. R., Copperwheat, C., Bezawada, N., Ives, D., Vick, A., and O'Brien, K. 2008. ULTRASPEC: High-speed spectroscopy with zero readout noise. Pages 132–139 of: D. Phelan, O. Ryan, & A. Shearer (eds.), *High Time Resolution Astrophysics: The Universe at Sub-Second Timescales*. American Institute of Physics Conference Series, vol. 984.

Dubus, G., Hameury, J.-M., and Lasota, J.-P. 2001. The disc instability model for X-ray transients: evidence for truncation and irradiation. A&A, **373**, 251–271.

Dubus, G., Lasota, J.-P., Hameury, J.-M., and Charles, P. 1999. X-ray irradiation in low-mass binary systems. MNRAS, **303**, 139–147.

Durant, M., Gandhi, P., Shahbaz, T., Fabian, A. P., Miller, J., Dhillon, V. S., and Marsh, T. R. 2008. Swift J1753.5-0127: a surprising optical/X-ray cross-correlation function. ApJL, **682**, L45–L48.

Edelson, R. A., and Krolik, J. H. 1988. The discrete correlation function – a new method for analyzing unevenly sampled variability data. ApJ, **333**, 646–659.

Esin, A. A., McClintock, J. E., and Narayan, R. 1997. Advection-dominated accretion and the spectral states of black hole X-ray binaries: application to Nova Muscae 1991. ApJ, **489**, 865–889.

Fabian, A. C., Guilbert, P. W., Motch, C., Ricketts, M., Ilovaisky, S. A., and Chevalier, C. 1982. GX 339-4 – Cyclotron radiation from an accretion flow. A&A, **111**, L9–L10.

Fender, R. P., Gallo, E., and Jonker, P. G. 2003. Jet-dominated states: an alternative to advection across black hole event horizons in "quiescent" X-ray binaries. MNRAS, **343**, L99–L103.

Fitzpatrick, E. L. 1999. Correcting for the effects of interstellar extinction. PASP, **111**, 63–75.

Frank, J., King, A., and Raine, D. J. 2002. *Accretion Power in Astrophysics*, Third ed. Cambridge University Press.

Froning, C. S., Robinson, E. L., and Bitner, M. A. 2007. Near-infrared spectra of the black hole X-ray binary A0620-00. ApJ, **663**, 1215–1224.

Fryer, C. L., and Kalogera, V. 2001. Theoretical black hole mass distributions. ApJ, **554**, 548–560.

Gallo, E., Fender, R. P., and Hynes, R. I. 2005. The radio spectrum of a quiescent stellar mass black hole. MNRAS, **356**, 1017–1021.

Gallo, E., Migliari, S., Markoff, S., Tomsick, J. A., Bailyn, C. D., Berta, S., Fender, R., and Miller-Jones, J. C. A. 2007. The spectral energy distribution of quiescent black hole X-ray binaries: new constraints from spitzer. ApJ, **670**, 600–609.

Gandhi, P., Makishima, K., Durant, M., Fabian, A. C., Dhillon, V. S., Marsh, T. R., Miller, J. M., Shahbaz, T., and Spruit, H. C. 2008. Rapid optical and X-ray timing observations of GX 339-4: flux correlations at the onset of a low/hard state. MNRAS, **390**, L29–L33.

Gaskell, C. M., and Peterson, B. M. 1987. The accuracy of cross-correlation estimates of quasar emission-line region sizes. ApJS, **65**, 1–11.

Gelino, D. M., Harrison, T. E., and Orosz, J. A. 2001. A multiwavelength, multiepoch study of the soft X-ray transient prototype, V616 Monocerotis (A0620-00). AJ, **122**, 2668–2678.

Gerend, D., and Boynton, P. E. 1976. Optical clues to the nature of Hercules X-1/HZ Herculis. ApJ, **209**, 562–573.

Giacconi, R., Gursky, H., Paolini, F. R., and Rossi, B. B. 1962. Evidence for X-rays from sources outside the solar system. Physical Review Letters, **9**, 439–443.

Gottlieb, E. W., Wright, E. L., and Liller, W. 1975. Optical studies of Uhuru sources. XI. A probable period for Scorpius X-1 = V818 Scorpii. ApJL, **195**, L33–L35.

Grindlay, J. E., McClintock, J. E., Canizares, C. R., Cominsky, L., Li, F. K., Lewin, W. H. G., and van Paradijs, J. 1978. Discovery of optical bursts from an X-ray burst source, MXB 1735-44. Nature, **274**, 567–568.

Hackwell, J. A., Grasdalen, G. L., Gehrz, R. D., Cominsky, L., Lewin, W. H. G., and van Paradijs, J. 1979. The detection of an optical burst coincident with an X-ray burst from MXB 1837+05 (Ser X-1). ApJL, **233**, L115–L119.

Hameury, J.-M., Lasota, J.-P., and Warner, B. 2000. The zoo of dwarf novae: illumination, evaporation and disc radius variation. A&A, **353**, 244–252.

Haswell, C. A., Hynes, R. I., King, A. R., and Schenker, K. 2002. The ultraviolet line spectrum of the soft X-ray transient XTE J1118+480: a CNO-processed core exposed. MNRAS, **332**, 928–932.

Haswell, C. A., King, A. R., Murray, J. R., and Charles, P. A. 2001. Superhumps in low-mass X-ray binaries. MNRAS, **321**, 475–480.

Haswell, C. A., Robinson, E. L., Horne, K., Stiening, R. F., and Abbott, T. M. C. 1993. On the mass of the compact object in the black hole binary A0620-00. ApJ, **411**, 802–812.

Hellier, C. 2001. On echo outbursts and ER UMa supercycles in SU UMa-type cataclysmic variables. PASP, **113**, 469–472.

Horne, K. 1994. Echo mapping problems, maximum entropy solutions. Pages 23–25 of: P. M. Gondhalekar, K. Horne, and B. M. Peterson (eds.), *Reverberation Mapping of the Broad-Line Region in Active Galactic Nuclei*. Astronomical Society of the Pacific Conference Series, vol. 69.

Horne, K., and Marsh, T. R. 1986. Emission line formation in accretion discs. MNRAS, **218**, 761–773.

Hynes, R. I. 2005. The optical and ultraviolet spectral energy distributions of short-period black hole X-ray transients in outburst. ApJ, **623**, 1026–1043.

Hynes, R. I., and Haswell, C. A. 1999. Hubble Space Telescope observations of the black hole X-ray transient GRO J0422+32 near quiescence. MNRAS, **303**, 101–106.

Hynes, R. I., Bradley, C. K., Rupen, M., Gallo, E., Fender, R. P., Casares, J., and Zurita, C. 2009b. The quiescent spectral energy distribution of V404 Cyg. MNRAS, **399**, 2239–2248.

Hynes, R. I., Brien, K. O., Mullally, F., and Ashcraft, T. 2009a. Echo mapping of Swift J1753.5-0127. MNRAS, **399**, 281–286.

Hynes, R. I., Charles, P. A., Casares, J., Haswell, C. A., Zurita, C., and Shahbaz, T. 2003b. Fast photometry of quiescent soft X-ray transients with the Acquisition Camera on Gemini-South. MNRAS, **340**, 447–456.

Hynes, R. I., Charles, P. A., Haswell, C. A., Casares, J., Zurita, C., and Serra-Ricart, M. 2001. Optical studies of the X-ray transient XTE J2123-058 – II. Phase-resolved spectroscopy. MNRAS, **324**, 180–190.

Hynes, R. I., Haswell, C. A., Chaty, S., Shrader, C. R., and Cui, W. 2002. The evolving accretion disc in the black hole X-ray transient XTE J1859+226. MNRAS, **331**, 169–179.

Hynes, R. I., Haswell, C. A., Cui, W., Shrader, C. R., O'Brien, K., Chaty, S., Skillman, D. R., Patterson, J., and Horne, K. 2003c. The remarkable rapid X-ray, ultraviolet, optical and infrared variability in the black hole XTE J1118+480. MNRAS, **345**, 292–310.

Hynes, R. I., Horne, K., O'Brien, K., Haswell, C. A., Robinson, E. L., King, A. R., Charles, P. A., and Pearson, K. J. 2006b. Multiwavelength observations of EXO 0748-676. I. Reprocessing of X-ray bursts. ApJ, **648**, 1156–1168.

Hynes, R. I., Mauche, C. W., Haswell, C. A., Shrader, C. R., Cui, W., and Chaty, S. 2000. The X-ray transient XTE J1118+480: multiwavelength observations of a low-state minioutburst. ApJL, **539**, L37–L40.

Hynes, R. I., O'Brien, K., Horne, K., Chen, W., and Haswell, C. A. 1998. Echoes from an irradiated disc in GRO J1655-40. MNRAS, **299**, L37–L41.

Hynes, R. I., Robinson, E. L., and Bitner, M. 2005. Observational constraints on cool disk material in quiescent black hole binaries. ApJ, **630**, 405–412.

Hynes, R. I., Robinson, E. L., Pearson, K. J., Gelino, D. M., Cui, W., Xue, Y. Q., Wood, M. A., Watson, T. K., Winget, D. E., and Silver, I. M. 2006a. Further evidence for variable synchrotron Emission in XTE J1118+480 in outburst. ApJ, **651**, 401–407.

Hynes, R. I., Steeghs, D., Casares, J., Charles, P. A., and O'Brien, K. 2003a. Dynamical evidence for a black hole in GX 339-4. ApJL, **583**, L95–L98.

Ilovaisky, S. A., Chevalier, C., White, N. E., Mason, K. O., Sanford, P. W., Delvaille, J. P., and Schnopper, H. W. 1980. Simultaneous X-ray and optical observations of rapid variability in Scorpius X-1. MNRAS, **191**, 81–93.

Johnston, H. M., Kulkarni, S. R., and Oke, J. B. 1989. The black hole A0620-00 and its accretion disk. ApJ, **345**, 492–497.

Jonker, P. G., Steeghs, D., Nelemans, G., and van der Klis, M. 2005. The radial velocity of the companion star in the low-mass X-ray binary 2S 0921-630: limits on the mass of the compact object. MNRAS, **356**, 621–626.

Kanbach, G., Straubmeier, C., Spruit, H. C., and Belloni, T. 2001. Correlated fast X-ray and optical variability in the black-hole candidate XTE J1118+480. Nature, **414**, 180–182.

King, A. R., and Ritter, H. 1998. The light curves of soft X-ray transients. MNRAS, **293**, L42–L48.

Koen, C. 2003. The analysis of indexed astronomical time-series – VIII. Cross-correlating noisy autoregressive series. MNRAS, **344**, 798–808.

Kong, A. K. H., McClintock, J. E., Garcia, M. R., Murray, S. S., and Barret, D. 2002. The X-ray spectra of black hole x-ray novae in quiescence as measured by Chandra. ApJ, **570**, 277–286.

Kuulkers, E. 1998. A0620-00 revisited: a black-hole transient case-study. New Astronomy Review, **42**, 1–22.

Lasota, J.-P. 2001. The disc instability model of dwarf novae and low-mass X-ray binary transients. New Astronomy Review, **45**, 449–508.

Lawrence, A., Cominsky, L., Engelke, C., Jernigan, G., Lewin, W. H. G., Matsuoka, M., Mitsuda, K., Oda, M., Ohashi, T., Pedersen, H., and van Paradijs, J. 1983. Simultaneous U, B, V, and X-ray measurements of a burst from 4U/MXB 1636-53. ApJ, **271**, 793–803.

Lewin, W. H. G., and van der Klis, M. (eds). 2006. *Compact Stellar X-Ray Sources*. Cambridge Astrophysics Series, vol. 39. Cambridge University Press.

Lewin, W. H. G., van Paradijs, J., and van den Heuvel, E. P. J. (eds). 1995. *X-ray Binaries*. Cambridge Astrophysics Series, vol. 26. Cambridge University Press.

Lynden-Bell, D. 1969. Galactic nuclei as collapsed old quasars. Nature, **223**, 690–694.

Markoff, S., Falcke, H., and Fender, R. 2001. A jet model for the broadband spectrum of XTE J1118+480. Synchrotron emission from radio to X-rays in the Low/Hard spectral state. A&A, **372**, L25–L28.

Marsh, T. R. 2001. Doppler tomography. Pages 1–26 of: H. M. J. Boffin, D. Steeghs, and J. Cuypers (eds.), *Astrotomography, Indirect Imaging Methods in Observational Astronomy*. Lecture Notes in Physics, Springer Verlag, vol. 573.

Marsh, T. R., Robinson, E. L., and Wood, J. H. 1994. Spectroscopy of A0620-00 – the mass of the black-hole and an image of its accretion disc. MNRAS, **266**, 137–154.

Mason, K. O., Seitzer, P., Tuohy, I. R., Hunt, L. K., Middleditch, J., Nelson, J. E., and White, N. E. 1980. A 5.57 hr modulation in the optical counterpart of 2S 1822-371. ApJL, **242**, L109–L113.

McClintock, J. E., and Remillard, R. A. 1986. The black hole binary A0620-00. ApJ, **308**, 110–122.

McClintock, J. E., and Remillard, R. A. 2000. HST/STIS UV spectroscopy of two quiescent X-ray novae: A0620-00 and Centaurus X-4. ApJ, **531**, 956–962.

McClintock, J. E., Canizares, C. R., and Tarter, C. B. 1975. On the origin of 4640-4650 A emission in X-ray stars. ApJ, **198**, 641–652.

McClintock, J. E., Canizares, C. R., Cominsky, L., Li, F. K., Lewin, W. H. G., van Paradijs, J., and Grindlay, J. E. 1979. A 3-s delay in an optical burst from X-ray burst source MXB 1735-44. Nature, **279**, 47–49.

McClintock, J. E., Horne, K., and Remillard, R. A. 1995. The dim inner accretion disk of the quiescent black hole A0620-00. ApJ, **442**, 358–365.

McClintock, J. E., Narayan, R., Garcia, M. R., Orosz, J. A., Remillard, R. A., and Murray, S. S. 2003. Multiwavelength spectrum of the black hole XTE J1118+480 in quiescence. ApJ, **593**, 435–451.

McGowan, K. E., Charles, P. A., O'Donoghue, D., and Smale, A. P. 2003. Correlated optical and X-ray variability in LMC X-2. MNRAS, **345**, 1039–1048.

Migliari, S., Tomsick, J. A., Maccarone, T. J., Gallo, E., Fender, R. P., Nelemans, G., and Russell, D. M. 2006. Spitzer reveals infrared optically thin synchrotron emission from the compact jet of the neutron star X-ray binary 4U 0614+091. ApJL, **643**, L41–L44.

Migliari, S., Tomsick, J. A., Miller-Jones, J. C. A., Heinz, S., Hynes, R. I., Fender, R. P., Gallo, E., Jonker, P. G., and Maccarone, T. J. 2010. The complete spectrum of the neutron star X-ray binary 4U0614+091. ApJ, **710**, 117–124.

Mirabel, I. F., and Rodríguez, L. F. 1994. A superluminal source in the Galaxy. Nature, **371**, 46–48.

Motch, C., Ilovaisky, S. A., and Chevalier, C. 1982. Discovery of fast optical activity in the X-ray source GX 339-4. A&A, **109**, L1–L4.

Motch, C., Ricketts, M. J., Page, C. G., Ilovaisky, S. A., and Chevalier, C. 1983. Simultaneous X-ray/optical observations of GX339-4 during the May 1981 optically bright state. A&A, **119**, 171–176.

Muno, M. P., and Mauerhan, J. 2006. Mid-infrared emission from dust around quiescent low-mass X-ray binaries. ApJL, **648**, L135–L138.

Muñoz-Darias, T., Casares, J., and Martínez-País, I. G. 2005. The "K-correction" for irradiated emission lines in LMXBs: evidence for a massive neutron star in X1822-371 (V691 CrA). ApJ, **635**, 502–507.

Muñoz-Darias, T., Martínez-País, I. G., Casares, J., Dhillon, V. S., Marsh, T. R., Cornelisse, R., Steeghs, D., and Charles, P. A. 2007. Echoes from the companion star in Sco X-1. MNRAS, **379**, 1637–1646.

Narayan, R., and McClintock, J. E. 2005. Inclination effects and beaming in black hole X-ray binaries. ApJ, **623**, 1017–1025.

Narayan, R., and Raymond, J. 1999. Thermal X-ray line emission from accreting black holes. ApJL, **515**, L69–L72.

Narayan, R., Barret, D., and McClintock, J. E. 1997. Advection-dominated accretion model of the black hole V404 Cygni in quiescence. ApJ, **482**, 448–464.

Narayan, R., McClintock, J. E., and Yi, I. 1996. A new model for black hole soft X-ray transients in quiescence. ApJ, **457**, 821–833.

Neil, E. T., Bailyn, C. D., and Cobb, B. E. 2007. Infrared monitoring of the microquasar GRS 1915+105: detection of orbital and superhump signatures. ApJ, **657**, 409–414.

Nelemans, G., Jonker, P. G., and Steeghs, D. 2006. Optical spectroscopy of (candidate) ultra-compact X-ray binaries: constraints on the composition of the donor stars. MNRAS, **370**, 255–262.

Nelemans, G., Jonker, P. G., Marsh, T. R., and van der Klis, M. 2004. Optical spectra of the carbon-oxygen accretion discs in the ultra-compact X-ray binaries 4U 0614+09, 4U 1543-624 and 2S 0918-549. MNRAS, **348**, L7–L11.

O'Brien, K., Horne, K., Hynes, R. I., Chen, W., Haswell, C. A., and Still, M. D. 2002. Echoes in X-ray binaries. MNRAS, **334**, 426–434.

O'Donoghue, D., and Charles, P. A. 1996. Have superhumps been seen in black hole soft X-ray transients? MNRAS, **282**, 191–205.

Ogilvie, G. I., and Dubus, G. 2001. Precessing warped accretion discs in X-ray binaries. MNRAS, **320**, 485–503.

Orosz, J. A. 2001. The spectroscopic mass ratio of the black hole binary XTE J1118+480. The Astronomer's Telegram, **67**.

Orosz, J. A., and Kuulkers, E. 1999. The optical light curves of Cygnus X-2 (V1341 Cyg) and the mass of its neutron star. MNRAS, **305**, 132–142.

Orosz, J. A., McClintock, J. E., Narayan, R., Bailyn, C. D., Hartman, J. D., Macri, L., Liu, J., Pietsch, W., Remillard, R. A., Shporer, A., and Mazeh, T. 2007. A 15.65-solar-mass black hole in an eclipsing binary in the nearby spiral galaxy M 33. Nature, **449**, 872–875.

Osaki, Y., Meyer, F., and Meyer-Hofmeister, E. 2001. Repetitive rebrightening of EG Cancri: evidence for viscosity decay in the quiescent disk? A&A, **370**, 488–495.

Özel, F., and Psaltis, D. 2009. Reconstructing the neutron-star equation of state from astrophysical measurements. Phys. Rev. D, **80**, 103003.

Parmar, A. N., White, N. E., Giommi, P., Haberl, F., Pedersen, H., and Mayor, M. 1985. EXO 0748-676. IAU Circ., **4039**.

Patterson, J., Kemp, J., Skillman, D. R., Harvey, D. A., Shafter, A. W., Vanmunster, T., Jensen, L., Fried, R., Kiyota, S., Thorstensen, J. R., and Taylor, C. J. 1998. Superhumps in cataclysmic binaries. XV. EG Cancri, king of the echo outbursts. PASP, **110**, 1290–1303.

Pearson, K. J., Hynes, R. I., Steeghs, D., Jonker, P. G., Haswell, C. A., King, A. R., O'Brien, K., Nelemans, G., and Méndez, M. 2006. Multiwavelength observations of EXO 0748-676. II. Emission-line behavior. ApJ, **648**, 1169–1180.

Pedersen, H., Lub, J., Inoue, H., Koyama, K., Makishima, K., Matsuoka, M., Mitsuda, K., Murakami, T., Oda, M., Ogawara, Y., Ohashi, T., Shibazaki, N., Tanaka, Y., Hayakawa, S., Kunieda, H., Makino, F., Masai, K., Nagase, F., Tawara, Y., Miyamoto, S., Tsunemi, H., Yamashita, K., Kondo, I., Jernigan, J. G., van Paradijs, J., Beardsley, A., Cominsky, L., Doty, J., and Lewin, W. H. G. 1982. Simultaneous optical and X-ray bursts from 4U/MXB 1636-53. ApJ, **263**, 325–339.

Peterson, B. M., and Horne, K. 2006. Reverberation mapping of active galactic nuclei. Page 89 of: M. Livio and S. Casertano (eds.), *Planets to Cosmology: Essential Science in the Final Years of the Hubble Space Telescope*. Space Telescope Science Institute Symposium Series. Cambridge University Press.

Petro, L. D., Bradt, H. V., Kelley, R. L., Horne, K., and Gomer, R. 1981. Rapid X-ray and optical flares from Scorpius X-1. ApJL, **251**, L7–L11.

Pfahl, E., Rappaport, S., and Podsiadlowski, P. 2003. The galactic population of low- and intermediate-mass x-ray binaries. ApJ, **597**, 1036–1048.

Pietsch, W., Steinle, H., and Gottwald, M. 1983. 4U 2129+47 = V1727 Cygni. IAU Circ., **3887**.

Quataert, E., and Narayan, R. 1999. Spectral models of advection-dominated accretion flows with winds. ApJ, **520**, 298–315.

Reynolds, M. T., Callanan, P. J., and Filippenko, A. V. 2007. Keck infrared observations of GRO J0422+32 in quiescence. MNRAS, **374**, 657–663.

Sandage, A., Osmer, P., Giacconi, R., Gorenstein, P., Gursky, H., Waters, J., Bradt, H., Garmire, G., Sreekantan, B. V., Oda, M., Osawa, K., and Jugaku, J. 1966. On the optical identification of Sco X-1. ApJ, **146**, 316–321.

Shahbaz, T., Bandyopadhyay, R., Charles, P. A., and Naylor, T. 1996. Infrared spectroscopy of V404 Cygni: limits on the accretion disc contamination. MNRAS, **282**, 977–981.

Shahbaz, T., Casares, J., Watson, C. A., Charles, P. A., Hynes, R. I., Shih, S. C., and Steeghs, D. 2004. The Massive neutron star or low-mass black hole in 2S 0921-630. ApJL, **616**, L123–L126.

Shahbaz, T., Dhillon, V. S., Marsh, T. R., Casares, J., Zurita, C., Charles, P. A., Haswell, C. A., and Hynes, R. I. 2005. ULTRACAM observations of the black hole X-ray transient XTE J1118+480 in quiescence. MNRAS, **362**, 975–982.

Shahbaz, T., Dhillon, V. S., Marsh, T. R., Zurita, C., Haswell, C. A., Charles, P. A., Hynes, R. I., and Casares, J. 2003. Multicolour observations of V404 Cyg with ULTRACAM. MNRAS, **346**, 1116–1124.

Shahbaz, T., Fender, R. P., Watson, C. A., and O'Brien, K. 2008a. The first polarimetric signatures of infrared jets in X-ray binaries. ApJ, **672**, 510–515.

Shahbaz, T., Ringwald, F. A., Bunn, J. C., Naylor, T., Charles, P. A., and Casares, J. 1994. The mass of the black hole in V404 Cygni. MNRAS, **271**, L10–L14.

Shahbaz, T., Watson, C. A., Zurita, C., Villaver, E., and Hernandez-Peralta, H. 2008b. Time-resolved optical photometry of the ultracompact binary 4U 0614+091. PASP, **120**, 848–851.

Shakura, N. I., and Sunyaev, R. A. 1973. Black holes in binary systems. Observational appearance. A&A, **24**, 337–355.

Smith, A. J., Haswell, C. A., Murray, J. R., Truss, M. R., and Foulkes, S. B. 2007. Comprehensive simulations of superhumps. MNRAS, **378**, 785–800.

Steeghs, D. 2003. Extending emission-line Doppler tomography: mapping-modulated line flux. MNRAS, **344**, 448–454.

Steeghs, D., and Casares, J. 2002. The mass donor of Scorpius X-1 revealed. ApJ, **568**, 273–278.

Strohmayer, T., and Bildsten, L. 2006. New views of thermonuclear bursts. Pages 113–156 of: Lewin, W. H. G., and van der Klis, M. (eds.), *Compact Stellar X-ray Sources*. Cambridge University Press.

Thorstensen, J., Charles, P., Bowyer, S., Briel, U. G., Doxsey, R. E., Griffiths, R. E., and Schwartz, D. A. 1979. A precise position and optical identification for 4U 2129+47 – X-ray heating and a 5.2 hour binary period. ApJL, **233**, L57–L61.

Truss, M. R., Wynn, G. A., Murray, J. R., and King, A. R. 2002. The origin of the rebrightening in soft X-ray transient outbursts. MNRAS, **337**, 1329–1339.

Uemura, M., Kato, T., Matsumoto, K., Honkawa, M., Cook, L., Martin, B., Masi, G., Oksanen, A., Moilanen, M., Novak, R., Sano, Y., and Ueda, Y. 2000. XTE J1118+480. IAU Circ., **7418**.

van der Klis, M. 2006. Rapid x-ray variability. Pages 39–112 of: Lewin, W. H. G., and van der Klis, M. (eds.), *Compact Stellar X-Ray Sources*. Cambridge Astrophysics Series, vol. 39. Cambridge University Press.

von Zeipel, H. 1924. The radiative equilibrium of a rotating system of gaseous masses. MNRAS, **84**, 665–683.

Wade, R. A., and Horne, K. 1988. The radial velocity curve and peculiar TiO distribution of the red secondary star in Z Chamaeleontis. ApJ, **324**, 411–430.

White, N. E., and Swank, J. H. 1982. The discovery of 50 minute periodic absorption events from 4U 1915-05. ApJL, **253**, L61–L66.

Whitehurst, R., and King, A. 1991. Superhumps, resonances and accretion discs. MNRAS, **249**, 25–35.

Wijers, R. A. M. J., and Pringle, J. E. 1999. Warped accretion discs and the long periods in X-ray binaries. MNRAS, **308**, 207–220.

Wolff, M. T., Ray, P. S., Wood, K. S., and Hertz, P. L. 2009. Eclipse timings of the transient low-mass X-ray binary EXO 0748-676. IV. The Rossi X-Ray Timing Explorer eclipses. ApJS, **183**, 156–170.

Zurita, C., Casares, J., Shahbaz, T., Charles, P. A., Hynes, R. I., Shugarov, S., Goransky, V., Pavlenko, E. P., and Kuznetsova, Y. 2000. Optical studies of the X-ray transient XTE J2123-058 – I. Photometry. MNRAS, **316**, 137–142.

Zurita, C., Casares, J., Shahbaz, T., Wagner, R. M., Foltz, C. B., Rodríguez-Gil, P., Hynes, R. I., Charles, P. A., Ryan, E., Schwarz, G., and Starrfield, S. G. 2002. Detection of superhumps in XTE J1118+480 approaching quiescence. MNRAS, **333**, 791–799.

5. X-ray binary populations in galaxies
GIUSEPPINA FABBIANO

Abstract

X-ray binaries are responsible for the bulk of the X-ray emission of our own galaxy. A lot has been learned about these bright X-ray sources since the beginning of X-ray astronomy, but significant questions are still open. These questions are related to the origin and evolution of these sources, and to how their properties depend on those of the parent stellar population. The discovery of several populations of X-ray binaries in external galaxies with Chandra, and to a lesser extent with XMM-Newton, gives us tools to look at these sources in a new way. Not only can we reconsider long-standing questions of galactic studies, such as the origin of low-mass X-ray binaries, but also we can look at the entire gamut of X-ray binary properties in a range of environments, from actively star-forming galaxies to older stellar systems. These observations have led to the discovery of several ultraluminous X-ray sources, thereby introducing new interesting possibilities for our understanding of X-ray binaries and possibly opening new paths to the discovery of the elusive intermediate-mass black holes.

5.1 Introduction and chapter outline

X-ray astronomy began with the unexpected discovery of a very luminous source, Sco X-1 (Giacconi et al., 1962), the first galactic X-ray binary (XRB) ever to be observed. XRBs, the most common luminous X-ray sources in the Milky Way, are binary systems composed of an evolved stellar remnant (neutron star [NS], black hole [BH], or white dwarf [WD]), and a stellar companion (for reviews on XRBs, see Lewin et al., 1995; Lewin and van der Klis, 2006). If the companion is a massive star (mass > 10 $M_\odot$), the XRB is called a high-mass X-ray binary (HMXB); if it is a low-mass star (mass ≤ 1 $M_\odot$), the XRB is called a low-mass X-ray binary (LMXB).

HMXBs are short-lived X-ray sources, with lifetimes ∼10 million years, regulated by the evolution of the massive companion. LMXBs evolve at a much slower pace, with lifetimes of ∼10^8 to 10^9 years due to the slow-evolving low-mass donor star (see Verbunt and van den Heuvel, 1995), so they are generally old systems. This is strictly valid for LMXBs originating from the evolution of native binary systems in the stellar field. However, LMXBs may also form from dynamical interactions in globular clusters (see Clark, 1975; Grindlay, 1984); these GC LMXBs can be formed virtually at any time, and some of them may be short-lived systems (e.g., the NS-WD ultracompact binaries; Bildsten and Deloye, 2004).

Ever since the discovery of Sco X-1 in 1962, the study of Galactic XRBs has continued to be a thriving area of research. This work has contributed to the knowledge of both accretion processes onto a compact object and the extreme state of matter found in these compact objects. In the late 1970s and 1980s, following the advent of imaging X-ray astronomy with the Einstein Observatory, XRBs started to be detected in a number of nearby galaxies (reviewed in Fabbiano, 1989). While observations with other X-ray telescopes (ROSAT, ASCA, XMM-Newton) have contributed to these studies, the subarcsecond resolution of the Chandra X-ray Observatory (Weisskopf et al., 2000) has revolutionized this field (reviewed in Fabbiano, 2006).

With Chandra, populations of XRBs have been detected in virtually all galaxies within 20 to 30 Mpc, providing a new tool for constraining the formation and evolution of XRBs. Given that the XRBs in a given galaxy are all at the same distance, with these observations we can establish their luminosities accurately (which is not the case for galactic

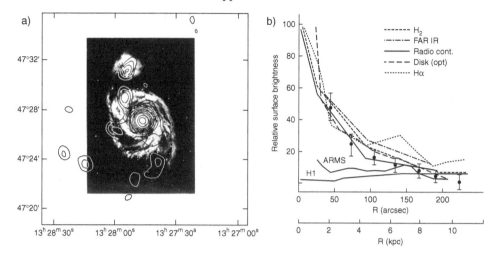

FIGURE 5.1. a) Einstein contours of M51 on the optical image; b) X-ray radial brightness profile for M51 (dots with error bars), compared with the disk and arm profiles (Palumbo et al., 1985). Reproduced by permission of the AAS.

XRBs), and we can correlate their properties with those of the parent stellar population. By detecting an ever-growing sample of XRBs, we are also sensitive to extreme objects that may be missing in the Milky Way, such as the ultraluminous X-ray sources (ULXs) that emit in excess of the Eddington luminosity for a NS or ∼10 solar-mass stellar BH system; ULXs have been suggested to be intermediate-mass BHs, bridging the gap between the supermassive BHs of AGNs and the stellar BHs found in some XRBs (Fabbiano, 1989, 2006).

In this chapter, I briefly review the outcome of the early pre-Chandra observations, and the questions raised by these results (Section 5.2); discuss the methods we are now using to study XRB populations (Section 5.3) and the application of these methods to characterize populations of XRBs (Section 5.4); review the current landscape for both HMXB (Section 5.5) and LMXB (Section 5.6) populations; revisit the issue of LMXB formation and evolution in the light of the Chandra results (Section 5.7); and conclude with a brief review of ULXs (Section 5.8).

5.2 Discovery and first conclusions: from Einstein to Chandra

The Einstein Observatory gave us the first X-ray images and spatially resolved spectra of galaxies. A review of these results can be found in Fabbiano (1989); the Catalog and Atlas of Galaxies (Fabbiano et al., 1992) gives a compendium of the images and integrated properties of normal galaxies observed with Einstein; the spectral properties of this emission are compiled in Kim et al. (1992a).

5.2.1 XRBs in spiral and irregular galaxies

Several nearby late-type galaxies were detected with Einstein (see Fabbiano, 1989). In these nearby galaxies, the X-ray images revealed a number of luminous X-ray sources, some variable, with hard emission spectra, properties consistent with those of the galactic XRBs (see Fabbiano and Trinchieri, 1987; the Einstein Catalog and Atlas of Galaxies, Fabbiano et al. 1992; Fig. 5.1). The unresolved X-ray emission also detected in these galaxies typically followed the radial distribution of the galaxy disks, with some enhancements in the spiral arms, suggesting that this emission was due to the integrated contributions of fainter unresolved XRBs and establishing a connection with the stellar population of the parent galaxies (see preceding references and Palumbo et al., 1985; Fig. 5.1).

These early observations also revealed a connection between the integrated emission of late-type galaxies and their star formation rate (SFR), in the sense of more actively star-forming galaxies having enhanced X-ray emission (relative to the stellar optical emission). This suggested either a more numerous or a more luminous population of X-ray sources in high-SFR galaxies, and a link between HMXB populations and SFR (Fabbiano et al., 1982, 1984).

More systematic multiwavelength correlation studies of spiral and irregular galaxy samples were performed once larger samples of galaxies had been observed (e.g., Fabbiano and Trinchieri, 1985; Fabbiano et al., 1988), culminating with the final study of the entire sample of 234 S0/a Irr galaxies observed with Einstein (Shapley et al., 2001; Fabbiano and Shapley, 2002). It is worth noting that these samples were analyzed with censored data analysis techniques, using both detections and upper limits to maximize the information available and avoid flux limit biases. These studies typically included, besides the X-ray, optical B, near and far-infrared, and radio continuum integrated emission. Significant differences were detected in the correlations of bulge (old stellar population) and disk/arm (younger stellar population) dominated galaxies, which were related to the different LMXB and HMXB populations.

These studies concluded that the integrated X-ray luminosity is correlated with the far-infrared luminosity (therefore SFR) in star-forming galaxies; it is instead correlated with the stellar luminosity (therefore stellar mass) in bulge-dominated galaxies. These results identified HMXB and LMXB populations as the major contributors to the X-ray emission of normal spiral and irregular galaxies. As discussed later in this chapter, these conclusions have been confirmed by Chandra observations.

5.2.2 The LMXB X-ray emission of E and S0 galaxies

When Einstein first surveyed the Virgo Cluster, it revealed extended X-ray sources associated with elliptical galaxies (Forman et al., 1979). While there was evidence pointing to extended hot gaseous halos as the origin of this emission (displacement from the stellar body, softer spectra than typical XRB spectra), these data could not exclude a sizeable LMXB contribution; after all, LMXB should be associated with old stellar systems, such as these galaxies. A vigorous debate ensued on the relative importance of these components of the X-ray emission (see Fabbiano, 1989). LMXBs could not be detected individually with Einstein in galaxies at the distance of Virgo (16 Mpc), so the debate was based on the average properties of the integrated X-ray emission of these galaxies.

Trinchieri and Fabbiano (1985), comparing the X-ray to optical ratios of X-ray-faint elliptical galaxies with the integrated emission of the bulge of M31 – where LMXBs could be easily detected – pointed out that the X-ray emission of these elliptical galaxies could be dominated by LMXBs. This conclusion was later strengthened by the X-ray spectra of these galaxies. The spectrum of the integrated X-ray emission of the bulge of M31 is hard, typical of LMXBs (Fabbiano et al., 1987). The same spectral signature was found with Einstein in a sample of elliptical and S0 galaxies with relatively low X-ray to optical ratios, consistent with those of the bulges of nearby spirals, and thus having little chance of harboring gaseous hot halo emission (Kim et al., 1992b). This hard LMXB component was later detected, in addition to a gaseous emission component, in X-ray luminous halo-dominated ellipticals with the CCD spectrometers on ASCA (Matsushita et al., 1994). With Chandras sub-arcsecond resolution imaging, we now have direct proof that LMXB populations are ubiquitous in elliptical and S0 galaxies (see Fabbiano, 2006).

5.3 Methods

With Chandra (and XMM-Newton for the most nearby galaxies; see Fabbiano, 2006), we have now ample evidence of the existence of XRB populations in all types of galaxy.

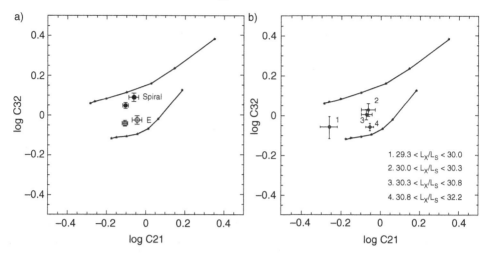

FIGURE 5.2. a) Average Einstein X-ray colors of spiral and elliptical galaxies; b) X-ray colors of X-ray–faint ellipticals are closer to those of spirals than the colors of halo-dominated galaxies (Kim et al., 1992b). Reproduced by permission of the AAS.

We can now begin to study how the characteristics of XRBs (e.g., spectra, luminosities) correlate with the properties of their parent stellar populations. To this end, X-ray astronomy is adapting methods long used in other wavelengths, such as photometry and luminosity functions. Here these methods are described; the results of their use are discussed in subsequent sections.

5.3.1 X-ray photometry

To constrain the physical parameters of the emission, X-ray spectra are typically fitted (e.g., using minimum χ^2) to models convolved with the response of the instrument. To yield meaningful results, this procedure, however, requires a minimum amount of ∼200 detected source counts. X-ray photometry can give us information for fainter sources, even detected with as few as ∼30 counts. X-ray photometry was first used to interpret the results of the Einstein observations of galaxies. By dividing the energy range into three energy bands and plotting the results in color–color diagrams, Kim et al. (1992b) showed that galaxies of different morphological types have different average X-ray spectra (Fig. 5.2) and that halo-dominated X-ray–bright ellipticals can be separated this way from their LMXB-dominated X-ray–faint counterparts.

This approach is now being used to interpret the result of Chandra and XMM observations. Chandra, in particular, can detect faint sources (typically, a 10-count source is highly significant with Chandra) because there is virtually no background in its small resolution element. The majority of the XRBs detected in galaxies are too faint for individual spectral analysis, but their position in color–color or color–luminosity diagrams can provide useful information. For example, Prestwich et al. (2003) (see Section 5.4; for other examples, see also Fabbiano, 2006) demonstrated that X-ray colors can be used to discriminate among different types of X-ray sources in galaxies (Fig. 5.3).

There is, however, a drawback of X-ray photometry of which one needs to be aware. This is the lack of a standard photometric system (such as UBV in optical astronomy). Different definitions of photometric boundaries have been used in different studies; moreover, the resulting X-ray colors or flux ratios are typically not corrected for the different responses of X-ray telescopes and instruments. As a result colors are very much instrument specific (for example, one cannot directly compare Chandra and XMM colors). Moreover, one has to be careful comparing colors obtained with the same instrument at

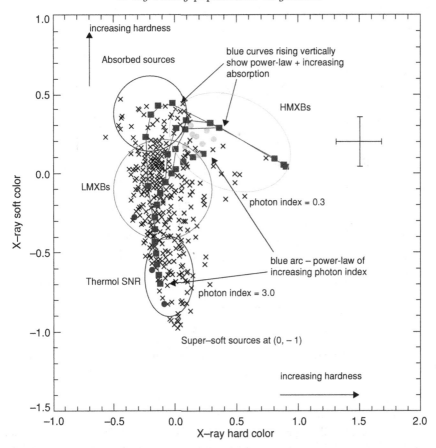

FIGURE 5.3. Chandra color-color diagram from Prestwich *et al.* (2003), identifying the loci of different types of galactic X-ray sources. Reproduced by permission of the AAS.

different epochs, because of changes in the instrument response with time (e.g., Chandra ACIS). Nevertheless, photometry is a powerful tool, and it is likely that we will achieve X-ray standards (and software to apply them) in the not too distant future. A beginning to this process can be found in Grimm et al. (2009).

5.3.2 *X-ray luminosity functions*

Luminosity functions are a well-known tool of astronomy. This tool is now being applied to the study of XRB population in galaxies. Given a population of XRBs, the X-ray luminosity function (XLF) can be constructed in either differential or cumulative form. The differential XLF gives the number of sources found at a given X-ray luminosity (dN/dLX) as a function of the luminosity. The cumulative luminosity function is the integral of this distribution and gives the number of X-ray sources detected with luminosity greater than a given value. Typically XLFs are parameterized in terms of power-law slopes, breaks, and overall normalization. These parameters can provide astrophysical information on the XRB populations, as shown in Sections 5.4 through 5.7. An early example of this approach for the study of XRB populations in galaxies can be found in the comparison of the Einstein XLFs of the two optically similar galaxies M31 and M81 (Fabbiano, 1988; Fig. 5.4), which showed an excess of luminous XRBs in M81. However, the limitation of those data did not allow discrimination between a difference in slope (an intrinsically larger number of higher luminosity sources in M81), and normalization (increased XRB formation at all luminosities in M81).

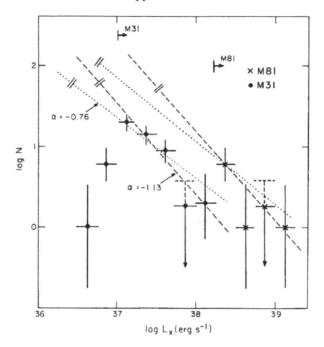

FIGURE 5.4. Einstein XLFs of M31 and M81 (Fabbiano, 1988). Reproduced by permission of the AAS.

5.3.3 Building the XLF

There are four steps in building and modeling the XLF: source detection, incompleteness correction, background AGN contamination correction, and parameter estimation.

(i) *Source detection.* One can apply one's favorite detection algorithm to the observations of a galaxy and obtain a sample of sources above a defined detection threshold or statistical significance (see, e.g., Kim and Fabbiano, 2003). At this stage, if the aim is to obtain an XLF that is as representative as possible of the XRB population of a galaxy, it is important to exclude obvious interlopers, such as extended sources and sources associated with bright nearby stars.

(ii) *Incompleteness correction.* For the XLF to be a true representation of the XRB population under study, it is important to correct biases in the source detection. These biases are mostly important at the lower luminosity end of the XLF. If not corrected, they may lead to the spurious detection of low-luminosity breaks (Fig. 5.5; see the detailed discussion in the appendix of Kim and Fabbiano, 2003).

In the case of Chandra, there are two major effects leading to incompleteness at the low luminosities. One is telescope specific and derives from the dependence of the point response function both on the energy of the incident photon and on the off-axis angle. The latter, in particular, leads to a significantly diminishing sensitivity a few arcminutes away from the optical axis. The second effect has to do with the presence of diffuse emission in the field (e.g., emission from a hot interstellar medium or from unresolved faint sources in the observed galaxy). These effects can be calculated a priori and corrections tailored to a given observation ("forward correction"; see Kim and Fabbiano, 2003). However, there are other effects that the forward correction cannot account for. These are the Eddington bias – that is, excess source detection at the lower luminosities resulting from positive statistical fluctuations of sources just below the detection threshold – and the effect of source confusion in crowded fields. While it is better to exclude the very crowded centermost regions of the galaxies, both effects, as well as those

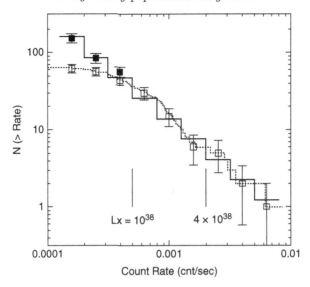

FIGURE 5.5. Observed and corrected (the three filled points at the low-luminosity end) XLF of NGC 1316 (Kim and Fabbiano, 2003). Reproduced by permission of the AAS.

included in the "forward correction," can be in principle minimized with simulations ("backward correction"; see Kim and Fabbiano, 2003). The approach here is to start with the observation itself, adding simulated sources of different count rate and at different positions in the field, running the detection algorithm, and comparing the results with the input data.

(iii) *Background AGN contamination.* This contamination, noted by many practitioners in the field, is particularly important in the case of very deep observations and/or observations covering a large area of the sky. It can be estimated using the results of deep sky X-ray surveys (see Gilfanov, 2004, for a clear explanation of this effect).

(iv) *Functional parameters.* Once the corrected XLF has been derived, it can be fitted to models to derive functional parameters. It is important to use the differential XLF to have a clean estimation of the statistical errors. Different methods have been used to fit the XLF, including χ^2, C statistics, and maximum likelihood (e.g., Kim and Fabbiano, 2004; Schmitt and Maccacaro, 1986; Zezas et al., 2007). XRBs are time variable, so in principle source variability could affect the XLF. However, in cases where multiple observations of a given galaxy are available, no significant differences have been seen in the shape of the individual XLFs (e.g., in the XLF of the Antennae galaxies; Zezas et al., 2007). This result justifies the practice of co-adding observations to achieve the sensitivity needed to measure the faintest part of the XLF. Zezas et al. (2007) point out that this approach may produce a bias, if several transient sources are present in the XRB population. In this case, the luminosity of the transients would not be representative of the "on" luminosity but would be the lower averaged value over the entire observing period.

5.4 Characterizing XRB populations with XLFs and X-ray colors

The XLFs of galactic HMXBs and LMXBs are different (Grimm et al., 2002). Whereas the HMXB XLF can be fitted with a single unbroken power law, the LMXB XLF follows a steeper power law at high luminosities ($> 5 \times 10^{37}$ erg s^{-1}) and then flattens up at the lower luminosities. Similarly, high-resolution observations of nearby galaxies show

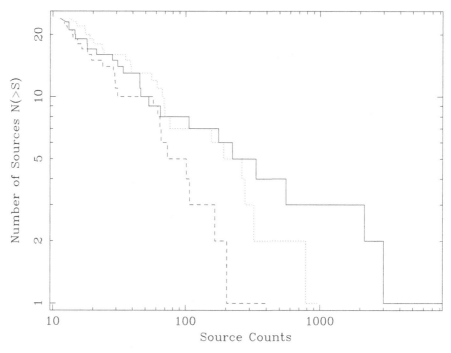

FIGURE 5.6. XLFs of sources in M81: solid arms; dotted moving toward the disk; dashed disk (Swartz et al., 2003). Reproduced by permission of the AAS.

differences in the XLFs of sources associated with different stellar populations (for a review, see Fabbiano, 2006). For example, in M31 the XLFs of sources in different stellar fields differ (Kong et al., 2003). In M81, a first report (Tennant et al., 2001) indicated that the XLF of the sources associated with the old stellar bulge is steeper than that of the sources associated with the younger stellar population of the disk and spiral arms, echoing the behavior of the galactic LMXB and HMXB XLFs. A further study of M81 (Swartz et al., 2003) noticed that the XLFs become gradually steeper, moving from the youngest actively star-forming regions of the spiral arms toward the interarm regions and the old stellar disk (Fig. 5.6). This suggests an aging of the XRB population, with younger and more luminous HMXBs fast fading away. Similar age-related differences are seen in the XLFs of different regions of the face-on spiral M83 (Soria and Wu, 2003), and in the XLFs of a sample of spiral and late-type galaxies (Kilgard et al., 2002, 2006).

The X-ray colors of sources can also be used to investigate their nature. In particular, Prestwich et al. (2003) (see Fig. 5.3) demonstrated that different types of well-known galactic sources have colors that fall in different loci of a Chandra color–color diagram and used this diagram to discriminate among various types of sources detected in external galaxies with Chandra. Soria and Wu (2003) used X-ray photometry together with XLFs and imaging in M83 to identify the most luminous sources with XRBs and the softer lower-luminosity sources in the spiral arms with SNR candidates. Using a similar approach, Prestwich et al. (2009) identify populations of hard and soft sources in eight spiral galaxies and find that these populations follow different XLFs. The hard sources have a steep XLF, typical of LMXB populations (see Section 5.6, e.g., Kim and Fabbiano, 2004), while the soft sources follow the same XLF as the "classical" wind-fed galactic HMXBs, which, however, occupy a different locus of their color–color diagram. These soft HMXBs tend to be more luminous than the wind-fed HMXBs, and Prestwich et al. (2009) suggest that they may be HMXBs in a very high–accretion–rate state (e.g., steep power law state; see the review of Remillard and McClintock, 2006, for a discussion of accretion states in XRB). In the ringed galaxy NGC 4736, these high-luminosity soft sources concentrate in the star-forming ring, while the hard sources with steep XLFs are found in the old bulge.

5.5 HMXB populations

HMXBs are short-lived X-ray binaries, with lifetimes limited to $\sim 10^7$ years by the fast evolution of the massive donor star (see Verbunt and van den Heuvel, 1995). These populations are therefore confined to young stellar regions, and, as demonstrated by the observations of M81 (Swartz et al., 2003; see Section 5.4), tend to disappear in older stellar populations. HMXBs are therefore a good indicator of the SFR of the parent galaxy.

5.5.1 The L_X–SFR connection

The connection between the integrated X-ray emission of star-forming galaxies (L_X) and their integrated far-infrared (FIR) emission was first noticed with the analysis of the sample of spiral and irregular galaxies observed with the Einstein Observatory (Fabbiano et al., 1988; Fabbiano and Shapley, 2002); see Subsection 5.2.1. Fabbiano and Shapley (2002) in particular point out that in the sample of actively star-forming galaxies observed with Einstein, L_X and FIR appear to be linearly correlated, and therefore the integrated X-ray luminosity of a galaxy is a reasonable proxy for its star formation rate. Grimm et al. (2003) confirmed this link by demonstrating that the normalizations of the Chandra XLFs of star-forming galaxies are proportional to the SFR of the galaxy. This result shows that these XLFs are dominated by short-lived HMXBs. They all follow similar power-law dependences with cumulative slope of about -0.5 to -0.6, which is also consistent with the slope of the XLF of galactic HMXBs (Grimm et al., 2002).

Grimm et al. (2003) noted that the XLFs of the galaxies with the highest SFR observed with Chandra, such as M82 and the Antennae (see also Zezas and Fabbiano, 2002) extend to luminosities of a few 10^{40} erg s^{-1}. These galaxies contain a number of ULXs, sources with luminosities exceeding the Eddington luminosity of typical galactic black-hole binaries (BHBs, which host BHs of $\sim$10–15 M$_\odot$; see Remillard and McClintock, 2006). These ULXs are discussed more extensively in Section 5.8; here we remark that the similarity of HMXB XLFs associates them strongly with the young stellar population.

The L_X–SFR correlation has been exploited in studies of high-redshift galaxies found in deep surveys (e.g., Lehmer et al., 2005; Ranalli et al., 2005), where it was concluded that the integrated X-ray luminosity L_X is a useful observational proxy for the integrated SFR of these distant galaxies and could then be used to study the history of star formation of the universe. Gilfanov et al. (2004) demonstrate that in low-SFR regimes the L_X–SFR correlation deviates from linearity, because of Poisson statistics affecting the number of luminous X-ray sources present in a given galaxy population, whereas linearity is approached for high enough SFR, where sources at the "natural" high-luminosity cutoff of the HMXB XLF can be found in large enough numbers. The break into a linear regime is in fact a function of this cutoff luminosity. Comparing the integrated galaxy L_X detected in nearby galaxies with Chandra with the results from Chandra observations of the Hubble Deep Field North (Fig. 5.7a), Gilfanov et al. (2004) find that the cutoff luminosity is near 10^{40} erg s^{-1}, consistent with the largest ULX luminosity detected in M82 and the Antennae. This luminosity is then suggested as a limiting luminosity for HMXBs, supporting the notion that the majority of ULXs are normal accreting stellar BHs. The same authors suggest that if a hypothetical population of intermediate-mass BHs is present in galaxies, it should appear as a further steepening of the L_X–SFR dependence at the extremely high SFR that may be present in young distant galaxies detected in future very deep X-ray surveys (Fig. 5.7b).

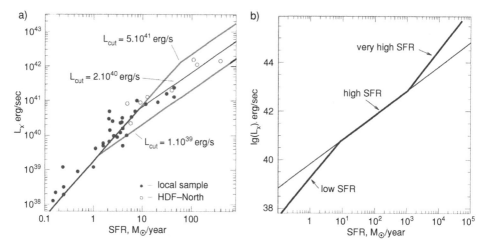

FIGURE 5.7. *a)* L_X SFR relation compared with data point; *b)* expected L_X SFR for an additional population of sources appearing at the highest SFR (Gilfanov *et al.*, 2004).

5.5.2 *Examples of high-luminosity XLFs in high-SFR galaxies: the Antennae and the Cartwheel*

The Antennae galaxies were observed several times over a couple of years with Chandra, resulting in a deep and valuable data set (Fabbiano *et al.*, 2004). The Antennae is a major merger of two similar-sized spirals, which results in tidal distortion and very active star formation, with SFR about 10 times larger than in our own Galaxy. In this dataset, 120 sources were detected, resulting in an XLF extending down to 2×10^{37} erg s^{-1} with a cumulative power-law slope of -0.5 (Zezas *et al.*, 2007). Ten of these sources are ULXs, giving us the ability to observe in detail the properties of a whole family of these intriguing sources. The ULXs vary, with correlated luminosity and spectra variability (see Fabbiano *et al.*, 2003b), consistent with being accreting binary systems. Only three of these ULXs are associated with a stellar cluster, although all of them are in the vicinity of a cluster, suggesting possible formation kicks (Zezas *et al.*, 2002). These characteristics, and the uninterrupted slope of the XLF, suggest that the ULXs of the Antennae are part of the HMXB population and may be binary systems in extremely high accretion states (King *et al.*, 2001).

An even more extreme young XRB population is that of the Cartwheel galaxy (Gao *et al.*, 2003), which hosts more than 20 ULXs. Based on the XLF of this population, Wolter and Trinchieri (2004) estimate an SFR between 12 and 20 solar masses per year.

5.5.3 *The low-luminosity XLF*

M33 is a small Sc galaxy in the Local Group that is more actively star forming than either the galaxy or M31. Grimm *et al.* (2005) analyzed the first Chandra survey of M33, which reached limiting luminosities near 2×10^{34} erg s^{-1}. A deeper study is now in progress by a different team. Grimm *et al.* (2005) report the detection of 261 sources. The color–color diagram (Fig. 5.8) suggests that only a few LMXBs are found in this population, which is instead dominated by wind-fed HMXBs and by SNRs. A significant contamination by background AGN is also found, consistent with the large angular area and depth of the survey. After correction for incompleteness and background contamination, the resulting XLF is consistent with the HMXB XLF. The spectra of these luminous sources and their time variability further suggest an HMXB population (Grimm *et al.*, 2007).

The Chandra survey of the SMC has pushed source detection to an even fainter limit, corresponding to $L_X \sim 10^{33}$ erg s^{-1}. The XLF built with these data gives convincing

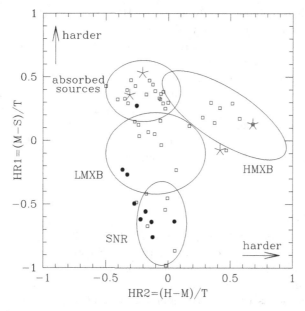

FIGURE 5.8. Chandra color–color diagram of the X-ray source in M33 (Grimm et al., 2005), showing a lack of "old" LMXB sources. Note that the cluster of "absorbed" sources is consistent with the expected contribution of background AGNs that will contaminate the X-ray source population of M33, given its large angular scale in the sky. Reproduced by permission of the AAS.

evidence of a flattening below 5×10^{34} erg s^{-1}, consistent with low accretion rates and the onset of the propeller effect in NS binaries (Zezas et al., in preparation; see Shtykovskiy and Gilfanov, 2005). Using this survey, Antoniou et al. (2009) extend the relation between the number of accreting binaries and the SFR of the parent population to the very low luminosities of quiescent Be XRBs.

5.5.4 HMXB and current star formation

The evolution of a massive binary system into a HMXB requires $\sim 10^6$ to 10^7 years. Therefore, although the normalization of the XLF is proportional to the SFR, more properly it should be said that the number of HMXBs represents the star formation that took place in a galaxy 5–60 $\sim$ Myr ago.

There are reports in the literature of star-forming regions too young for HMXB formation. In a survey of 12 nearby galaxies, Tyler et al. (2004) find that actively star-forming regions (based on Hα, molecular gas, and IR emission) lack X-ray emission; Shtykovskiy and Gilfanov (2005) find the same effect in the XMM-Newton survey of the LMC. More recently, Shtykovskiy and Gilfanov (2007) compare the distributions of X-ray emission and H II regions in the spiral arms of M51 and suggest that the distribution of X-ray emission is not as peaked as that of the H II region, reflecting earlier occurrences of star formation that have been left behind by the compression wave.

5.5.5 Population synthesis (PS) modeling of the HMXB XLF

The XLFs of sources in a given system reflect the formation, evolution, and physical properties of the X-ray source population. These differences are evident in different regions of the same galaxy, as discussed previously, and also in the comparison of different galaxy populations. Although very young populations may follow the "universal" 0.5 power-law cumulative XLF, aging of these populations results in the loss of the most luminous XRBs (fast-evolving, highly accreting HMXBs), and therefore in a steepening of the XLF slope, at least at high luminosities (see Section 5.4).

Early theoretical work attempted to interpret the XLFs using ad hoc power-law models and accounting for aging and impulsive birth of XRB populations (Wu, 2001; Kaaret, 2002; Kilgard et al., 2002). More recent work by Belczynski et al. (2004) instead developed synthetic XLFs for a range of stellar populations, showing that the XLF slopes depend on both the age of the starburst and the metallicity of the stellar population. Comparison of these models with the observed XLF of NGC 1569, a well-studied star-forming galaxy, shows that the best-fitting model is the one developed for a synthetic population with stellar age and metallicity closest to the stellar population of this galaxy.

5.6 LMXB populations

As discussed in Section 5.2.2, the presence of LMXB populations in elliptical galaxies was inferred 20 to 30 yr ago from the analysis of Einstein data (e.g., Trinchieri and Fabbiano, 1985; Kim et al., 1992b) but was controversial at the time (see Fabbiano, 1989). Chandra images now have shown populations of point sources in all elliptical galaxies at least as far as the Virgo cluster (Sarazin et al., 2000; see Fabbiano, 2006). These sources have luminosities consistent with those of luminous XRBs. They also have typically hard spectra, with possibly softer spectra in the most high-luminosity sources, suggesting the presence of disk emission in BH binaries (Irwin et al., 2003). Variability detected in some cases in these early Chandra observations is also consistent with the presence of compact accreting objects (e.g., Kraft et al., 2001; Sivakoff et al., 2005).

5.6.1 The L_X-stellar-mass and GC connections

Most of the earlier Chandra observations were relatively short (although typically with exposures as long as half a day) and could detect only the most luminous LMXBs, with $L_X \geq 5 \times 10^{37}$ erg s^{-1}. The XLFs of these high-luminosity populations tend to be remarkably similar and follow steeper slopes (cumulative slope ~ -1) than those of HMXB XLFs of corresponding luminosity. They also show a high-luminosity break near 4.5×10^{38} erg s^{-1}. Although there are uncertainties, this break luminosity is higher than the Eddington luminosity of a "standard" NS (Kim and Fabbiano, 2004; Gilfanov, 2004; see other references in Fabbiano, 2006). Gilfanov (2004) suggested that the normalization of these XLFs (i.e., the number of LMXBs in a given population) is a function of the stellar mass of the galaxy. This suggestion agrees with the idea that the LMXB population is a slowly evolving population of native binary systems. However, a second parameter, besides stellar luminosity/mass, is also important: the GC content of the galaxy (Kim and Fabbiano, 2004). Galaxies with a larger GC specific frequency (SGC; number of GC per unit stellar luminosity) tend to have a large number of LMXBs for a given total optical luminosity of the galaxy (Fig. 5.9). This result shows that both (i) evolution of native binary systems and (ii) dynamical LMXB formation and evolution in GCs contribute to the LMXB populations. Both mechanisms have been suggested and debated in the case of galactic LMXBs (see Clark, 1975; Grindlay, 1984; Verbunt and van den Heuvel, 1995; see also review in Fabbiano, 2006).

To better understand the properties of LMXB populations, two nearby elliptical galaxies with similar old stellar populations were observed with deep Chandra exposures (several days of observation): NGC 3379 and NGC 4278. The most conspicuous difference between these two galaxies is in their SGC: NGC 3379 is GC poor, whereas NGC 4278 is GC rich. Since both galaxies have very little hot gaseous emission, the detection of faint LMXBs is optimized. The aims of this project, of which I was principal investigator, were several:

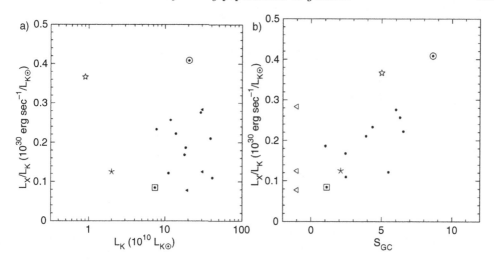

FIGURE 5.9. *a)* L_X L_K correlation, where L_X is the total integrated LMXB luminosity of each galaxy in the sample; note the scatter; *b)* The scatter is correlated with the GC specific frequency SGC (Kim and Fabbiano, 2004). Reproduced by permission of the AAS.

(i) to obtain samples of LMXBs reaching luminosities more representative of galactic LMXBs (with detection thresholds in the 10^{36} erg s^{-1} range), to explore the low-luminosity XLF;
(ii) to observe in time monitoring mode, with several observations within 1 year, in order to learn about the time and spectral variability of these sources;
(iii) to correlate the LMXBs with GCs in these galaxies (from HST observations), in order to compare the properties of field and GC subsamples.

These observations resulted in 98 sources detected within the optical D_{25} of NGC 3379, 10 of which were in GCs (Brassington *et al.*, 2008), and 180 sources within the optical D_{25} of NGC 4278, 39 of which were in GCs (Brassington *et al.*, 2009). In both cases, the overall radial distribution of the number density of these sources follows the optical stellar surface brightness, associating them with the parent galaxies. The majority of these sources are variable, both in flux and in spectral properties, and several sources are strong transient candidates. The Chandra X-ray colors of these populations cluster around the colors expected for typical LMXB spectral parameters. The number of detected sources correlates with the GC content of the parent galaxy (Fig. 5.10; Kim *et al.*, 2006a), highlighting the importance of GC formation for LMXBs. Some of the results of these observations are discussed more fully later in this paper.

5.6.2 *LMXBs and GCs: properties and correlations*

Several studies have attempted to explore the LMXB-GC connection by combining the samples of LMXBs from Chandra observations with samples of GCs detected with HST. These works have led to the identification of "field" and GC LMXB samples, which have then been compared in the attempt to set observational constraints to the evolution and nature of these LMXBs (see, e.g., Fabbiano, 2006; Brassington *et al.*, 2008, 2009; Sivakoff *et al.*, 2007).

Three key factors have emerged that favor the presence of a LMXB in a GC: GC luminosity, metallicity, and compactness (e.g., Kundu *et al.*, 2002; Maccarone *et al.*, 2003; Kim *et al.*, 2006b; Sivakoff *et al.*, 2007). The connections with luminosity and compactness link directly the presence of LMXBs with an environment that fosters their dynamical formation. The reason for the preference of red, high-metallicity GCs as LMXB hosts is less clear. Stellar winds speeding up the LMXB evolution in blue metal-poor GCs

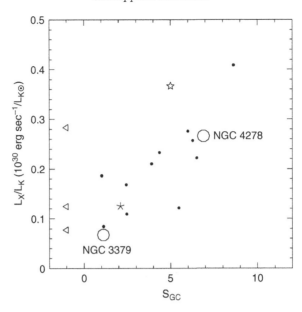

FIGURE 5.10. The number of detected LMXBs in NGC 3379 and NGC 4278 correlate with their GC content (SGC) (Kim et al., 2006a). Reproduced by permission of the AAS.

have been suggested as a way to eliminate LMXBs in these clusters (Maccarone et al., 2004). Another mechanism, proposed by Ivanova (2006), is the lack of outer convective zones in metal-poor main sequence stars resulting in no magnetic braking, and thus impeding the formation of a compact binary system in blue GCs.

In the Milky Way, where one can probe the X-ray emission of GCs down to a very low luminosity ($\sim 10^{30}$ erg s^{-1} with Chandra, e.g., Grindlay et al., 2001), denser and more compact GCs tend preferentially to host LMXBs; these are the GCs where dynamical interactions would be favored (Verbunt and Lewin, 2006; Bregman et al., 2006). These more compact GCs are found at smaller galactocentric radii (R_G), independent of their luminosity (van den Bergh et al., 1991). In their Chandra and HST survey of Virgo galaxies, Sivakoff et al. (2007) did not find any R_G dependence of the association of an X-ray source ($L_X > 5 \times 10^{37}$ erg s^{-1}) with a GC. They instead report at all radii strong dependences on GC luminosity, compactness, and color. Using our deeper Chandra survey of NGC 4278, we find a significant radial effect not on the presence, but on the luminosity of the X-ray sources (Fabbiano et al., 2010): the more luminous X-ray sources tend to be found at smaller R_G in this galaxy (Fig. 5.11). A similar effect is not found in the control sample of field LMXBs in NGC 4278, ruling out analysis and sampling biases and pointing to an intrinsic property of GC-LMXBs.

As observed in the Milky Way (van den Bergh et al., 1991), there is no R_G dependence of the GC luminosity in NGC 4278, whereas there is a very weak link between GC luminosity and X-ray source luminosity. If GCs at smaller R_G are more compact, as in the Milky Way, the NGC 4278 result may be explained with a higher probability of LMXB formation in these GCs, resulting in multiple luminous LMXBs in a single cluster. The higher X-ray luminosities of the more central GCs could be due to the presence of multiple LMXBs. In the Milky Way the number of GC X-ray sources increases with a higher formation rate parameter (a function of both GC luminosity and compactness; see Pooley et al., 2003); lacking a radial dependence of GC luminosity, GC compactness would be mostly responsible for the larger X-ray luminosity of more central GCs in NGC 4278.

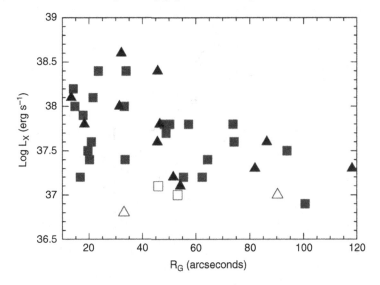

FIGURE 5.11. X-ray luminosity of GC sources in NGC 4278 versus R_G. Red GCs are squares and blue GCs are triangles. The Spearman Rank probability of no correlation is P = 0.0008 (Fabbiano et al., 2010). Reproduced by permission of the AAS.

5.6.3 Radial distributions of LMXBs

The radial distributions of LMXB populations (both in GCs and in the stellar field) have been compared with the overall stellar light distribution and the distribution of GCs. These comparisons were done in an attempt to shed light on the origin of these sources. If, for example, LMXBs were predominantly formed in GCs and then dispersed in the stellar field (Grindlay, 1984), their spatial distribution may look less centrally concentrated than that of the stellar field, following the GC distribution; if the majority of field LMXBs derived from the evolution of native binary systems (Verbunt and van den Heuvel, 1995), it may be expected instead that their spatial distribution would more closely resemble that of the stellar light. If GC and field LMXB have distinct origins, their distributions should differ, resembling those of the parent populations, unless there is a radial dependence of the intrinsic properties of GC systems that may affect LMXB formation, as discussed previously. Although results have been somewhat controversial in the past (Kim et al., 2006b; Kundu et al., 2007), it appears that if one excludes the centermost areas, where source crowding may bias the results, the overall distribution of LMXBs follows that of the stellar light, at least in the two galaxies for which a sizeable population was derived, NGC 3379 and NGC 4278 (Brassington et al., 2008, 2009). The same is true from NGC 4697, which was surveyed to a nearly similar depth (Fig. 5.12, from Kim, 2009, private communication).

D.-W. Kim (2009, private communication) finds that, in NGC 3379 and NGC 4278, the radial distributions of the number densities of X-ray sources in the field and in GC are statistically consistent, with the exclusion of the central $10''$ (a few 100 pc), which were not used because of source confusion. Although it suggests that GCs at smaller R_G may be more prone to LMXB formation, this result may be in contradiction with the conclusions of the Virgo survey (Sivakoff et al. 2007, which, however, consider a higher-luminosity population of LMXBs), unless it is indicative of a prevalence of more compact GC at smaller radii. Also, red GC, which are prevalently associated with LMXBs, tend to be relatively more centrally concentrated than blue GCs.

As discussed earlier, the LMXB populations of the innermost regions of elliptical galaxies cannot be studied even with Chandra because of source confusion. However, this is

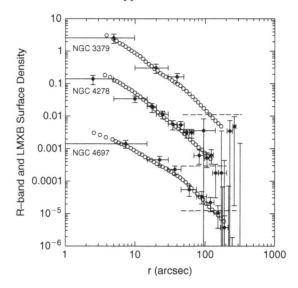

FIGURE 5.12. Radial distributions of the number count of LMXBs in three galaxies compared with the stellar surface brightness distribution (Kim, 2009, private communication; see also Brassington et al., 2008, 2009; Kim et al., 2009). Reproduced by permission of the AAS.

not true for the nearby bulge of M31. In this galaxy, Voss and Gilfanov (2007) report a central excess of LMXBs in the inner bulge (within 150 pc) over what would be expected if the LMXB distribution followed that of the stellar light. They suggest that this result may be indicative of a third LMXB population in addition to the GC and field LMXBs. These sources at the center of M31 may have been formed via dynamical interactions of stars in the inner bulge.

5.6.4 X-ray spectra of LMXBs in elliptical galaxies

The deep observations of NGC 3379 and NGC 4278 offer us the opportunity to study in details the spectra and the spectral variability of the most luminous X-ray sources in these galaxies and to compare these results with the wealth of information that is available for galactic LMXBs. Given their high luminosities, these sources are likely to be BHBs; does their spectral variability follow the patterns seen in galactic BHBs (for a review, see Remillard and McClintock, 2006)? In particular, can we distinguish among disk-dominated (thermal, high/soft) states, power-law-dominated (possibly jet-dominated) hard/low states, and steep power-law (very high; possibly Comptonized) states?

The spectra of galactic BHBs are typically well represented with two spectral components, with varying flux ratios: a power law and an accretion disk (see Remillard and McClintock, 2006). The statistics of LMXB detection in external galaxies does not give us enough counts to attempt meaningful double-component fits in most cases. Fitting the data with single-component models, either power-law or accretion disk (multitemperature blackbody) typically yields comparable fit statistics. Therefore, we cannot discriminate between these models based on goodness-of-fit arguments alone. Is there some way of looking at the parameter spaces and using these results as a discriminator?

To explore how much it is possible to learn from the spectral analysis of these data, with single-component model fits, Brassington et al. (2010) resorted to an extensive set of simulations. They simulated a range of two-component spectra and fitted the results to single-component models. The simulated spectra were power-law and accretion disk combinations, with the power-law photon index covering the range $\Gamma = 1.5$ to 2.3 in steps of 0.2 and the disk temperature covering the range $kT = 0.5$ to 2.25 keV in steps of 0.25 keV. The flux ratios of the two components were varied between 0% and 100% for each

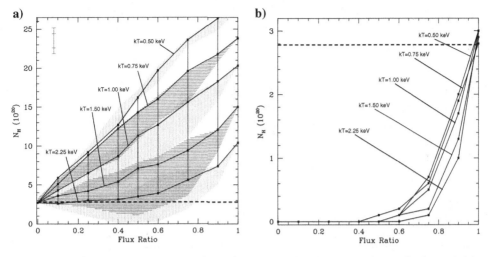

FIGURE 5.13. a) N_H versus the ratio of disk/power-law flux for power-law–only fits; solid lines are for $\Gamma = 1.7$ power laws; the horizontal dashed line is at the line of sight N_H; b) same for disk-only fits (Brassington et al., 2010). Reproduced by permission of the AAS.

combination of power-law photon index Γ and disk temperature kT; in all cases, only a galactic line-of-sight neutral hydrogen absorption column, N_H, was assumed. This N_H is broadly speaking appropriate for LMXBs where extensive stellar winds are not present; moreover, we would not expect significant absorption within the elliptical galaxies, given the general lack of cold ISM in these galaxies. For each combination of parameters, 100 spectra were simulated with 500 and 1,000 detected counts, with noise added.

This exercise demonstrates that if the data are fitted with a single absorbed power-law model, the greater the contribution to the spectrum of a disk component, the larger (above the line-of-sight value) the N_H that is returned by the fit. This effect is significant for the typical range of temperatures of galactic LMXB accretion disks. Conversely, the presence of even a small power-law component in the spectrum yields unphysically small N_H, well below the line-of-sight value, when the spectrum is fitted with a single absorbed disk component (Fig. 5.13). These simulations demonstrate that the value of the best-fit N_H can give indications on the relative prominence of the power-law or thermal emission in the spectra.

With the results of these simulations in mind, evaluating the spectra and spectral variability of the luminous LMXBs of NGC 3379 and NGC 4278 shows that the most luminous sources, and the variable sources when at the highest luminosity in these samples, favor more prominent accretion disk components (Fig. 5.14; Brassington et al., 2010; Fabbiano et al., 2010).

In disk-dominated spectra (selected according to the results of the simulations), flux-related temperature variations of the disk component are seen that may be connected with variations in the accretion rate. Plotting these results in the BHB L_X–T_{in} diagram, where L_X is the source luminosity and T_{in} is the temperature of the inner accretion disk, we find trends consistent with those observed in galactic XRBs (Fig. 5.15). Moreover, the majority of these sources are also consistent with BH masses in the range of those measured in galactic BHBs (see Remillard and McClintock, 2006). These results confirm that these luminous sources are typical LMXB systems.

NGC 4278 yields sizeable samples of both field and GC LMXBs, allowing the comparison of the average spectral properties of selected subsamples (Fabbiano et al., 2010). Using a single power-law model, the jointly fitted X-ray spectrum of the most luminous field LMXBs ($L_X > 1.5 \times 10^{38}$ erg s^{-1}) requires substantially larger N_H than the jointly-fitted spectrum of less luminous field LMXBs, suggesting on average disk-dominated

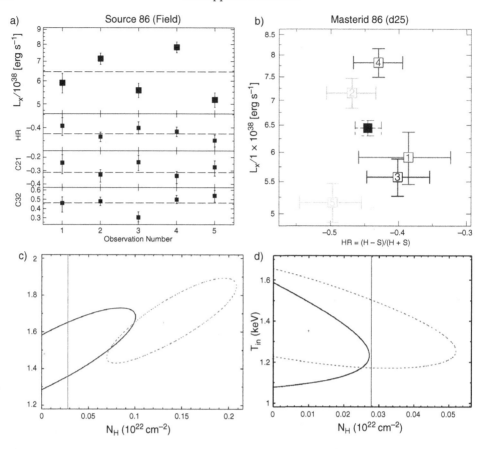

FIGURE 5.14. *a)* Luminosity of source 86 in NGC 3379 (Brassington *et al.*, 2008) for separate Chandra observations; *b)* luminosity versus hardness ratio of this source, suggesting luminosity-correlated spectral variability. *c)* Γ–N_H parameter space for single power-law fit; *d)* kT–N_H parameter space for single–disk fit; in both cases, the contours are 2σ, solid observations $1 + 3$, dashed observations $2 + 4$; the vertical line represents the line-of-sight N_H. In the power-law fit, the lower-luminosity spectra are consistent with galactic line-of-sight N_H, whereas the high-luminosity spectra exclude it; conversely, the disk fit requires unphysically low N_H in the low-luminosity spectrum, whereas the high-luminosity parameter space is consistent with the line-of-sight N_H (from Brassington *et al.*, 2010). Reproduced by permission of the AAS.

emission in these luminous sources. Comparison between average field and GC LMXB spectra show that high-luminosity sources have compatible spectra, suggesting in both cases disk-dominated emission. However, lower-luminosity sources ($L_X < 6 \times 10^{37}$ erg s^{-1}) behave in an unexpected way. While field sources have an average power-law dominated spectrum with line of sight N_H, consistent with intrinsic power-law emission spectra, the GC average spectrum is either an absorbed power-law or a disk-dominated spectrum. This result opens the possibility that the populations of lower luminosity LMXBs in the field and in GCs may be intrinsically different.

We also readdressed the issue of possible spectral differences between LMXBs in red and blue GCs. Harder spectra in blue-GC LMXBs were suggested in NGC 4472 by Maccarone *et al.* (2003) and explained as resulting from absorption by radiatively induced winds in the metal-poor stars of the blue GC. These winds would also speed up the evolution of the binary, effectively reducing the number of active LMXBs in blue GCs, and so explain the relative excess of LMXBs in red clusters (Maccarone *et al.*, 2004). However, no difference in X-ray colors could be seen between red and blue GC LMXBs in a larger sample (Kim *et al.*, 2006b), and a slight – but not significant – effect was

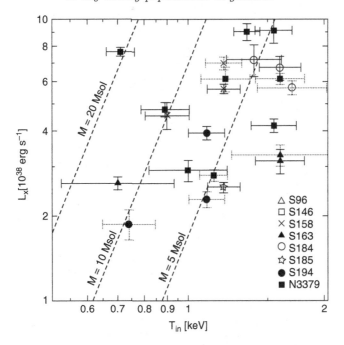

FIGURE 5.15. Luminosity disk temperature diagram for sources in NGC 3379 and NGC 4278 (Fabbiano et al., 2010; see also Brassington et al., 2010). Reproduced by permission of the AAS.

reported in NGC 4697 (Sivakoff et al., 2008). Similarly, in NGC 4278 we find a slight but not statistical significant difference in the sense of a possible excess N_H in blue metal-poor GC spectra. It is worth noting that the alternative explanation proposed for the lack of LMXBs in blue GC by Ivanova (2006; Subsection 5.6.2) does not predict any spectral effect.

5.6.5 The deeper LMXB XLF

As discussed earlier, the high-luminosity ($L_X > 5 \times 10^{37}$ erg s^{-1}) LMXB XLF can be approximated with a steep power law with a high-luminosity break (Kim and Fabbiano, 2004; Gilfanov, 2004). How does the XLF behave at the lower X-ray luminosities more typical of galactic LMXBs? In particular, is there a low luminosity break, as suggested by the galactic LMXB XLF (Grimm et al., 2002)? Moreover, in the high-luminosity regime, no differences were found between field and GC LMXB XLFs (Kim et al., 2006b). This similarity is consistent with (but does not prove) a similar evolutionary path for the two LMXB populations. Does this similarity extend to lower LMXB luminosities?

Observations of the bulge of M31 (Voss and Gilfanov, 2007) and of the nearby radio galaxy Cen A (Voss et al., 2009) suggest a low-luminosity break near 10^{37} erg s^{-1} in the XLF, and a more dramatic lack of GC LMXBs at these lower luminosities (Fig. 5.16).

The deep Chandra observations of the three elliptical galaxies (NGC 3379, NGC 4278, and NGC 4697) confirm this effect (Kim et al., 2009). Kim and colleagues also point out a possible alternative explanation for the low-luminosity break in the field LMXB XLF, consistent with the statistics of the data: not a real break, but a localized excess of sources at that luminosity (bump).

These data, providing source counts and optical identification down to similarly deep detection thresholds, were also used to revisit the question of field evolution versus GC formation of LMXBs. Kim et al. (2009) find that the number of GC-LMXBs increases with the galaxy GC-specific frequency (SN), as expected. However, the number of field

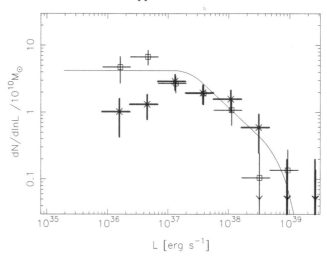

FIGURE 5.16. XLFs of field (squares) and GC (cross) LMXBs in Cen A (from Voss *et al.*, 2009). Reproduced by permission of the AAS.

LMXBs also increases with SN, albeit with a weaker dependence (Fig. 5.17). This result suggests that although there is a fraction of LMXBs that derive from the evolution of native field binaries (see also Irwin, 2005; Juett, 2005; Kim *et al.*, 2006b), the field-LMXB population contains also sources that were formed in GCs.

5.6.6 *Symptoms of rejuvenation?*

Although the stellar population of elliptical galaxies is old, there are galaxies where either morphological or spectral indicators suggest relatively recent interaction and rejuvenation (see Schweizer and Seitzer, 1992). Early reports of rejuvenation in the LMXB population noted asymmetrically distributed and luminous sources in NGC 720

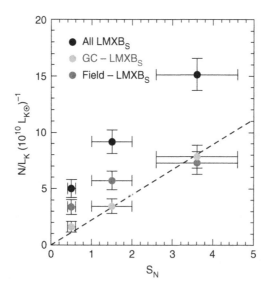

FIGURE 5.17. Number counts (per unit galaxy K-band luminosity) of LMXBs detected in three galaxies versus the GC specific frequency. The number of GC-LMXBs increases with the GC specific frequency, as expected, but there is also a similar – although smaller – effect in the field-LMXBs (Kim *et al.*, 2009). Reproduced by permission of the AAS. (A color version of this figure is available in the online version of the original paper.)

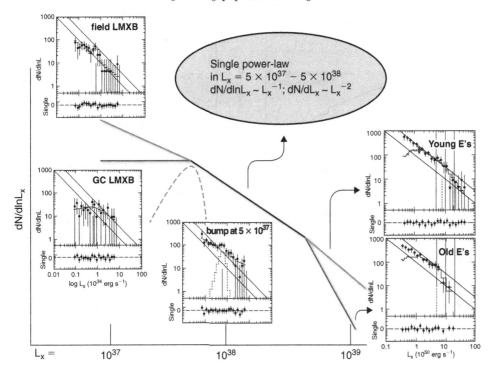

FIGURE 5.18. Schematic summary of the LMXB XLF features (Kim and Fabbiano, 2010). Reproduced by permission of the AAS.

(Jeltema et al., 2003) and NGC 4261 (Zezas et al., 2003). In a recent work, Kim and Fabbiano (2010) have looked at two samples of "rejuvenated" and "old" elliptical galaxies observed with Chandra, finding a difference in the luminosity functions that demonstrates an excess of very luminous sources ($L_X > 5 \times 10^{38}$ erg s^{-1}) in the "rejuvenated" galaxies. Figure 5.18 gives a summary of the LMXB XLF findings.

5.7 GC and field LMXB formation and evolution

Several papers have provided theoretical predictions and interpretations relevant to the understanding of LMXB populations and their evolution. A sample of these papers is listed here:

- Piro and Bildsten (2002) point out that field LMXB populations are likely to be dominated by transient sources with Red Giant donors.
- King (2002) further points out that these transient LMXBs should resemble galactic microquasars.
- Bildsten and Deloye (2004) suggest that the bulk of the luminous sources in elliptical galaxies ($L_X > 10^{37}$ erg s^{-1}) may be ultracompact (UC) 8- to 10-min.–period NS + WD binaries formed in GCs. UCs would be short-lived but continuously generated. This paper predicts an XLF in agreement with the observations.
- Postnov and Kuranov (2005) instead explore the evolution of field LMXBs to explain the XLF.
- Ivanova and Kalogera (2006) model the high end of the XLF with transient BH sources and use these results to constrain the BH mass function.
- Fragos et al. (2008) model the NGC 3379 and NGC 4278 XLFs with field LMXB population synthesis (PS).
- Ivanova et al. (2008) present PS studies of compact NS binaries in GCs.

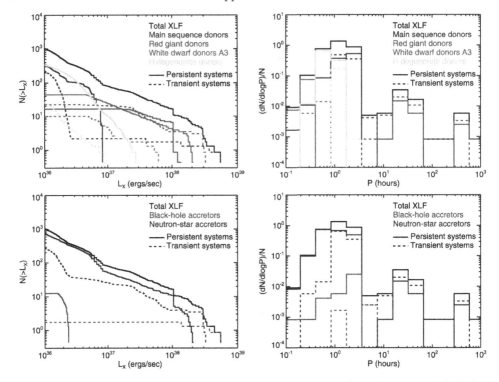

FIGURE 5.19. Synthetic field LMXB XLFs and their components from Fragos et al. (2008). See Fragos et al. (2008) in order to see the graphics color code. Reproduced by permission of the AAS.

- Fragos et al. (2009) apply the PS approach of Fragos et al. (2008) to constrain the properties of X-ray transients in NGC 3379 and NGC 4278.

In the following we discuss some of the work that bears more directly on the interpretation of the observations.

5.7.1 Population synthesis modeling of the XLFs

Now that we have a well-defined "deep" LMXB XLF, what can we learn from it? In particular, can we explain its shape, the low-luminosity flattening, and the relative dearth of low-luminosity GC sources?

Fragos et al. (2008) developed a suite of synthetic LMXB populations, matched to the stellar populations of NGC 3379 and NGC 4278. These models followed the evolution of native field binaries in galaxies of appropriate stellar mass, age and metallicity using the StarTrack code (Belczynski et al., 2002). Model populations were derived for different initial mass functions, binary fractions, magnetic braking laws, and common envelope efficiency. Both persistent and transient systems were included in these synthetic populations. XLFs were derived from these models and compared with the observed LMXB XLF. A set of models was found that fit the slope of the observed XLF (Fig. 5.19). Although this result demonstrated that the PS approach provides models consistent with the data, it does not by any means prove that this particular set of PS models correctly represent the LMXB population. In particular, these models are derived for field LMXBs only, so although they may be relevant for the overall LMXB population of the GC-poor galaxy NGC 3379, they certainly do not represent the GC-LMXBs, which are a large component of the LMXB population of the GC-rich galaxy NGC 4278.

A check to the model is given by comparison with the observed number of transients in these galaxies, since transients are an important component of the field LMXB population

(Piro and Bildsten, 2002; King, 2002). Fragos et al. (2009) used two best-fit PS models from Fragos et al. (2008), one with variable and other with constant transient duty cycle, to predict the number of transients that should be detected in each population, given the Chandra monitoring patterns of NGC 3379 and NGC 4278. The variable duty cycle model predicts a number of transients in agreement with the observations, when the results are normalized by the stellar mass of the galaxy. Interestingly, for NGC 3379, where the GC contribution is small, normalizing the model by the total number of detected LMXBs gives a reasonable agreement with the observations of Brassington et al. (2008). Instead, the latter normalization grossly overpredicts the number of transients in NGC 4278 (Brassington et al., 2009), where a large fraction of the LMXBs must be of dynamical origin, for which the model does not apply.

5.7.2 Explaining the low-luminosity XLF break

In subsection 5.6.4 we discussed the detection of a low-luminosity break (at a few 10^{37} erg s^{-1}) in the LMXB XLF. There are at least three papers in the literature attempting to explain this break. Postnov and Kuranov (2005) suggest that the break may be related to the change of the binary braking mechanism from magnetic stellar winds at high luminosity, to gravitational radiation at low luminosity. Revnivtsev et al. (2008) instead propose that the break may reflect the transition from the high/soft disk-dominated emission, where the luminosity is proportional to the accretion rate, to the low/hard state regime, where for advection-dominated flows, the luminosity follows a steeper function of the accretion rate; assuming an intrinsic relation between the number of detected LMXBs and the accretion rate, they can reproduce the XLF break. A different approach was followed by Kim et al. (2009), who compared the XLF of field LMXBs with the PS model of Fragos et al. (2008), suggesting that the break may be a "bump" related to the contribution of the red giant LMXB population.

Kim et al. (2009) also sought to explain the dearth of GC-LMXBs at $L_X < 5 \times 10^{37}$ erg s^{-1}, relative to the field LMXB population (see Figs. 5.16 and 5.18). It is unlikely that this result is due to underestimating the number of low-luminosity GC-LMXBs because of source confusion, resulting in "spurious" detections of high-L_X LMXBs. Even if some crowding may occur (see Section 5.6.2), the variability pattern of GC sources (Brassington et al., 2008, 2009) suggests that it is unlikely that this is an overwhelming effect. Another possibility is that transient sources may appear prevalently in the field population, thereby spuriously steepening the field-LMXB XLF at the high luminosity (so that the perceived lack of low-luminosity GC sources would instead be an excess of high-luminosity GC sources). However, some instances of transients (or possible transients) have also been reported in GCs (Brassington et al., 2008, 2009). The possibility that the break may be related to the onset of transient behavior in UC binaries (NS+WD) that could be responsible for the GC X-ray emission (Bildsten and Deloye, 2004) has been debated, but not conclusively, because the break in that case should appear at even lower luminosity. It is clear that this is an area where more theoretical work will be needed to explain the observations satisfactorily.

5.8 Ultraluminous X-ray sources (ULXs)

ULXs are sources detected in external galaxies, outside the nuclear regions, with luminosity $L_X > 10^{39}$ up to 10^{41} erg s^{-1}, well exceeding the Eddington luminosity of a NS or of a BH of a few M$_\odot$. These sources have excited a lot of interest, and many papers have been written about them, because they could be the counterparts of BHs too massive to originate from the evolution of normal massive stars. I do not attempt here to give a complete review of all the work on this subject (see Fabbiano, 2006, and references therein for some reviews); I give just a summary of where I think we are in this field.

The reason why ULXs are so exciting is that they could be the elusive IMBHs, filling the gap between the supermassive BHs at the nuclei of galaxies and the stellar-mass BHs in galactic BHBs. These IMBHs could be the remnant of Pop III stars and subsequent mergers early in the history of the universe (Madau and Rees, 2001; Volonteri et al., 2003) or could be the result of core collapse in GCs (Portegies Zwart et al., 1999; Portegies Zwart and McMillan, 2002). But are ULXs really IMBHs? To convincingly prove that we have found IMBHs, we must first demonstrate that other less exotic possibilities are excluded. In particular, could ULXs be just a particular state of otherwise normal XRBs? Several alternative possibilities have been advanced: first of all, stellar black holes could be as massive as 30 $M_\odot$, and possibly more in a low-metallicity environment (e.g., see Belczynski et al., 2004); evolutionary calculations can produce X-ray binaries as luminous as 10^{40} erg s^{-1} (Rappaport et al., 2005); the emission may be enhanced because of moderate or strong beaming (King et al., 2001; Körding et al., 2002) or of super-Eddington disk emission (Begelman, 2002); or, finally, microquasars such as SS433 (see Mirabel and Rodríguez, 1999), seen face-on, would appear as luminous ULXs (Begelman et al., 2006). Basically, alternative explanations can be found for most ULXs.

5.8.1 ULXs and the underlying stellar population

As discussed in Sections 5.5.1 and 5.5.2, ULXs are usually found in actively star-forming galaxies. A recent study, comparing the positions of ULXs with the Sloan survey results, concludes that ULXs tend to be associated with OB associations (Swartz et al., 2009). No association is reported with super-star-clusters, consistent with the results on the Antennae galaxies (Zezas et al., 2002). Interestingly, the colors of the stellar regions associated with ULXs tend to be redder than those of H II regions, consistent with an aging of the stellar population commensurable with the evolution time of a massive X-ray binary.

Although a few sources with ULX luminosities have been detected in elliptical galaxies, these occurrences are rare. Irwin et al. (2004) demonstrated that the number of sources with apparent luminosities in excess of 2×10^{39} erg s^{-1}, detected with Chandra in these galaxies, is consistent with the expected number of background AGNs. However, Liu et al. (2006), using the ROSAT HRI survey, suggest that there may be a number of ULXs associated with the LMXB population. A variable ULX, consistent with a 30 $M_\odot$ BH, was found in the elliptical galaxy NGC 3379, with no optical counterpart that could suggest an interloper (Fabbiano et al., 2006).

5.8.2 Spectra and variability of ULXs: can they constrain the BH mass?

Widespread flux and spectral variability has been observed in ULXs, consistent with the presence of compact accretion-fuelled sources in binary systems (e.g., in M82, Strohmayer and Mushotzky, 2003; Dewangan et al., 2006a; Kaaret et al., 2006; Feng and Kaaret, 2007; in IC342, Kubota et al., 2001; and in the Antennae, Fabbiano et al., 2003b).

Considerable debate has originated from the interpretation of the X-ray spectra of these sources. XMM-Newton has provided the highest signal-to-noise-ratio spectra of ULXs, so the most recent debate has focused on the model fitting of these data. The emission of these sources has typically been modeled with the accretion disk and corona model of BH binaries. If the disk is a typical "thin" Shakura and Sunyaev (1973) disk, then the mass of the BH is related to the temperature of the inner disk (e.g., Makishima et al., 2000), because for a thin accretion disk:
(i)

$$L_{\text{disk}} \sim 4\pi R_{\text{in}}^2 \sigma T_{\text{in}}^4, \qquad (5.1)$$

(ii)
$$R_{in} \sim R_{ISCO} = \alpha GM_{BH} c^{-2}, \qquad (5.2)$$

where R_{ISCO} is the innermost stable orbit around the BH. $\alpha = 6$ Schwarzschild, $\alpha = 1$ Kerr.

It follows that

(iii)
$$M_{BH} \sim L_{disk}^{1/2} T_{in}^{-2}. \qquad (5.3)$$

If $L \sim M_{BH}$ (near Eddington), then

(iv)
$$M_{BH} \sim T_{in}^{-4}. \qquad (5.4)$$

In this approximation, cooler disk temperatures would indicate larger BH masses. In particular, Miller et al. (2003) first reported a cool disk ($T_{in} \sim 150$ eV) in a ULX in NGC 1313, concluding that the BH mass should be in excess of 1000 $M_\odot$, and therefore claiming the detection of an IMBH. A subsequent paper (Miller et al., 2004) suggested that ULXs occupy a different locus than stellar BH binaries in the L_X–disk kT plane, consistent with cool disks in the former, thus reinforcing the IMBH interpretation.

However, the underlying assumption in this work is that the temperature measured from the X-ray spectra is indicative of the emission at the innermost accretion disk radius (R_{in}) and that this radius is the innermost stable circular orbit (ISCO) around the BH. But does $R_{in} = R_{ISCO}$? If this is not true, the disk temperature cannot be used as a proxy for the BH mass. Other work followed, showing that indeed R_{in} may be larger than R_{ISCO} (e.g., Stobbart et al., 2006; Gonçalves and Soria, 2006).

A key observation against the uniqueness of the IMBH hypothesis is the behavior of a galactic BHB with a measured mass of 10 $M_\odot$, XTEJ1550-564. This source has been found to occupy different loci in the L_X–kT diagram at different times, which taken at face values would indicate different BH masses. Clearly this is impossible: Kubota and Done (2004) show that the same effect could be obtained in the presence of a Comptonized corona in the inner part of the accretion disk. In this case the disk temperature is not the temperature at the ISCO, but a larger radius. Therefore, one cannot simply infer masses from the disk temperature, unless it is clear that the emission is purely from a thin disk component (Done and Kubota, 2006). Gonçalves and Soria (2006), Soria et al. (2007), Soria and Kuncic (2008), and Kajava and Poutanen (2009) propose a semiempirical ULX–BH binary analogy, suggesting a spectral evolution linked to increasing accretion rate that in the most extreme high-accretion case leads to the ULX state. This ULX state would be characterized by Comptonized emission and possibly outflows from the central regions of the accretion disk (see King and Pounds, 2003; Begelman et al., 2006). In this state the innermost (undisrupted) radius of the accretion disk (R_{in}) is larger than the R_{ISCO}. Direct spectral evidence of high-luminosity outflows is found in the XMM spectra of the ULX NGC 1365 X-1 (Soria et al., 2007). In this ULX, Eddington limit considerations set an upper limit of 200 $M_\odot$ on the mass of the BH; the BH mass estimate could be further reduced by mild beaming (King et al., 2001) to values consistent with the upper range of stellar BH masses.

Quasi-periodic oscillations (QPOs) of 10 to 100 mHz have been detected in some ULXs (e.g., in M82, Strohmayer and Mushotzky, 2003; Fiorito and Titarchuk, 2004; Dewangan et al., 2006a; in Ho IX X-1, Dewangan et al., 2006b; in NGC 5408 X-1, Strohmayer et al., 2007; Strohmayer and Mushotzky, 2009). Using the established scaling relationships of galactic BHBs between QPO frequency, spectral power-law index, and BH mass (see Vignarca et al., 2003; Shaposhnikov and Titarchuk, 2009), these QPOs have been used to infer IMBHs of $\sim$1,000 solar masses (see Strohmayer and Mushotzky, 2009, and references therein). However, the assumption underlying these measurements is the same as that for the measurements based on the disk temperature, that is, that $R_{in} = R_{ISCO}$. As in the case of the spectral measurements, if this is not true, because of outflows or inner Comptonized regions, the argument is no longer valid (see Soria and Kuncic, 2008).

Winter et al. (2007) take a different approach to constrain ULX masses. Using a sample of ULXs observed with XMM, they find that their spectra suggest hard power-law components ($\Gamma \sim 1.2$–1.8), analogous to those measured in galactic BHBs in low-state (Remillard and McClintock, 2006). Using this analogy, Winter et al. suggest that these ULXs are in "low-state"; if the BHB model is scalable, this would imply IMBH masses. Looking at the disk components of these spectra, Winter et al. identify two regimes of temperature; 1 and 0.1 keV, suggesting that there may be both stellar BH and IMBH ULXs. However, these conclusions assume the presence of stable accretion disks. Soria and Ghosh (2009) reiterate that ULXs are likely to have super-Eddington accretion rates that always impede the formation of stable thin disks, and therefore these spectral measurements cannot pose direct constraints on the BH mass.

5.8.3 Supersoft ULXs

A few ULXs have supersoft (SS) X-ray colors; they also tend to be variable sources. Examples include Antennae X-13 (Fabbiano et al., 2003a), M101 SS ULX (Mukai et al., 2003, 2005; Kong et al., 2004), and NGC 4631 X-1 (Soria and Ghosh, 2009). The SS X-ray spectrum of these ULXs, if fitted with a thin disk model, would imply very cold disks and large (IMBH) masses. However, considering the variability patterns, other solutions are more convincing, including stellar mass BHs and even white dwarf accretors (e.g., Fabbiano et al., 2003a; Soria and Ghosh, 2009). A good example of the controversy surrounding these sources is given by the SS ULX in M101. This source was discovered with Chandra by Mukai et al. (2003). These authors noticed that the spectral variability observed during three data intervals could be explained with a model of an expanding BH photosphere, with constant bolometric luminosity, but diminishing temperature with increasing radius. They inferred a BH mass of 15 to 25 $M_\odot$. Additional Chandra and XMM observations revealed an outburst and a spectrum reminiscent of the low/hard high/soft behavior of galactic BHBs. These data were fitted by Kong et al. (2004) with a composite power-law + disk model, from which they inferred IMBH masses. However, Mukai et al. (2005), using the same data, noticed that the highest ULX luminosity (a few 10^{41} erg s^{-1}) inferred by Kong et al. (2004) occurred when the source had the lowest count rate, and was due to a best-fit model with both low temperature and very high intrinsic absorption. Mukai et al. (2005), adopting a different disk model, with relativistic emission lines seen in galactic sources, demonstrated that these high luminosities are not required. They found instead that the luminosity is correlated with the disk temperature, as would be expected for increasing accretion rates; they estimate the BH mass to be 20 to 40 $M_\odot$, well within the stellar BH mass range.

5.8.4 So what are these ULXs?

The controversy is ongoing, and the number of papers on ULXs increasing. Summarized here are some of the proposed models and the related interpretation on the nature of ULXs:

- **IMBH.** Phenomenological model: power law with soft component
  - Thermal emission from cool optically thick thin accretion disk (BB) + power-law from Comptonized emission from corona (used by Miller et al., 2003)
- **XRB.** Slim disk models (hot thicker disk at super-E accretion rates) (e.g., Ebisawa et al., 2003; Stobbart et al., 2006; Vierdayanti et al., 2006).
- **XRB.** Self-consistent Comptonized models, with disk-corona interaction (Svensson and Zdziarski, 1994; Done and Kubota, 2006; Socrates and Davis, 2006)
  - Dissipates accretion power in corona, lower flux and kT in inner disk
  - Gives reasonable results for very–high–state BH XRBs

- **XRB.** Semiphenomenological models with Comptonization, ionized absorber, outflow (e.g., Gonçalves and Soria, 2006; Soria et al., 2007).
- **IMBH.** QPO BHB analogy (e.g., Shaposhnikov and Titarchuk, 2009)
- **XRB.** SS-ULXs: White dwarf or BH photospheric outflow? (e.g., Fabbiano et al., 2003b, Antennae SS-ULX; Mukai et al., 2005, M101 SS-ULX; Soria and Ghosh, 2009, NGC 4631 X-1).
- **XRB.** SS433 type model: Jet, Eddington columnar outflow, outer disk.

It is clear that there is no single solution, and it could well be that the ULX category includes different types of X-ray sources, perhaps even some IMBHs. How can we solve the issue of the nature of the ULXs? As shown by the examples given in Sections 5.8.3 and 5.8.4, it is unlikely that X-ray spectra or variability measurements can unequivocally demonstrate the presence of IMBHs in these sources: consistency is not demonstration. I believe that an ironclad solution to this problem will come from optical follow-up of these sources. In particular, if the mass function of the binary system can be measured, there will be a definite constraint on the mass of the BH. HST may be able to provide these data for a few nearby ULXs; future large-area and high-resolution optical telescopes will be needed for a more definite answer.

5.9 Acknowledgments

I thank several colleagues for discussions and providing figures for this article and the Power Point presentations: Nicky Brassington, Dong-Woo Kim, Andrea Prestwich, Doug Swartz, Marat Gilfanov, Tassos Fragos, and Roberto Soria. I thank the IAC winter school for their warm hospitality.

REFERENCES

Antoniou, V., Zezas, A., and Hatzidimitriou, D. 2009 (Mar.). A comprehensive study of the link between star-formation history and X-ray source populations in the SMC. Pages 355–360 of: J. T. van Loon and J. M. Oliveira (eds.), *IAU Symposium*. IAU Symposium, vol. 256.

Begelman, M. C. 2002. Super-Eddington fluxes from thin accretion disks? ApJ, **568**(Apr.), L97–L100.

Begelman, M. C., King, A. R., and Pringle, J. E. 2006. The nature of SS433 and the ultraluminous X-ray sources. MNRAS, **370**(July), 399–404.

Belczynski, K., Bulik, T., and Kluźniak, W. 2002. Population synthesis of neutron stars, strange (quark) stars, and black holes. ApJ, **567**(Mar.), L63–L66.

Belczynski, K., Kalogera, V., Zezas, A., and Fabbiano, G. 2004. X-ray binary populations: the luminosity function of NGC 1569. ApJ, **601**(Feb.), L147–L150.

Bildsten, L., and Deloye, C. J. 2004. Ultracompact binaries as bright X-ray sources in elliptical galaxies. ApJ, **607**(June), L119–L122.

Brassington, N. J., Fabbiano, G., Blake, S., Zezas, A., Angelini, L., Davies, R. L., Gallagher, J., Kalogera, V., Kim, D.-W., King, A. R., Kundu, A., Trinchieri, G., and Zepf, S. 2010. The X-ray spectra of the luminous LMXBs in NGC 3379: field and globular cluster sources. ApJ, **725**(Dec.), 1805–1823.

Brassington, N. J., Fabbiano, G., Kim, D.-W., Zezas, A., Zepf, S., Kundu, A., Angelini, L., Davies, R. L., Gallagher, J., Kalogera, V., Fragos, T., King, A. R., Pellegrini, S., and Trinchieri, G. 2008. Deep Chandra monitoring observations of NGC 3379: catalog of source properties. ApJS, **179**(Nov.), 142–165.

Brassington, N. J., Fabbiano, G., Kim, D.-W., Zezas, A., Zepf, S., Kundu, A., Angelini, L., Davies, R. L., Gallagher, J., Kalogera, V., Fragos, T., King, A. R., Pellegrini, S., and Trinchieri, G. 2009. Deep Chandra monitoring observations of NGC 4278: catalog of source properties. ApJS, **181**(Apr.), 605–626.

Bregman, J. N., Irwin, J. A., Seitzer, P., and Flores, M. 2006. Galactic globular clusters with luminous X-ray binaries. ApJ, **640**(Mar.), 282–287.

Clark, G. W. 1975. X-ray binaries in globular clusters. ApJ, **199**(Aug.), L143–L145.

Dewangan, G. C., Griffiths, R. E., and Rao, A. R. 2006b. Quasi-periodic oscillations and strongly Comptonized X-ray emission from Holmberg IX X-1. APJ, **641**(Apr.), L125–L128.

Dewangan, G. C., Titarchuk, L., and Griffiths, R. E. 2006a. Black hole mass of the ultraluminous X-ray source M82 X-1. ApJ, **637**(Jan.), L21–L24.

Done, C., and Kubota, A. 2006. Disc-corona energetics in the very high state of galactic black holes. MNRAS, **371**(Sept.), 1216–1230.

Ebisawa, K., Życki, P., Kubota, A., Mizuno, T., and Watarai, K.-Y. 2003. Accretion disk spectra of ultraluminous X-ray sources in nearby spiral galaxies and galactic superluminal jet sources. ApJ, **597**(Nov.), 780–797.

Fabbiano, G. 1988. The X-ray emission of M81 and its nucleus. ApJ, **325**(Feb.), 544–562.

Fabbiano, G. 1989. X rays from normal galaxies. ARA&A, **27**, 87–138.

Fabbiano, G. 2006. Populations of X-ray sources in galaxies. ARA&A, **44**(Sept.), 323–366.

Fabbiano, G., and Shapley, A. 2002. A multivariate statistical analysis of spiral galaxy luminosities. II. Morphology-dependent multiwavelength emission properties. ApJ, **565**(Feb.), 908–920.

Fabbiano, G., and Trinchieri, G. 1985. A statistical analysis of the Einstein normal galaxy sample. I – Spiral and irregular galaxies. ApJ, **296**(Sept.), 430–457.

Fabbiano, G., and Trinchieri, G. 1987. X-ray observations of spiral galaxies. II – Images and spectral parameters of 13 galaxies. ApJ, **315**(April), 46–67.

Fabbiano, G., Baldi, A., King, A. R., Ponman, T. J., Raymond, J., Read, A., Rots, A., Schweizer, F., and Zezas, A. 2004. X-raying chemical evolution and galaxy formation in the Antennae. ApJ, **605**(Apr.), L21–L24.

Fabbiano, G., Brassington, N. J., Lentati, L., Angelini, L., Davies, R. L., Gallagher, J., Kalogera, V., Kim, D.-W., King, A. R., Kundu, A., Pellegrini, S., Richings, A. J., Trinchieri, G., Zezas, A., and Zepf, S. 2010. Field and globular cluster low-mass X-ray binaries in NGC 4278. ApJ, **725**(Dec.), 1824–1847.

Fabbiano, G., Feigelson, E., and Zamorani, G. 1982. X-ray observations of peculiar galaxies with the Einstein Observatory. ApJ, **256**(May), 397–409.

Fabbiano, G., Gioia, I. M., and Trinchieri, G. 1988. A five-band study of spiral galaxies – X-ray, optical, near- and far-infrared, and radio continuum correlations. ApJ, **324**(Jan.), 749–766.

Fabbiano, G., Kim, D.-W., and Trinchieri, G. 1992. An X-ray catalog and atlas of galaxies. ApJS, **80**(June), 531–644.

Fabbiano, G., Kim, D.-W., Fragos, T., Kalogera, V., King, A. R., Angelini, L., Davies, R. L., Gallagher, J. S., Pellegrini, S., Trinchieri, G., Zepf, S. E., and Zezas, A. 2006. The modulated emission of the ultraluminous X-ray source in NGC 3379. ApJ, **650**(Oct.), 879–884.

Fabbiano, G., King, A. R., Zezas, A., Ponman, T. J., Rots, A., and Schweizer, F. 2003a. A variable ultraluminous supersoft X-ray source in "the Antennae": stellar-mass black hole or white dwarf? ApJ, **591**(July), 843–849.

Fabbiano, G., Trinchieri, G., and MacDonald, A. 1984. X-ray observations of spiral galaxies. I – Integrated properties. ApJ, **284**(Sept.), 65–74.

Fabbiano, G., Trinchieri, G., and van Speybroeck, L. S. 1987. The X-ray spectral properties of the bulge of M31. ApJ, **316**(May), 127–131.

Fabbiano, G., Zezas, A., King, A. R., Ponman, T. J., Rots, A., and Schweizer, F. 2003b. The time-variable ultraluminous X-ray sources of "the Antennae." ApJ, **584**(Feb.), L5–L8.

Feng, H., and Kaaret, P. 2007. Origin of the X-ray quasi-periodic oscillations and identification of a transient ultraluminous X-ray source in M82. ApJ, **669**(Nov.), 106–108.

Fiorito, R., and Titarchuk, L. 2004. Is M82 X-1 really an intermediate-mass black hole? X-ray spectral and timing evidence. ApJ, **614**(Oct.), L113–L116.

Forman, W., Schwarz, J., Jones, C., Liller, W., and Fabian, A. C. 1979. X-ray observations of galaxies in the Virgo cluster. ApJ, **234**(Nov.), L27–L31.

Fragos, T., Kalogera, V., Belczynski, K., Fabbiano, G., Kim, D.-W., Brassington, N. J., Angelini, L., Davies, R. L., Gallagher, J. S., King, A. R., Pellegrini, S., Trinchieri, G., Zepf, S. E., Kundu, A., and Zezas, A. 2008. Models for low-mass X-ray binaries in the elliptical galaxies NGC 3379 and NGC 4278: comparison with observations. ApJ, **683**(Aug.), 346–356.

Fragos, T., Kalogera, V., Willems, B., Belczynski, K., Fabbiano, G., Brassington, N. J., Kim, D.-W., Angelini, L., Davies, R. L., Gallagher, J. S., King, A. R., Pellegrini, S., Trinchieri,

G., Zepf, S. E., and Zezas, A. 2009. Transient low-mass X-ray binary populations in elliptical galaxies NGC 3379 and NGC 4278. ApJ, **702**(Sept.), L143–L147.

Gao, Y., Wang, Q. D., Appleton, P. N., and Lucas, R. A. 2003. Nonnuclear hyper/ultraluminous X-ray Sources in the starbursting Cartwheel ring galaxy. ApJ, **596**(Oct.), L171–L174.

Giacconi, R., Gursky, H., Paolini, F. R., and Rossi, B. B. 1962. Evidence for X-rays from sources outside the solar system. Physical Review Letters, **9**(Dec.), 439–443.

Gilfanov, M. 2004. Low-mass X-ray binaries as a stellar mass indicator for the host galaxy. MNRAS, **349**(Mar.), 146–168.

Gilfanov, M., Grimm, H.-J., and Sunyaev, R. 2004. Statistical properties of the combined emission of a population of discrete sources: astrophysical implications. MNRAS, **351**(July), 1365–1378.

Gonçalves, A. C., and Soria, R. 2006. On the weakness of disc models in bright ULXs. MNRAS, **371**(Sept.), 673–683.

Grimm, H.-J., Gilfanov, M., and Sunyaev, R. 2002. The Milky Way in X-rays for an outside observer. Log(N)–Log(S) and luminosity function of X-ray binaries from RXTE/ASM data. A&A, **391**(Sept.), 923–944.

Grimm, H.-J., Gilfanov, M., and Sunyaev, R. 2003. High-mass X-ray binaries as a star formation rate indicator in distant galaxies. MNRAS, **339**(Mar.), 793–809.

Grimm, H.-J., McDowell, J., Fabbiano, G., and Elvis, M. 2009. An X-ray photometry system. I. Chandra ACIS. ApJ, **690**(Jan.), 128–142.

Grimm, H.-J., McDowell, J., Zezas, A., Kim, D.-W., and Fabbiano, G. 2005. The X-ray binary population in M33. I. Source list and luminosity function. ApJS, **161**(Dec.), 271–303.

Grimm, H.-J., McDowell, J., Zezas, A., Kim, D.-W., and Fabbiano, G. 2007. The X-ray binary population in M33. II. X-ray spectra and variability. ApJS, **173**(Nov.), 70–84.

Grindlay, J. E. 1984. Globular cluster origin of X-ray bursters. Advances in Space Research, **3**, 19–27.

Grindlay, J. E., Heinke, C., Edmonds, P. D., and Murray, S. S. 2001. High-resolution X-ray imaging of a globular cluster core: compact binaries in 47Tuc. Science, **292**(June), 2290–2295.

Irwin, J. A. 2005. The birthplace of low-mass X-ray binaries: field versus globular cluster populations. ApJ, **631**(Sept.), 511–517.

Irwin, J. A., Athey, A. E., and Bregman, J. N. 2003. X-ray spectral properties of low-mass X-ray binaries in nearby galaxies. ApJ, **587**(Apr.), 356–366.

Irwin, J. A., Bregman, J. N., and Athey, A. E. 2004. The lack of very ultraluminous X-ray sources in early-type galaxies. ApJ, **601**(Feb.), L143–L146.

Ivanova, N. 2006. Low-mass X-ray binaries and metallicity dependence: story of failures. ApJ, **636**(Jan.), 979–984.

Ivanova, N., and Kalogera, V. 2006. The brightest point X-ray sources in elliptical galaxies and the mass spectrum of accreting black holes. ApJ, **636**(Jan.), 985–994.

Ivanova, N., Heinke, C. O., Rasio, F. A., Belczynski, K., and Fregeau, J. M. 2008. Formation and evolution of compact binaries in globular clusters – II. Binaries with neutron stars. MNRAS, **386**(May), 553–576.

Jeltema, T. E., Canizares, C. R., Buote, D. A., and Garmire, G. P. 2003. X-ray source population in the elliptical galaxy NGC 720 with Chandra. ApJ, **585**(Mar.), 756–766.

Juett, A. M. 2005. On the nature of X-ray sources in early-type galaxies. ApJ, **621**(Mar.), L25–L28.

Kaaret, P. 2002. A Chandra high resolution camera observation of X-ray point sources in M31. ApJ, **578**(Oct.), 114–125.

Kaaret, P., Simet, M. G., and Lang, C. C. 2006. A 62 day X-ray periodicity and an X-ray flare from the ultraluminous X-ray source in M82. ApJ, **646**(July), 174–183.

Kajava, J. J. E., and Poutanen, J. 2009. Spectral variability of ultraluminous X-ray sources. MNRAS, **398**(Sept.), 1450–1460.

Kilgard, R. E., Cowan, J. J., Garcia, M. R., Kaaret, P., Krauss, M. I., McDowell, J. C., Prestwich, A. H., Primini, F. A., Stockdale, C. J., Trinchieri, G., Ward, M. J., and Zezas, A. 2006. Erratum: "A Chandra survey of nearby spiral galaxies. I. Point source catalogs" ApJS, **159**, 214 (2005); ApJS, **163**(Apr.), 424–425.

Kilgard, R. E., Kaaret, P., Krauss, M. I., Prestwich, A. H., Raley, M. T., and Zezas, A. 2002. A minisurvey of X-ray point sources in starburst and nonstarburst galaxies. ApJ, **573**(July), 138–143.

Kim, D.-W., and Fabbiano, G. 2003. Chandra X-ray observations of NGC 1316 (Fornax A). ApJ, **586**(Apr.), 826–849.

Kim, D.-W., and Fabbiano, G. 2004. X-ray luminosity function and total luminosity of low-mass X-ray binaries in early-type galaxies. ApJ, **611**(Aug.), 846–857.

Kim, D.-W., and Fabbiano, G. 2010. X-ray properties of young early-type galaxies. I. X-ray luminosity function of low-mass X-ray binaries. ApJ, **721**(Oct.), 1523–1530.

Kim, D.-W., Fabbiano, G., and Trinchieri, G. 1992a. The X-ray spectra of galaxies. I – Spectral FITS of individual galaxies and X-ray colors. ApJS, **80**(June), 645–681.

Kim, D.-W., Fabbiano, G., and Trinchieri, G. 1992b. The X-ray spectra of galaxies. II – Average spectral properties and emission mechanisms. ApJ, **393**(July), 134–148.

Kim, D.-W., Fabbiano, G., Brassington, N. J., Fragos, T., Kalogera, V., Zezas, A., Jordán, A., Sivakoff, G. R., Kundu, A., Zepf, S. E., Angelini, L., Davies, R. L., Gallagher, J. S., Juett, A. M., King, A. R., Pellegrini, S., Sarazin, C. L., and Trinchieri, G. 2009. Comparing GC and field LMXBs in elliptical galaxies with Deep Chandra and Hubble data. ApJ, **703**(Sept.), 829–844.

Kim, D.-W., Fabbiano, G., Kalogera, V., King, A. R., Pellegrini, S., Trinchieri, G., Zepf, S. E., Zezas, A., Angelini, L., Davies, R. L., and Gallagher, J. S. 2006a. Probing the low-luminosity X-ray luminosity function in normal elliptical galaxies. ApJ, **652**(Dec.), 1090–1096.

Kim, E., Kim, D.-W., Fabbiano, G., Lee, M. G., Park, H. S., Geisler, D., and Dirsch, B. 2006b. Low-mass X-ray binaries in six elliptical galaxies: connection to globular clusters. ApJ, **647**(Aug.), 276–292.

King, A. R. 2002. The brightest black holes. MNRAS, **335**(Sept.), L13–L16.

King, A. R., and Pounds, K. A. 2003. Black hole winds. MNRAS, **345**(Oct.), 657–659.

King, A. R., Davies, M. B., Ward, M. J., Fabbiano, G., and Elvis, M. 2001. Ultraluminous X-ray sources in external galaxies. ApJ, **552**(May), L109–L112.

Kong, A. K. H., DiStefano, R., Garcia, M. R., and Greiner, J. 2003. Chandra studies of the X-ray point source luminosity functions of M31. ApJ, **585**(Mar.), 298–304.

Kong, A. K. H., Di Stefano, R., and Yuan, F. 2004. Evidence of an intermediate-mass black hole: Chandra and XMM-Newton observations of the ultraluminous supersoft X-ray source in M101 during its 2004 outburst. ApJ, **617**(Dec.), L49–L52.

Körding, E., Falcke, H., and Markoff, S. 2002. Population X: are the super-Eddington X-ray sources beamed jets in microblazars or intermediate mass black holes? A&A, **382**(Jan.), L13–L16.

Kraft, R. P., Kregenow, J. M., Forman, W. R., Jones, C., and Murray, S. S. 2001. Chandra observations of the X-ray point source population in Centaurus A. ApJ, **560**(Oct.), 675–688.

Kubota, A., and Done, C. 2004. The very high state accretion disc structure from the galactic black hole transient XTE J1550-564. MNRAS, **353**(Sept.), 980–990.

Kubota, A., Mizuno, T., Makishima, K., Fukazawa, Y., Kotoku, J., Ohnishi, T., and Tashiro, M. 2001. Discovery of spectral transitions from two ultraluminous compact X-ray sources in IC 342. ApJ, **547**(Feb.), L119–L122.

Kundu, A., Maccarone, T. J., and Zepf, S. E. 2002. The low-mass X-ray binary globular cluster connection in NGC 4472. ApJ, **574**(July), L5–L9.

Kundu, A., Maccarone, T. J., and Zepf, S. E. 2007. Probing the formation of low-mass X-ray binaries in globular clusters and the field. ApJ, **662**(June), 525–543.

Lehmer, B. D., Brandt, W. N., Alexander, D. M., Bauer, F. E., Conselice, C. J., Dickinson, M. E., Giavalisco, M., Grogin, N. A., Koekemoer, A. M., Lee, K.-S., Moustakas, L. A., and Schneider, D. P. 2005. X-ray properties of Lyman break galaxies in the Great Observatories Origins Deep Survey. AJ, **129**(Jan.), 1–8.

Lewin, W. H. G., and van der Klis, M. (eds.) 2006. *Compact Stellar X-Ray Sources*. Cambridge Astrophysics Series No. 39, Cambridge University Press.

Lewin, W. H. G., van Paradijs, J., and van den Heuvel, E. P. J. (eds.). 1995. *X-Ray Binaries*. Cambridge Astrophysics Series No. 26, Cambridge University Press.

Liu, J.-F., Bregman, J. N., and Irwin, J. 2006. Ultraluminous X-ray sources in nearby galaxies from ROSAT HRI observations. II. Statistical properties. ApJ, **642**(May), 171–187.

Maccarone, T. J., Kundu, A., and Zepf, S. E. 2003. The low-mass X-ray binary–globular cluster connection. II. NGC 4472 X-ray source properties and source catalogs. ApJ, **586**(Apr.), 814–825.

Maccarone, T. J., Kundu, A., and Zepf, S. E. 2004. An explanation for metallicity effects on X-ray binary properties. ApJ, **606**(May), 430–435.

Madau, P., and Rees, M. J. 2001. Massive black holes as Population III remnants. ApJ, **551**(Apr.), L27–L30.

Makishima, K., Kubota, A., Mizuno, T., Ohnishi, T., Tashiro, M., Aruga, Y., Asai, K., Dotani, T., Mitsuda, K., Ueda, Y., Uno, S., Yamaoka, K., Ebisawa, K., Kohmura, Y., and Okada, K. 2000. The nature of ultraluminous compact X-ray sources in nearby spiral galaxies. ApJ, **535**(June), 632–643.

Matsushita, K., Makishima, K., Awaki, H., Canizares, C. R., Fabian, A. C., Fukazawa, Y., Loewenstein, M., Matsumoto, H., Mihara, T., Mushotzky, R. F., Ohashi, T., Ricker, G. R., Serlemitsos, P. J., Tsuru, T., Tsusaka, Y., and Yamazaki, T. 1994. Detections of hard X-ray emissions from bright early-type galaxies with ASCA. ApJ, **436**(Nov.), L41–L45.

Miller, J. M., Fabbiano, G., Miller, M. C., and Fabian, A. C. 2003. X-ray spectroscopic evidence for intermediate-mass black holes: cool accretion disks in two ultraluminous X-ray sources. ApJ, **585**(Mar.), L37–L40.

Miller, J. M., Fabian, A. C., and Miller, M. C. 2004. A comparison of intermediate-mass black hole candidate ultraluminous X-ray sources and stellar-mass black holes. ApJ, **614**(Oct.), L117–L120.

Mirabel, I. F., and Rodríguez, L. F. 1999. Sources of relativistic jets in the galaxy. ARA&A, **37**, 409–443.

Mukai, K., Pence, W. D., Snowden, S. L., and Kuntz, K. D. 2003. Chandra observation of luminous and ultraluminous X-ray binaries in M101. ApJ, **582**(Jan.), 184–189.

Mukai, K., Still, M., Corbet, R. H. D., Kuntz, K. D., and Barnard, R. 2005. The X-ray properties of M101 ULX-1 = CXOKM101 J140332.74+542102. ApJ, **634**(Dec.), 1085–1092.

Palumbo, G. G. C., Fabbiano, G., Trinchieri, G., and Fransson, C. 1985. An X-ray study of M51 (NGC 5194) and its companion (NGC 5195). ApJ, **298**(Nov.), 259–267.

Piro, A. L., and Bildsten, L. 2002. Transient X-ray binaries in elliptical galaxies. ApJ, **571**(June), L103–L106.

Pooley, D., Lewin, W. H. G., Anderson, S. F., Baumgardt, H., Filippenko, A. V., Gaensler, B. M., Homer, L., Hut, P., Kaspi, V. M., Makino, J., Margon, B., McMillan, S., Portegies Zwart, S., van der Klis, M., and Verbunt, F. 2003. Dynamical formation of close binary systems in globular clusters. ApJ, **591**(July), L131–L134.

Portegies Zwart, S. F., and McMillan, S. L. W. 2002. The runaway growth of intermediate-mass black holes in dense star clusters. ApJ, **576**(Sept.), 899–907.

Portegies Zwart, S. F., Makino, J., McMillan, S. L. W., and Hut, P. 1999. Star cluster ecology. III. Runaway collisions in young compact star clusters. A&A, **348**(Aug.), 117–126.

Postnov, K. A., and Kuranov, A. G. 2005. The luminosity function of low-mass X-ray binaries in galaxies. Astronomy Letters, **31**(Jan.), 7–14.

Prestwich, A. H., Irwin, J. A., Kilgard, R. E., Krauss, M. I., Zezas, A., Primini, F., Kaaret, P., and Boroson, B. 2003. Classifying X-ray sources in external galaxies from X-ray colors. ApJ, **595**(Oct.), 719–726.

Prestwich, A. H., Kilgard, R. E., Primini, F., McDowell, J. C., and Zezas, A. 2009. The luminosity function of X-ray sources in spiral galaxies. ApJ, **705**(Nov.), 1632–1636.

Ranalli, P., Comastri, A., and Setti, G. 2005. The X-ray luminosity function and number counts of spiral galaxies. A&A, **440**(Sept.), 23–37.

Rappaport, S. A., Podsiadlowski, P., and Pfahl, E. 2005. Stellar-mass black hole binaries as ultraluminous X-ray sources. MNRAS, **356**(Jan.), 401–414.

Remillard, R. A., and McClintock, J. E. 2006. X-ray properties of black-hole binaries. ARA&A, **44**(Sept.), 49–92.

Revnivtsev, M., Lutovinov, A., Churazov, E., Sazonov, S., Gilfanov, M., Grebenev, S., and Sunyaev, R. 2008. Low-mass X-ray binaries in the bulge of the Milky Way. A&A, **491**(Nov.), 209–217.

Sarazin, C. L., Irwin, J. A., and Bregman, J. N. 2000. Resolving the mystery of X-ray–faint elliptical galaxies: Chandra X-ray observations of NGC 4697. ApJ, **544**(Dec.), L101–L105.

Schmitt, J. H. M. M., and Maccacaro, T. 1986. Number-counts slope estimation in the presence of Poisson noise. ApJ, **310**(Nov.), 334–342.

Schweizer, F., and Seitzer, P. 1992. Correlations between UBV colors and fine structure in E and S0 galaxies – a first attempt at dating ancient merger events. AJ, **104**(Sept.), 1039–1067.

Shakura, N. I., and Sunyaev, R. A. 1973. Black holes in binary systems. Observational appearance. A&A, **24**, 337–355.

Shapley, A., Fabbiano, G., and Eskridge, P. B. 2001. A multivariate statistical analysis of spiral galaxy luminosities. I. Data and results. ApJS, **137**(Nov.), 139–199.

Shaposhnikov, N., and Titarchuk, L. 2009. Determination of black hole masses in galactic black hole binaries using scaling of spectral and variability characteristics. ApJ, **699**(July), 453–468.

Shtykovskiy, P., and Gilfanov, M. 2005. High mass X-ray binaries in the LMC: dependence on the stellar population age and the "propeller" effect. A&A, **431**(Feb.), 597–614.

Shtykovskiy, P. E., and Gilfanov, M. R. 2007. High-mass X-ray binaries and the spiral structure of the host galaxy. Astronomy Letters, **33**(May), 299–308.

Sivakoff, G. R., Jordán, A., Juett, A. M., Sarazin, C. L., and Irwin, J. A. 2008. Deep Chandra X-ray observations of low mass X-ray binary candidates in the early-type galaxy NGC 4697. ArXiv e-prints, June.

Sivakoff, G. R., Jordán, A., Sarazin, C. L., Blakeslee, J. P., Côté, P., Ferrarese, L., Juett, A. M., Mei, S., and Peng, E. W. 2007. The low-mass X-ray binary and globular cluster connection in Virgo Cluster early-type galaxies: optical properties. ApJ, **660**(May), 1246–1263.

Sivakoff, G. R., Sarazin, C. L., and Jordán, A. 2005. Luminous X-ray flares from low-mass X-ray binary candidates in the early-type galaxy NGC 4697. ApJ, **624**(May), L17–L20.

Socrates, A., and Davis, S. W. 2006. Ultraluminous X-ray sources powered by radiatively efficient two-phase super-Eddington accretion onto stellar-mass black holes. ApJ, **651**(Nov.), 1049–1058.

Soria, R., and Ghosh, K. K. 2009. Different types of Ultraluminous X-ray sources in NGC 4631. ApJ, **696**(May), 287–297.

Soria, R., and Kuncic, Z. 2008. Black hole mass estimates from soft X-ray spectra. Advances in Space Research, **42**(Aug.), 517–522.

Soria, R., and Wu, K. 2003. Properties of discrete X-ray sources in the starburst spiral galaxy M 83. A&A, **410**(Oct.), 53–74.

Soria, R., Baldi, A., Risaliti, G., Fabbiano, G., King, A., La Parola, V., and Zezas, A. 2007. New flaring of an ultraluminous X-ray source in NGC1365. MNRAS, **379**(Aug.), 1313–1324.

Stobbart, A.-M., Roberts, T. P., and Wilms, J. 2006. XMM-Newton observations of the brightest ultraluminous X-ray sources. MNRAS, **368**(May), 397–413.

Strohmayer, T. E., and Mushotzky, R. F. 2003. Discovery of X-ray quasi-periodic oscillations from an ultraluminous X-ray source in M82: evidence against beaming. ApJ, **586**(Mar.), L61–L64.

Strohmayer, T. E., and Mushotzky, R. F. 2009. Evidence for an intermediate-mass black hole in NGC 5408 X-1. ApJ, **703**(Oct.), 1386–1393.

Strohmayer, T. E., Mushotzky, R. F., Winter, L., Soria, R., Uttley, P., and Cropper, M. 2007. Quasi-periodic variability in NGC 5408 X-1. ApJ, **660**(May), 580–586.

Svensson, R., and Zdziarski, A. A. 1994. Black hole accretion disks with coronae. ApJ, **436**(Dec.), 599–606.

Swartz, D. A., Ghosh, K. K., McCollough, M. L., Pannuti, T. G., Tennant, A. F., and Wu, K. 2003. Chandra X-ray observations of the spiral galaxy M81. ApJS, **144**(Feb.), 213–242.

Swartz, D. A., Tennant, A. F., and Soria, R. 2009. Ultraluminous X-ray source correlations with star-forming regions. ApJ, **703**(Sept.), 159–168.

Tennant, A. F., Wu, K., Ghosh, K. K., Kolodziejczak, J. J., and Swartz, D. A. 2001. Properties of the Chandra sources in M81. ApJ, **549**(Mar.), L43–L46.

Trinchieri, G., and Fabbiano, G. 1985. A statistical analysis of the Einstein Normal Galaxy Sample – part two – Elliptical and S0 galaxies. ApJ, **296**(Sept.), 447+.

Tyler, K., Quillen, A. C., LaPage, A., and Rieke, G. H. 2004. Diffuse X-ray emission in spiral galaxies. ApJ, **610**(July), 213–225.

van den Bergh, S., Morbey, C., and Pazder, J. 1991. Diameters of Galactic globular clusters. ApJ, **375**(July), 594–599.

Verbunt, F., and Lewin, W. H. G. 2006. *Globular cluster X-ray sources*. Pages 341–379. Cambridge Astrophysics Series No. 39, Cambridge University Press.

Verbunt, F., and van den Heuvel, E. P. J. 1995. Formation and evolution of neutron stars and black holes in binaries. Pages 457–494 of: W. H. G. Lewin, J. van Paradijs, and E. P. J. van den Heuvel (eds.), *X-Ray Binaries*. Cambridge Astrophysics Series No. 26, Cambridge University Press.

Vierdayanti, K., Mineshige, S., Ebisawa, K., and Kawaguchi, T. 2006. Do ultraluminous X-ray sources really contain intermediate-mass black holes? PASJ, **58**(Oct.), 915–923.

Vignarca, F., Migliari, S., Belloni, T., Psaltis, D., and van der Klis, M. 2003. Tracing the power-law component in the energy spectrum of black hole candidates as a function of the QPO frequency. A&A, **397**(Jan.), 729–738.

Volonteri, M., Haardt, F., and Madau, P. 2003. The assembly and merging history of supermassive black holes in hierarchical models of galaxy formation. ApJ, **582**(Jan.), 559–573.

Voss, R., and Gilfanov, M. 2007. The dynamical formation of LMXBs in dense stellar environments: globular clusters and the inner bulge of M31. MNRAS, **380**(Oct.), 1685–1702.

Voss, R., Gilfanov, M., Sivakoff, G. R., Kraft, R. P., Jordán, A., Raychaudhury, S., Birkinshaw, M., Brassington, N. J., Croston, J. H., Evans, D. A., Forman, W. R., Hardcastle, M. J., Harris, W. E., Jones, C., Juett, A. M., Murray, S. S., Sarazin, C. L., Woodley, K. A., and Worrall, D. M. 2009. Luminosity functions of LMXBs in Centaurus a: globular clusters versus the field. ApJ, **701**(Aug.), 471–480.

Weisskopf, M. C., Tananbaum, H. D., Van Speybroeck, L. P., and O'Dell, S. L. 2000 (July). Chandra X-ray Observatory (CXO): overview. Pages 2–16 of: J. E. Truemper and B. Aschenbach (eds.), *Society of Photo-Optical Instrumentation Engineers (SPIE) Conference Series*. Society of Photo-Optical Instrumentation Engineers (SPIE) Conference Series, vol. 4012.

Winter, L. M., Mushotzky, R. F., and Reynolds, C. S. 2007. Elemental abundances of nearby galaxies through high signal-to-noise ratio XMM-Newton observations of ultraluminous X-ray sources. ApJ, **655**(Jan.), 163–178.

Wolter, A., and Trinchieri, G. 2004. A thorough study of the intriguing X-ray emission from the Cartwheel ring. A&A, **426**(Nov.), 787–796.

Wu, K. 2001. Populations of X-ray binaries and the dynamical history of their host galaxies. PASA, **18**, 443–450.

Zezas, A., and Fabbiano, G. 2002. Chandra observations of "the Antennae" galaxies (NGC 4038/4039). IV. The X-ray source luminosity function and the nature of ultraluminous X-ray sources. ApJ, **577**(Oct.), 726–737.

Zezas, A., Fabbiano, G., Baldi, A., Schweizer, F., King, A. R., Rots, A. H., and Ponman, T. J. 2007. Chandra monitoring observations of the Antennae galaxies. II. X-ray luminosity functions. ApJ, **661**(May), 135–148.

Zezas, A., Fabbiano, G., Rots, A. H., and Murray, S. S. 2002. Chandra observations of "the Antennae" galaxies (NGC 4038/4039). III. X-ray properties and multiwavelength associations of the X-ray source population. ApJ, **577**(Oct.), 710–725.

Zezas, A., Hernquist, L., Fabbiano, G., and Miller, J. 2003. NGC 4261 and NGC 4697: rejuvenated elliptical galaxies. ApJ, **599**(Dec.), L73–L77.

6. Observational characteristics of accretion onto black holes I

CHRISTINE DONE

Abstract

These notes resulted from a series of lectures at the IAC winter school. They are designed to help students, especially those just starting in subject, to get hold of the fundamental tools used to study accretion powered sources. As such, the references give a place to start reading, rather than representing a complete survey of work done in the field.

I outline Compton scattering and blackbody radiation as the two predominant radiation mechanisms for accreting black holes, producing the hard X-ray tail and disk spectral components, respectively. The interaction of this radiation with matter can result in photoelectric absorption and/or reflection. While the basic processes can be found in any textbook, here I focus on how these can be used as a toolkit to interpret the spectra and variability of black-hole binaries (hereafter BHB) and active galactic nuclei (AGN). I also discuss how to use these to physically interpret real data using the publicly available XSPEC spectral fitting package (Arnaud, 1996), and how this has led to current models (and controversies) of the accretion flow in both BHB and AGN.

6.1 Fundamentals of accretion flows: observation and theory

6.1.1 Plotting spectra

Spectra can often be (roughly) represented as a power law. This can be written as a differential photon number density (photons per second per square cm per energy band) as $N(E) = N_0 E^{-\Gamma}$, where Γ is photon index. The energy flux is then simply $F(E) = EN(E) = N_0 E^{-(\Gamma-1)} = N_0 E^{-\alpha}$, where $\alpha = \Gamma - 1$ is energy index.

Power law spectra are broadband: that is, the emission spans many decades in energy. Thus, in general we plot logarithmically, in $\log E$, with $d\log E$ rather than dE as the constant. The number of photons per bin is $N(E)dE = N(E)EdE/E = EN(E)d\log E = F(E)d\log E$. Therefore, somewhat counterintuitively, plotting $F(E)$ on a logarithmic energy scale shows the *number* of photons rather than flux.

Similarly, energy per bin is $F(E)dE = F(E)EdE/E = EF(E)d\log E$. Thus, to see the energy at which the source luminosity peaks on a logarithmic frequency scale means we have to plot $\log EF(E)$ $(= \nu F(\nu))$ versus $\log E$. In these units, hard spectra have $\Gamma < 2$ and so peak at high energies. Soft spectra have $\Gamma > 2$ and peak at low energies. Flat spectra with $\Gamma = 2$ means equal power per decade. These are illustrated in Figure 6.1.

One other very common way for spectra to be plotted is in counts per second per cm^2, $C(E)$. This is not the same as $N(E)$; the counts refer to number of photons detected, not emitted. Such spectra show more about the detector than the intrinsic spectrum, as these are convolved with the detector response, giving $C(E) = \int N(E_0)R(E, E_0)dE_0$, where $R(E, E_0)$ is the detector response, that is, the probability that a photon of input energy E_0 is detected at energy E. Instrument responses for X-rays are generally complex, so the spectra are generally analyzed in counts space, by convolving models of the intrinsic spectrum through the detector response and minimizing the difference between the predicted and detected counts to derive the best fit model parameters. Although such "counts spectra" constitute the basic observational data, they do not give a great deal of physical insight. Deconvolving these data using a model to plot them in $\nu f(\nu)$ space is strongly recommended that is, in XSPEC using the command IPLOT EEUF.

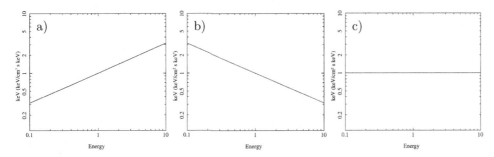

FIGURE 6.1. The $\nu f(\nu)$ spectra of a power law with photon index Γ of a) 1.5, b) 2.5, and c) 2.0. This representation makes it clear that the power output peaks at high energies for hard spectra ($\Gamma < 2$), whereas it peaks at low energies for soft spectra ($\Gamma > 2$).

6.1.2 Plotting variability

Plotting variability is analogous to plotting spectra. A light curve, $I(t)$, spanning time T, with points every Δt can be decomposed into a sum of sinusoids:

$$I(t) = I_0 + \Sigma_{i=1}^{N} A_i \sin(2\pi\nu_i t + \phi_i), \quad (6.1)$$

where I_0 is the average flux over that time scale, $\nu_i = i/T$ with $i = 1, 2, \ldots N$ and $N = T/(2\Delta t)$. This is more useful when normalized to the average, giving the fractional change in intensity $I(t)/I_0 = 1 + \Sigma(A_i/I_0)\sin(2\pi\nu_i t + \phi_i)$. The power spectrum $P(\nu)$ is $(A_i/I_0)^2$ versus ν_i, and the integral $\int P(\nu)d\nu = (\sigma/I_0)^2$ that is, the squared total r.m.s. variability of the light curve.

Similar to the energy spectra, the power spectrum is generally broadband and so is plotted logarithmically. Then the total variability power in a bin is $P(\nu)d\nu = P(\nu)\nu d\nu/\nu = \nu P(\nu)d\log\nu$. Thus, similarly to spectra, a peak in $\nu P(\nu)$ versus $\log\nu$ shows the frequency at which the variability power peaks, so this is the more physical way to plot power spectra.

The data often show power-law power spectra, with $P(\nu) \propto \nu^0$ on long time scales breaking to ν^{-1} at a low-frequency break ν_b and then breaking again to ν^{-2} at a high-frequency break ν_h. This is termed band-limited noise or flat-top noise, where the "flat top" has $P(\nu) \propto \nu^{-1}$, as this has equal variability power per decade.

Figure 6.2 illustrates this for data from a low/hard state (see Section 6.1.4) in Cyg X-1. The light curves are shown over different time scales, T. These are all normalized to their mean, so it is easy to see from the light curves that the fractional variability is very low on time scales shorter than 0.01 s. At longer time scales, the fractional variability increases, then remains constant for 1 to 10 s, and then drops again.

6.1.3 Spectra and variability of the Shakura-Sunyaev disk

The underlying physics of a Shakura-Sunyaev accretion disk can be illustrated in a very simple derivation just conserving energy (rather than the proper derivation, which conserves energy and angular momentum).

A mass accretion rate $\dot{M}$ spiraling inward from R to $R - dR$ liberates potential energy at a rate $dE/dt = L_{\rm pot} = (GM\dot{M}/R^2) \times dR$. The virial theorem says that only half of this can be radiated, so $dL_{\rm rad} = GM\dot{M}dR/(2R^2)$. If this thermalizes to a blackbody, then $dL = dA\sigma_{SB}T^4$, where σ_{SB} is the Stefan-Boltzmann constant and area of the annulus $dA = 2 \times 2\pi R \times dR$ (where the factor 2 comes from the fact that there is a top and bottom face of the ring). Then the luminosity from the annulus $dL_{\rm rad} = GM\dot{M}dR/(2R^2) = 4\pi R \times dR\sigma_{SB}T^4$ or $\sigma_{SB}T^4(R) = GM\dot{M}/8\pi R^3$. This is only out by a factor $3(1 - (R_{\rm in}/R)^{1/2})$, which comes from a full analysis including angular momentum (see Chapter 1 by H. Spruit in these proceedings).

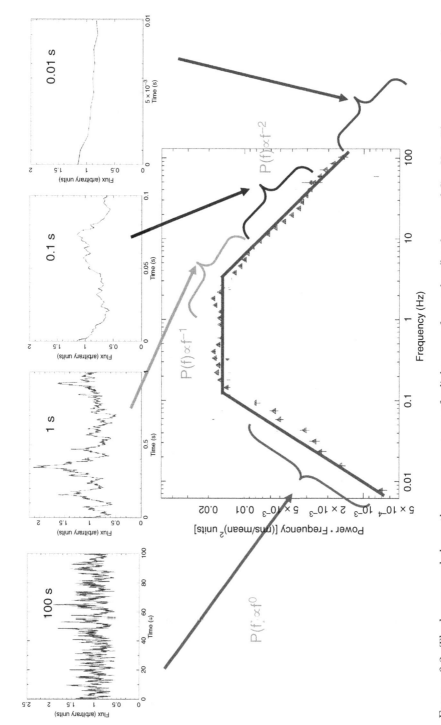

FIGURE 6.2. The lower panel shows the power spectrum of a light curve from a low/hard state of Cyg X-1. The upper panels show segments of the lght curve spanning time scales of 100 s, 1 s, 0.1 s, and 0.01 s, respectively. These are each normalized to their mean, so it is easy to see that the fractional variability is largest in the 1-s light curve and that this is what the power spectrum measures. Figure courtesy of P. Uttley.

Thus, the spectrum from a disk is a sum of blackbody components, with increasing temperature and luminosity emitted from a decreasing area as the radius decreases. The peak luminosity and temperature then comes from R_{in} (modulo the corrections for the inner boundary condition). Using the very approximate treatment just given, the total luminosity of the disc $L_{disc} = GM\dot{M}/(2R_{in})$ so substituting for $GM\dot{M}$ gives $\sigma_{SB}T^4(R) = R_{in}L_{disc}/4\pi R^3 \propto L_{disc} \times (R_{in}/R) \times 1/(4\pi R^2)$. So $\sigma_{SB}T^4_{max} = \sigma_{SB}T^4(R_{in}) = L_{disc}/(4\pi R_{in}^2)$, giving an observational constraint on R_{in} if we can measure T_{max} and L_{disc}. This is important because R_{in} is set by general relativity at the last stable orbit around the black hole, which is itself dependent on spin. Angular momentum, J, is typically a mass, times a velocity, times a size scale. The smallest size scale for a black hole is that of the event horizon, which is always larger than $R_g = GM/c^2$, and the fastest velocity is the speed of light. Thus $|J| < McGM/c^2$, or spin-per-unit mass, $|J|/M = a_*GM/c$, where $a_* \leq 1$. The last stable orbit is at 6 R_g for a zero spin ($a_* = 0$: Schwarzschild) black hole, decreasing to 1 R_g for a maximally rotating Kerr black hole for the disk corotating with the black hole spin ($a_* = 1$), or 9 R_g for a counterrotating disk ($a_* = -1$). Thus, to convert the observed emission area to spin, we need to know the mass of the black hole so we can put the observed inner radius into gravitational radii $r_{in} = R_{in}/R_g$, giving us a way to observationally measure black hole spin.

In XSPEC, a commonly used model for the disc is DISKBB. This assumes that $T^4 \propto r^{-3}$, that is, has no inner boundary condition. It is adequate to fit the high-energy part of the disk spectrum, that is, the peak and Wien tail, but the derived normalization needs to be corrected for the lack of boundary condition. It also assumes that each radius emits as a true blackbody, which is only true if the disk is effectively optically thick to absorption at all frequencies. Free-free (continuum) absorption drops as a function of frequency, so the highest energy photons from each radii are unlikely to thermalize. This forms instead a modified (or diluted) blackbody, with effective temperature that is a factor f_{col} (termed a color temperature correction) higher than for complete thermalization. The full disk spectrum is then a sum of these modified blackbodies, but this can likewise be approximately described by a single color temperature correction to a "sum of blackbodies" disk spectrum (Shimura and Takahara, 1995), giving rise to a further correction to the DISKBB normalization. The final factor is that the emission from each radius is smeared out by the combination of special and general relativistic effects that arise from the rapid rotation of the emitting material in a strong gravitational field (Cunningham, 1975; see Section 6.4.4). Again, these corrections can be applied to the DISKBB model (e.g., Kubota et al., 2001; Gierliński and Done, 2004a) but are only easily available as tabulated values for spin 0 and 0.998 in Zhang et al. (1997). Hence, a better approach is to use KERRBB, which incorporates the stress-free boundary condition and relativistic smearing for any spin (Li et al., 2005) for a given color temperature correction factor. An even better approach is to use BHSPEC, which calculates the intrinsic spectrum from each radius using full radiative transfer through the disk atmosphere, including partially ionized metal opacities, rather than assuming a color temperature – corrected blackbody form (Davis et al., 2005). This imprints atomic features onto the emission from each radius, distorting the spectrum from the smooth continuum as produced by KERRBB (Done and Davis, 2008; Kubota et al., 2010).

To zeroth order, the emitted spectrum does not require any assumptions about the nature of the viscosity, parameterized by α by Shakura and Sunyaev (1973). However, variability is dependent on this, as variability in the emitted spectrum requires that the mass accretion rate through the disk change. Material can only fall in if its angular momentum is transported outward via "viscous" stresses, now known to be due to the magneto rotational instability (MRI: see Chapter 8 by J. Hawley in these proceedings). The viscous time scale $t_{visc} \approx \alpha^{-1}(H/R)^{-2}t_{dyn}$, where H is the vertical scale height of the disk and $t_{dyn} = 2\pi R_g(r^{3/2} + a_*)/c$ is the dynamical (orbital) time scale, which is ~ 5 ms for a Schwarzschild black hole of 10 $M_\odot$.

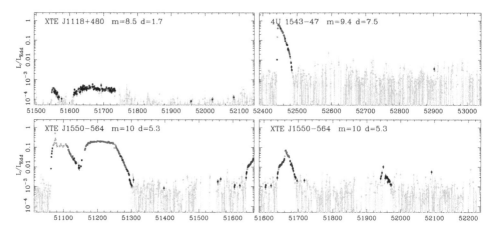

FIGURE 6.3. The long-term All Sky Monitor RXTE light curves of some transient BHB (DGK07).

Modeling the observed variability of the disk gives an estimate for $\alpha = 0.1$ (King et al., 2007), though current simulations of the MRI give stresses that are an order of magnitude lower than this (see Chapter 8 by J. Hawley in these proceedings).

The disk models give a geometrically thin solution $H/R \sim 0.01$, so the very fastest variability from changes in mass accretion rate at the innermost edge is 100,000 times longer than the dynamical time scale. Thus, accretion discs in BHB should only vary on time scales longer than a few hundred seconds.

6.1.4 Observed spectra and power spectra of black-hole binaries
Disk dominated states

The mass accretion rate through the entire disc in BHB can vary over weeks, months, or years, triggered by the disk instability (see Chapter 4 by R. Hynes in these proceedings, or Done et al., 2007, hereafter DGK07). This means that a single object (with constant distance, inclination, and spin) can map out how the spectrum changes as a function of luminosity as shown in Figure 6.3.

Figure 6.4a shows that these can indeed show spectra that look very like the simple accretion disk models described previously (Section 6.1.3). This disk emission also shows very little rapid variability on time scales of less than a few hundred seconds, as expected (Churazov et al., 2001). Collating disk spectra on longer time scales at different $\dot{m}$ gives $L_{\text{disc}} \propto T_{\max}^4$ (Fig. 6.4b) clear observational evidence for a constant–size-scale inner radius to the disk despite the large change in mass accretion rate. This is exactly as predicted for the behavior at the last stable orbit and is a key test of Einstein's gravity in the strong field limit. Indeed, given how close the last stable orbit is to the event horizon ($r = 6$ compared to the horizon at $r = 2$ for a Schwarzschild black hole), this represents almost the strongest gravitational field we could ever observe.

Folding in all the required corrections (see earlier discussion) allows us to measure this fixed size scale and translate it into a measure of spin if there are good system parameter estimates (see Chapter 4 by R. Hynes in these proceedings). To date, all objects that show convincing $L \propto T^4$ tracks give moderate spins, as opposed to extreme Kerr (Davis et al., 2006; Middleton et al., 2006; Shafee et al., 2006; Gou et al., 2010). This is as expected from current (probably quite uncertain) supernova collapse models (see Chapter 2 by P. Podsiadlowski in these proceedings; also Gammie et al., 2004; Kolehmainen and Done, 2010) and BHB in low-mass X-ray binaries should have a spin distribution that accurately reflects their birth spin. This is because the black hole mass must approximately

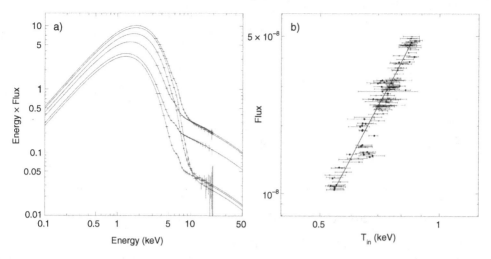

FIGURE 6.4. a) A selection of disk dominated spectra from the transient BHB GX339-4. b) The disc flux versus its temperature derived from fits to all the disk dominated data. The line shows the $L_{\text{disc}} \propto T_{\text{max}}^4$ relation expected from a constant size scale for the inner radius of the disk (Kolehmainen and Done, 2010).

double in order to significantly change the spin, which is not possible in an LMXB even if the black hole (which must be more than 3 $M_\odot$) completely accretes the entire low-mass ($\lesssim 1$ $M_\odot$) companion star (King and Kolb, 1999). However, high spins are derived for two objects that do not have good $L \propto T^4$ tracks, namely GRS1915+105 (McClintock et al., 2006; but see Middleton et al., 2006) and LMC X-1 (Gou et al., 2009).

The high-energy tail

However, even the most disk dominated (also termed high/soft state) spectra also have a tail extending out to higher energies with $\Gamma \sim 2$ (see Fig. 6.4a). This tail carries only a very small fraction of the power in the data discussed above, but it can be much stronger and even dominate the energetics. Where the tail coexists with a strong disk component (very high/intermediate or steep power law state), it has $\Gamma > 2$, so the spectra are soft. But where the disk is weak, the tail can dominate the total energy, with $\Gamma < 2$ so the spectra are hard (low/hard state), peaking above 100 keV. All these very different spectra are shown in Fig. 6.5a. The combination of these two plots shows that typically, hard spectra are only seen at low fractions of Eddington, whereas disk or disk-plus-soft-tail spectra are seen only at high fractions of Eddington. This is actually very surprising, as the classic Shakura-Sunyaev disk is unstable to a radiation pressure instability above $\sim 0.05 L_{\text{Edd}}$. Thus, we might expect that the disk is disrupted into some other type of flow at high fractions of Eddington, and that we see clean disk spectra only at low fractions of Eddington. This is entirely the reverse of what is seen (as first recognized by Nowak, 1995; see also Gierliński and Done, 2004a).

The tail has more rapid variability than the disc component (Churazov et al., 2001; Gierliński and Zdziarski, 2005, though the disk variability is enhanced on time scales longer than 1 in the low/hard state; Wilkinson and Uttley, 2009), typically with large fluctuations on time scales from 100 to 0.05 s. The dynamical (Keplerian orbit) time scale for a 10 $M_\odot$ black hole at $6R_g$ is 0.005 s, so these time scales are at least 10 × longer than Keplerian. If the variability is driven by changes in mass accretion rate, then the expected time scale is a factor $\alpha(H/R)^2$ longer, so this requires $H/R \sim 1$, that is, a geometrically thick flow.

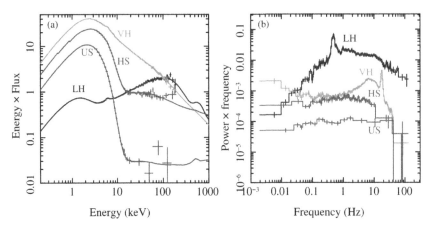

FIGURE 6.5. a) A selection of spectra from the transient BHB GRO J1655-40. Soft spectra are typically seen at high fractions of Eddington, whereas hard spectra are typically seen at low Eddington fractions. Spectra with both strong disk and a strong, soft tail are seen at intermediate luminosities and at the very highest luminosities (DGK07). b) The corresponding variability power spectra from 2 to 60 keV. The variability power on these short time scales drops dramatically when the disk contributes substantially to the spectrum, as the disk is very stable on these time scales, so it dilutes the variability of the tail.

6.1.5 Theory of geometrically thick flows: ADAFs

Pressure forces must be important in a geometrically thick flow, by contrast to a Shakura-Sunyaev disk. If these are from gas pressure, then the flow must be hot, with protons close to the virial temperature of 10^{12} K. The data require that the electrons be hot in the low/hard state in order to produce the high-energy tail via Compton scattering (see Section 6.2), but generally only out to 100 keV, that is, 10^9 K. These two very different requirements on the temperature can both be satisfied in a two-temperature plasma, where the ion temperature is much hotter than the electron temperature. This happens in plasmas that are not very dense, as the electrons can radiate much more efficiently than the protons, so the electrons lose energy much more rapidly than the protons. Even if the electrons and protons are heated at the same rate, the proton temperature will be hotter than that of the electrons if the protons and electrons do not interact enough to equilibrate their temperatures. This gives a further requirement that the optical depth of the flow needs to be low.

This leads to the idea of a hot, geometrically thick, optically thin flow replacing the cool, geometrically thin, optically thick Shakura-Sunyaev disk. The exact structure of this flow is not well known at present; advection-dominated accretion flows (ADAFs) are the most well known, but there can also be additional effects from convection, winds, and the jet (DGK07). Ultimately, magnetohydrodynamic simulations in full general relativity including radiative cooling are probably needed to fully explore the complex properties of these flows (J. Hawley, Chapter 8 in these proceedings).

When ADAFs (Narayan and Yi, 1995) were first proposed, a key issue was how to produce such flow from an originally geometrically thin cool disk (to the extent that one theoretician said "turbulence generated by theorists waving their hands"). However, there is now a mechanism to do this via an evaporation instability. If the cool disk is in thermal contact with the hot flow, then there is heat conduction between the two, which can lead to either evaporation of the disk into the hot flow or condensation of the hot flow onto the disk. At low mass accretion rates, evaporation predominates in the inner disk, giving rise to a radially truncated disk/hot inner flow geometry (Liu et al., 1999; Różańska and Czerny, 2000; Mayer and Pringle, 2007). This is exactly the geometry required in the phenomenological truncated disk/hot inner flow models described in the next section.

The hot flow can exist only if the electrons and protons do not interact often enough to thermalize their energy. This depends on optical depth, and the flow collapses when $\tau \gtrsim 2$ to 3, which occurs at $\dot{m} = 1.3\,\alpha^2 \sim 0.01$ for $\alpha = 0.1$ (Esin et al., 1997). This is very close to the luminosity of the transition from soft to hard spectra seen on the outburst decline (Maccarone, 2003), though more complicated behavior is seen on the hard-to-soft transition on the rise (hysteresis: Miyamoto et al., 1995; Yu and Yan, 2009), plausibly due to the rapid accretion rate changes pushing the system into nonequilibriuim states (Gladstone et al., 2007).

But there are issues. I stress again that the flow should be more complex than an ADAF, as these do not include other pieces of physics that are known to be present (convection, winds, the jet, changing advected fraction with radius: see, e.g., DGK07). Indeed, standard ADAF solutions are somewhat too optically thin and hot to match the observed Compton spectra (Malzac and Belmont, 2009), and this is especially an issue at the lowest L/L_{Edd}, where the observed X-ray spectra are far too smooth to be produced by the predicted very optically thin flow (Pszota et al., 2008).

6.1.6 Truncated disk/hot inner flow models

These two very different types of solution for the accretion flow can be put together into the truncated disk/hot inner flow model. At high fractions of Eddington, we typically see strong evidence for a disk down to the last stable orbit (see Fig. 6.4). At low fractions of Eddington, we can have one of these hot, optically thin geometrically thick flows. The only way to go from one to the other is for the disk to move inward. As it penetrates further into the flow, more seed photons from the disk are intercepted by the flow, so the Compton spectrum softens (see Section 6.2).

The MRI turbulence in the hot inner flow generates the rapid variability at each radius, modulating the mass accretion rate to the next radius. Thus, the total variability is the product (not the sum!) of variability from all radii within the hot flow (Lyubarskii, 1997; Kotov et al., 2001; Arévalo and Uttley, 2006). This rather naturally gives rise to a key observational requirement that the r.m.s. variability σ (see Section 6.1.2) in the light curve, as measured over a fixed set of frequencies (duration T and sampling Δt), is proportional to the mean intensity I_0. This is the r.m.s.–flux relation and cannot be produced by a superposition (addition) of uncorrelated events such as the phenomenological "shot noise" models. Instead this observation *requires* that the fluctuations be multiplicative (Uttley and McHardy, 2001; Uttley et al., 2005), as sketched in Figure 6.6.

As the disk extends progressively inward for softer spectra, the flow at larger radii cannot fluctuate on such large amplitudes, as the disk is underneath it. The large-amplitude fluctuations can only be produced from radii inward of the truncated disk. This gets progressively smaller as the disk comes in, so the longer time scale (lower frequency) fluctuations are progressively lost. Therefore, the power spectrum narrows, with ν_b increasing while the amount of high-frequency power stays approximately the same, as seen in the data (Fig. 6.7a). These models can be made quantitative and can match the major features of the correlated changes in both the energy spectra and the power spectra as the source makes a transition from the low/hard to high/soft states (Fig. 6.7b; Ingram and Done, 2010).

However, the power spectrum also contains a characteristic low-frequency QPO, which also moves along with ν_b (Fig. 6.7a). This can be very successfully modeled in both frequency and spectrum as Lense-Thirring (vertical) precession of the entire hot inner flow (Ingram et al., 2009), using the same transition radius as required for the low-frequency break in the broadband power spectrum (Ingram and Done, 2010; see Fig. 6.7c). Since

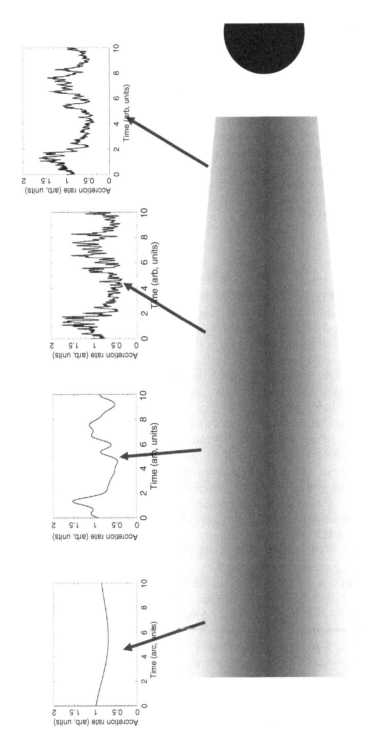

FIGURE 6.6. A propagating fluctuation model. Large radii produce long-time-scale fluctuations in mass accretion rate, which modulate faster variations in mass accretion rate produced by fluctuations at smaller radii. Figure courtesy of P. Uttley.

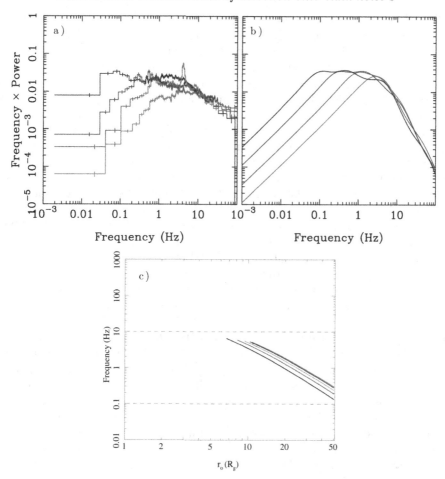

FIGURE 6.7. *a)* and *b)* show how the power spectra evolve as the source makes a transition from the low/hard to very high state. Low-frequency power is progressively lost as the spectrum softens, whereas the high-frequency power remains the same. *c)* shows how this can be modeled in a truncated disk/hot inner flow geometry, where the transition radius decreases, progressively losing the large radii (low-frequency variability) parts of the hot flow (DGK07). The strong low-frequency QPO also moves to higher frequencies, consistent with Lense-Thirring precession of the hot flow (Ingram et al., 2009).

this is a model involving vertical precession, the QPO should be strongest for more highly inclined sources, as observed (Schnittman et al., 2006).

The hot flow is also a good candidate for the base of the jet, in which case the collapse of the hot flow seen in the high/soft state spectra also triggers the collapse of the jet, as observed (see Chapter 7 by R. Fender in these proceedings).

6.1.7 *Scale up to AGN*

These models can be easily scaled up to AGN, keeping the same geometry as a function of $L/L_{\rm Edd}$ but changing the disk temperature as expected for a supermassive BH. The luminosity scales as M for mass accretion at the same fraction of Eddington, and the inner radius scales as M. Therefore, the emitting area scales as M^2, so $T^4 \propto L/A \propto 1/M$. Disk temperature *decreases*, peaking in the UV rather than soft X-rays. Interstellar absorption in the host galaxy and in our galaxy effectively screens this emission (see Section 6.3.1), so the disk peak cannot be directly observed in the same way as in BHB.

The strong UV flux from the disk also excites multiple UV line transitions (see Section 6.3.2) from any material around the nucleus. This environment is much less clean than in

LMXBs as there are many more sources of gas to be illuminated in the rich environment of a galaxy center (e.g., molecular clouds, the obscuring torus), giving strong line emission from the broad-line region (BLR) and narrow-line region (NLR).

Apart from these differences, we should otherwise see the same behavior in terms of the spectral and variability changes as the mass accretion rate changes, and in the jet power. Thus, these models predict intrinsic changes in the ionizing nuclear spectrum as a function of mass accretion rate as well as changes in the observed spectrum due to obscuration. This is in contrast to the original "unification models" of AGN in which Seyfert 1 and Seyfert 2 nuclei were intrinsically the same, but viewed at different orientations to a molecular torus.

There is growing evidence for intrinsic differences in nuclear spectra. The optical emission line ratios can be quite different in *unobscured* AGN of similar mass; for example, LINERS show different line ratios to Seyfert 1s, which are different from narrow-line Seyfert 1s. This is a clear indication that the ionizing spectrum is intrinsically different, as expected from their very different L_{bol}/L_{Edd}. This can be seen directly from compilations of the spectral energy distributions (SEDs) of these different types of AGN. The fraction of power carried by the X-rays drops for increasing L_{bol}/L_{Edd} in much the same way as for BHB. The soft tail at high L/L_{Edd} carries a smaller fraction of bolometric luminosity so L_x has to be multiplied by a larger factor to get L_{bol} (Vasudevan and Fabian, 2007).

The jet should also change with state as in BHB. This gives a clear potential explanation for the origin of radio loud/radio quiet dichotomy. This matches quite well to the observed radio populations (Körding et al., 2006), but there is a persistent suggestion that this is not all that is required, with the most powerful radio jets being found in the most massive AGN (e.g., Dunlop et al., 2003). Incorporating supermassive black-hole growth and its feedback onto galaxy formation into the semianalytic codes to model the growth of structures in the universe may give the answer to this. These show a correlation between supermassive black-hole mass and mass accretion rate such that largest black holes in massive ellipticals are now all accreting in gas-poor environments and so accrete via a hot flow with correspondingly strong radio jet (Fanidakis et al., 2009).

6.2 Compton scattering to make the high-energy tail

Compton scattering is just an energy exchange process between the photon and electron, and the energy exchange is completely analytic. The output photon energy, ϵ_{out}, is given by

$$\epsilon_{out} = \frac{\epsilon_{in}(1 - \beta \cos \theta_{ei})}{1 - \beta \cos \theta_{eo} + (\epsilon_{in}/\gamma)(1 - \cos \theta_{io})} \quad (6.2)$$

where θ_{ei}, θ_{eo}, and θ_{io} are the angles between the electron and input photon, electron and output photon, and input and output photon, respectively; $\gamma = (1 - \beta^2)^{-1/2}$ is the electron Lorentz factor so that its kinetic energy is $E = (\gamma^2 - 1)^{1/2} m_e c^2$; and $\epsilon = h\nu/m_e c^2$ is the photon energy relative to the rest mass energy of the electron. In general, this simply says that whichever of the input electron and photon has the most energy shares some of this with the other.

An electron at rest has $E = 0 < \epsilon$. The photon hits the electron, and momentum conservation means that the electron recoils from the collision, so the photon loses energy in Compton downscattering. If the photons and electrons are isotropic and $\epsilon_{in} \ll 1$, then the angle-averaged energy loss is $\epsilon_{out} = \epsilon_{in}/(1 + \epsilon_{in}) \approx \epsilon_{in}(1 - \epsilon_{in})$. Thus, the change in energy $\epsilon_{out} - \epsilon_{in} = \Delta\epsilon = -\epsilon_{in}^2$. Alternatively, for $\epsilon_{in} \gg 1$, the photon loses almost all its energy in the collision.

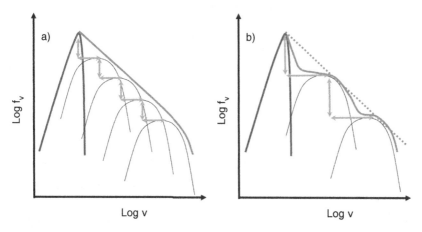

FIGURE 6.8. *a)* shows the spectrum built up from repeated thermal Compton upscattering events for optically thin ($\tau \lesssim 1$) material. A fraction τ of the seed photons (thick dark line) are boosted in energy by $1 + 4\Theta$, and then these form the seed photons for the next scattering, so each scattering order (thin light lines) is shifted down and to the right by the same factor (as indicated by the arrows), giving a power law (thick light line – envelope). *b)* shows that the same spectral index can be obtained by higher Θ and lower τ, but the wider separation of the individual scattering orders results in a bumpy spectrum (thick light line) rather than a smooth power law (dotted line).

6.2.1 Thermal Compton Upscattering: Theory

Since Comptonization conserves photon number, it is easiest to draw in $\log F(E)$ versus $\log E$ as $F(E) d \log E = $ photon number per bin on a logarithmic energy scale (see Section 6.1.1). See also the review by Gilfanov (2010).

In a thermal distribution of electrons, the typical random velocity is set by the electron temperature $\Theta = kT_e/m_e c^2$ as $v^2 \sim 3kT_e/m_e$ so $\beta^2 = 3\Theta$. Again, for isotropic electron and photon distributions, this can be averaged over angle to give $\epsilon_{\text{out}} = (1 + 4\Theta + 16\Theta^2 + \cdots)\epsilon_{\text{in}} \approx (1 + 4\Theta)\epsilon_{\text{in}}$ for $\Theta \ll 1$. So, in scattering we change energy by $\epsilon_{\text{out}} - \epsilon_{\text{in}} = \Delta\epsilon = 4\Theta\epsilon_{\text{in}}$, and photons are Compton upscattered. Obviously there is a limit to this, since the photon cannot gain more energy than the electron started out with, so this approximation only holds for $\epsilon_{\text{out}} \lesssim 3\Theta$.

Photons can only interact with electrons if they collide. The probability a photon will meet an electron can be calculated from the optical depth. An electron has a (Thomson) cross section σ_T for interaction with a photon. Thus it is probable that the photon will interact if there is one electron within the volume swept out by the photon where the length of the volume is simply the path length, R, and the cross-sectional area is the cross section for interaction of the photon with an electron, σ_T. Optical depth, τ, is defined as the number of electrons within this volume, so $\tau = nR\sigma_T$ where n is the electron number density. The scattering probability is $e^{-\tau} \approx 1 - \tau$ for $\tau \ll 1$.

The seed photons are initially at some energy, ϵ_{in}, so only a fraction, τ, of these are scattered in optically thin material to energy $\epsilon_{\text{out},1} = (1 + 4\Theta)\epsilon_{\text{in}}$. But these scattered photons themselves also can be scattered to $\epsilon_{\text{out},2} = (1 + 4\Theta)\epsilon_{\text{out},1} = (1 + 4\Theta)^2 \epsilon_{\text{in}}$. These photons can be scattered again to $\epsilon_{\text{out},3}$ and so on until they reach the limit of the electron energy after N scatterings, where $\epsilon_{\text{out},N} = (1 + 4\Theta)^N \epsilon_{\text{in}} \sim 3\Theta$. Since the energy boost and fraction of photons scattered is constant, this gives a power law of slope $\log f(\epsilon) \propto \ln(1/\tau)/\ln(1 + 4\Theta)$, that is, $f(\epsilon) \propto \epsilon^{-\alpha}$ with $\alpha = \ln\tau/\ln(1 + 4\Theta)$. This is a power law from the seed photon energy at ϵ_{in} up to 3Θ. Thus, the power law index is determined by both the temperature and optical depth of the electrons. These cannot be determined independently without observations at high energy to constrain Θ, as the same spectral index could be produced by making τ smaller while increasing Θ

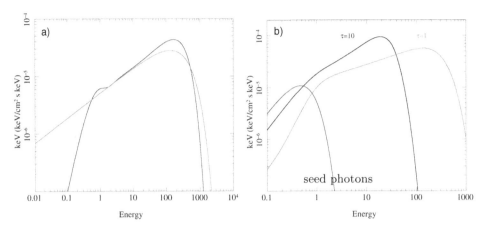

FIGURE 6.9. a) The dark line shows the thermal Comptonization spectrum computed by COMPPS for $\tau = 2$ and $kT_e = 100$ keV with seed photons at 0.2 keV (using geom = 0 and covering fraction = 1). The slight bump at the seed photon energy is from the fraction $\exp(-\tau)$ of the seed photons that escape without scattering. The Compton upscattered spectrum does not extend down below the seed photon energy, so there is an abrupt downturn below $3kT_{\rm seed} = 0.6$ keV that is not present in the cutoff power-law spectrum. At high energies, the Compton rollover at the electron temperature is much sharper than an exponential, giving a large difference in spectral parameters. b) shows an energetic approach to Comptonization for $\ell_h/\ell_s = 10$ with $\tau = 10$ and 1. Each of these two Comptonized spectra contains 10× more energy than the seed photons, but each has different spectral index and electron temperature.

(see Fig. 6.8a and b). However, there are some constraints, as the spectrum is only a smooth power law in the limit where the orders overlap – that is, τ not too small and Θ not too big – and the energy bandpass is not close to either the electron temperature or seed photon energy (see Fig. 6.8a and b).

This list of caveats means that often a power law is not a good approximation for a Comptonized spectrum. If the temperatures are nonrelativistic and the optical depth not too small, then COMPTT (Titarchuk, 1994) or NTHCOMP (Zdziarski et al., 1996, where the "N" in front of THCOMP denotes that it does not have the reflected component of Zycki et al., 1999) can be used, as these both include the downturn in the Compton emission close to the seed photon energy that affects the derived disk properties (Kubota and Done, 2004), as well as the rollover at the electron temperature.

However, for temperatures much above ∼100 keV with good high-energy data, relativistic corrections become important, and COMPPS (Poutanen and Svensson, 1996) or EQPAIR (Coppi, 1999) should be used. The Compton rollover at the electron temperature is rather sharper than an exponential, so using an exponentially cutoff power law is not a good approximation (see Fig. 6.9a) and will distort the derived reflected fraction (see Section 6.4).

In optically thick Comptonization, with $\tau \gg 1$, almost all the photons are scattered each time, so almost all of them end up at the electron temperature of 3Θ, forming a Wien peak (Fig. 6.10a). The average distance a photon travels before scattering is $\tau = 1$, that is, a mean free path of $\lambda = 1/(n\sigma_T)$. Thus, after one scattering, the distance traveled is $d_1^2 = \lambda^2 + \lambda^2 - 2\lambda^2 \cos\theta_{12}$, whereas after two scatterings, this is $d_2^2 = d_1^2 + \lambda^2 - 2\lambda d_1 \cos\theta_{2,3}$, and after N scatterings, $d_N^2 = d_{N-1}^2 + \lambda^2 - 2\lambda d_{N-1} \cos\theta_{N,N+1}$. Since the scattering randomizes the direction, the angles average out, leaving $d_N^2 = N\lambda^2 = N/(n\sigma_T)^2$. The photon can escape when $d_N = R$, so the average number of scatterings before escape is $N \sim \tau^2$ (see Fig. 6.10b).

The amount of energy exchange from the electrons to photons in Comptonization can be roughly characterized by the Compton y parameter. The fractional change in energy of

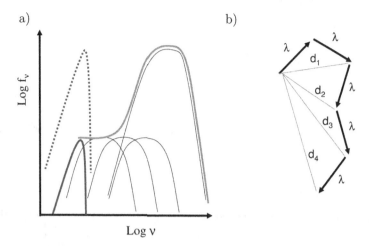

FIGURE 6.10. a) shows how the spectrum built up from repeated thermal Compton upscattering events for optically thick ($\tau \gtrsim 1$) material. Almost all ($\exp^{-\tau}$) the seed photons are scattered multiple times until they all pile up at the mean electron energy of $\sim 3\Theta$. b) shows the electron path, randomizing direction by scattering after an average distance $\lambda = 1/n\sigma_T$ (i.e., traveling $\tau = 1$).

the photon distribution, y, is the average number of scatterings times average fractional energy boost per scattering such that $y \approx (4\Theta + 16\Theta^2)(\tau + \tau^2) \approx 4\Theta\tau^2$ in the optically thick, low-temperature limit. If $y \ll 1$, then the electrons make very little difference to the spectrum, whereas for $y \gtrsim 1$, Comptonization is very important in determining the emergent spectrum.

6.2.2 Comptonization via energetics

Describing the spectrum in terms of τ and Θ is the "classical" way to talk about Compton scattering. However, the physical situation is better described by τ and energetics. There is some electron region with optical depth τ, heated by a power input ℓ_h, making a Comptonized spectrum from some seed photon luminosity ℓ_s. Now that we are doing this by luminosity instead of photon number, we should plot $\nu F(\nu)$ rather than $F(\nu)$. The seed photons peak at $\sim 3kT_{\text{seed}}$, with $\nu F(\nu) = \ell_s$. The "power law" Compton spectrum always points back to this point, forming a power law of energy index $\alpha \sim \log \tau / \log(1 + 4\Theta)$ that extends from here to $\sim 3kT_e$ and has total power ℓ_h. The resulting equation can then be solved for Θ (see Haardt and Maraschi, 1993). This is shown in Fig. 6.9b, using the EQPAIR model. The seed photons (red) are Comptonized by hot electrons that have 10× as much power as in the seed photons. For a large optical depth (blue: $\tau = 10$) this energy is shared among many particles, so the electron temperature is lower than for a smaller optical depth (green: $\tau = 1$). The spectrum cannot extend out to high energies and so has to be harder in order to contain the requisite amount of power.

This energetic approach gives more physical insight when we come to consider seed photons produced by reprocessing of the hard X-ray photons illuminating the disk (see Section 6.5.2).

6.2.3 Thermal Compton scattering: observations of low/hard state

Figure 6.11 shows two examples of low/hard state spectra from a BHB, one that rolls over at ~ 90 keV, and one that extends to ~ 200 keV. Both rollovers look like thermal Comptonization, but with different temperatures (~ 30 keV and ~ 100 keV, respectively, that is, $\Theta \sim 0.06$ and 0.2). The optical depth can then be derived from the spectral index, but not quite as easily as described previously, as τ is of order unity rather than $\ll 1$ as

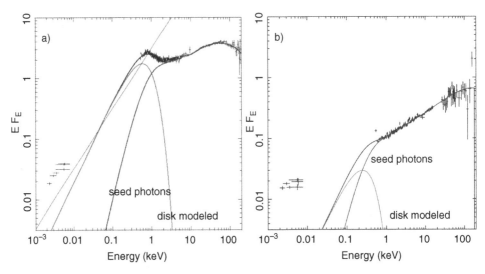

FIGURE 6.11. The unabsorbed data from the BHB transient 1753.5-0127 modeled with a disk and thermal Compton scattering of seed photons from the disk, both labeled in the graph. *a)* shows a bright low/hard state, where the optical/UV data clearly lie far below the extrapolated hard X-ray emission, indicating that the X-ray spectrum must break (i.e., have its seed photons) somewhere between the UV and soft X-ray bandpass. The disk is then the obvious source for these seed photons. The optical/UV points are some way below the extrapolated disk emission, but this leaves room for hard X-ray reprocessing, and possibly some component from the radio jet spectrum extrapolated up to this band (thin gray line). However, *b)* shows the very different dim low/hard state spectrum from the same source. There is still a very weak soft X-ray component, but the optical/UV now lie on the extrapolated hard X-ray spectrum, making it more likely that the seed photons for the Compton scattering are at energies below the optical, that is, they are probably cyclosynchrotron (Chiang et al., 2010).

required for the analytic expression. A more proper treatment gives $\tau \sim 0.6$. Then the fraction of unscattered seed photons should be only $1 - e^{-\tau}$, but we actually see more than this in Figure 6.11a. Thus we require that not all the seed photons go through the hot electron region. That is, either the geometry is a truncated disk, or the electron regions are small compared to the disk – either localized magnetic reconnection regions above the disk or a jet. See Section 6.5.2 for how the energetics of Compton scattering strongly favor the truncated disk.

The spectra shown earlier are fitted assuming that the observed soft X-ray component provides the seed photons for the Compton upscattering. This can be seen explicitly where there are multiwavelength observations, extending the bandpass down to the optical/UV where the outer parts of the disk can dominate the emission. Figure 6.11a shows this for a bright low/hard state in the transient BHB XTE J1753.5-0127 (Chiang et al., 2010). It is clear that extrapolating the hard X-ray power law down to the optical/UV will produce far more emission than is observed. Thus, the hard X-ray power law must break between the UV and soft X-ray, that is, the seed photons for the Compton upscattering should be somewhere in this range. Since there is an obvious soft X-ray thermal component, this is the obvious seed photon identification. Fitting the soft X-rays with a disk component slightly underproduces the optical/UV emission, but this can be enhanced by reprocessing of hard X-rays illuminating the outer disk (van Paradijs, 1996).

However, for dimmer low/hard states, the hard X-rays extrapolate directly onto the optical/UV emission (Fig. 6.11b, Chiang et al., 2010; also see Motch et al., 1985). This looks much more as if the seed photons are at lower energies than the optical. In the truncated disk picture, the disk can be so far away that it subtends a very small solid angle to the hot electron region, which is concentrated at small radii. Thus the number of seed photons from the disk illuminating the hot electrons can be very small and can be less important than seed photons produced by the hot flow itself. The same thermal electrons that make the Compton spectrum can make both bremsstrahlung (from interactions with protons) and cyclosynchrotron (from interactions with any magnetic field such as the tangled field produced by the MRI). The bremsstrahlung spectrum will peak at kT_e, so these seed photons have similar energies to the electrons and thus cannot gain much from Compton scattering. However, the cyclosynchroton typically peak in the IR/optical region, so these can be the seed photons for a power law that extends from the optical to the hard X-ray region (Narayan and Yi, 1995; Di Matteo et al., 1997; Wardziński and Zdziarski, 2000; Malzac and Belmont, 2009).

Evidence for a change in seed photons is also seen in the variability (see Chapter 4 by R. Hynes, this volume). In bright states (both bright low/hard and high/soft/very-high states), the optical variability is a lagged and smoothed version of the X-ray variability, showing that it is from reprocessed hard X-ray illumination of the outer accretion disk. However, this changes in the dim low/hard state, with the optical having more rapid variability than the hard X-rays, and often *leading* the hard X-ray variability (Kanbach et al., 2001; Gandhi et al., 2008; Durant et al., 2008; Hynes et al., 2009). This completely rules out a reprocessing origin, clearly showing the change in the optical emission mechanism.

6.2.4 Nonthermal Compton Scattering: High/Soft State

Whereas the low/hard state can be fairly well described by thermal Compton scattering, the same is not true for the tail seen in the high soft state. This has $\Gamma \sim 2$ and clearly extends out past 1 MeV, and probably past 10 MeV for Cyg X-1 (Gierliński et al., 1999; McConnell et al., 2002), as shown in Fig. 6.12a. If this were thermal Compton scattering, then the electron temperature must be $\Theta \gtrsim 1$, requiring optical depth $\tau \ll 1$ in order to produce this photon index. The separate Compton orders are then well separated, and the spectrum should be bumpy (see Fig. 6.8b) rather than the smooth power law seen in the data (the bump in the data is from reflection: see Section 6.4). Thus, this tail cannot be produced by thermal Comptonization.

Instead, it can be nonthermal Compton scattering, where the electron number density has a power-law distribution rather than a Maxwellian – that is, $n(\gamma) \propto \gamma^{-p}$ from $\gamma = 1$ to $\gamma_{\max}$. Going back to the original equation for the Compton-scattered energy boost, for $\gamma \gg 1$ the output photon is beamed into a cone of angle $1/\gamma$ along the input electron direction. This gives an angle-averaged output photon energy of $\epsilon_{\text{out}} = (4/3\gamma^2 - 1)\epsilon_{\text{in}} \approx \gamma^2 \epsilon_{\text{in}}$ for an isotropic distribution of input photons and electrons. Thus, the Compton-scattered spectrum extends from ϵ_{in} to $\gamma_{\max}^2 \epsilon_{\text{in}}$, forming a power law from a single scattering order.

The power law index of the resulting photon spectrum can be calculated from an energetic argument. The rate at which the electrons lose energy is the rate at which the photons gain energy, giving $F(\epsilon)d\epsilon \propto \dot\gamma n(\gamma)d\gamma$, where $\dot\gamma \propto \gamma^2$ is the rate at which a single electron of energy γ loses energy and $n(\gamma)$ is the number of electrons at that energy. Thus, $F(\epsilon) \propto \gamma^2 \gamma^{-p} d\gamma/d\epsilon$. Because $\epsilon \sim \gamma^2 \epsilon_i$, $d\epsilon/d\gamma = 2\gamma$, so $F(\epsilon) \propto \gamma^{-(p-1)} \propto \epsilon^{-(p-1)/2}$, that is, an energy spectral index of $\alpha = (p-1)/2$ (G. Ghisellini, private communication).

For an optically thin electron region, the electrons intercept only a fraction τ of the seed photons and scatter them to $\gamma_{\max}^2 \epsilon_{\text{in}}$ with an energy spectral index of $(p-1)/2$. These can themselves be scattered into a second-order Compton spectrum to $\gamma_{\max}^2(\gamma^2 \epsilon_{\text{in}}) = \gamma^4 \epsilon_{\text{in}}$, but very soon the large energy boost means that these hit the limit of the electron energy of $\epsilon_{\text{out}} = \gamma_{\max}$. The resulting spectrum is shown schematically in Figure 6.13.

Thus, the high/soft state of Cyg X-1 requires $\gamma_{\max} > 30$ to get seed photons from the disk at 1 keV upscattered to 1 MeV, or $\gamma_{\max} > 100$ to get to 10 MeV, whereas the energy

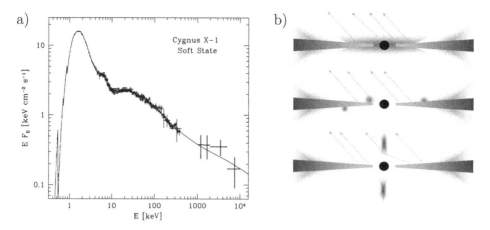

FIGURE 6.12. *a)* shows a composite spectrum of the high/soft state of Cyg X-1 from multiple instruments. Interstellar absorption is not removed, causing the drop below 1 keV. Nonetheless, the disk clearly dominates the low-energy data (BeppoSAX LECS), whereas the soft tail extends out to ∼10 MeV (OSSE COMPTEL). The tail itself is smooth, but there is curvature and spectral features seen from 5 to 20 keV (BeppoSAX HPGSPC) from reflection (see Section 6.4). Figure from McConnell *et al.* (2002). *b)* shows the potential source geometries in which the disk photons can dominate the spectrum at low energies as observed. This requires that the electron acceleration region must be optically thin, or that it must be localized, either over the disk or in the base of the jet. Reproduced by permission of the AAS.

spectral index of $1.2 = (p-1)/2$ implies $p \sim 3$. Such nonthermal Compton spectra can be modeled using either COMPPS or EQPAIR.

Figure 6.12a shows that the seed photons from the disk are clearly seen as distinct from the tail. This requires either that the optical depth is very low or the electron acceleration region does not intercept many of the seed photons from the disk, that is, localized acceleration regions as shown schematically in Figure 6.12b.

6.2.5 Thermal-nonthermal (hybrid) Compton scattering

The high/soft states can transition smoothly into the very high or intermediate state spectra, with the tail becoming softer and carrying a larger fraction of the total power (Fig. 6.14a). The disk then merges smoothly into the tail, showing that the hot electron region completely covers the inner disk emission and is optically thick. The tail still extends up to 1 MeV, so it clearly also contains nonthermal electrons and is rather soft but has a complex curvature (see Fig. 6.14b).

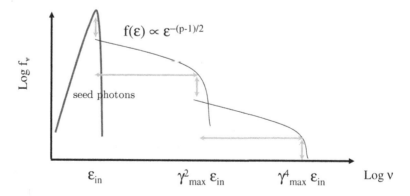

FIGURE 6.13. Schematic of nonthermal Compton scattering, where the seed photons form a power law from a single scattering due to the power law electron distribution from $\gamma = 1 - \gamma_{\max}$.

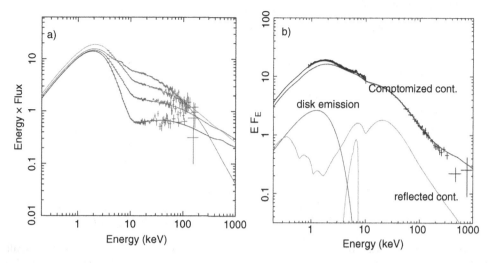

FIGURE 6.14. a) Transition from high/soft to extreme very high state in the BHB transient XTE J1550-564. b) Broader band spectrum of the extreme very high state. The emission is dominated at all energies by the Comptonized continuum, and there is very little of the disk emission that can be seen directly. This requires that the corona be optically thick and cover most of the inner disk, in contrast to the possible high/soft state geometries. The reflected continuum and its associated iron emission line are also shown (see Section 6.4).

The extent of the tail shows that there must be nonthermal Compton scattering, as in the high/soft state. However, the tail is softer, so the electron index must be more negative than in the high/soft state spectrum. Thus, the mean electron energy is lower. Yet the lack of direct disk emission requires an optical depth of $\gtrsim 1$. This means that there are multiple Compton scattering orders forming the spectrum in a similar way to thermal Compton scattering, but from a nonthermal distribution. The energetic limit to which photons can be scattered is γ_{max}, but because the energy boost on each scattering is small, the spectrum actually rolls over at $m_e c^2 = 511$ keV as the cross section for scattering drops at this point, where $\gamma \epsilon \sim 1$ (as the cross section transitions to Klein-Nishina rather than the constant Thomson cross section seen at lower collision energies). Thus, optically thick, nonthermal Comptonized spectra with a steep power law electron distribution do not produce a power-law spectrum. Instead there is a break at 511 keV (Ghisellini, 1989), as shown schematically in Fig. 6.15a, which means that this cannot fit the observed tail at high energies seen in the very high state spectrum, as shown in Fig. 6.15b (Gierliński and Done, 2003).

Thus, neither thermal nor nonthermal Compton scattering can produce the tail seen in these very high state data. Instead, the spectra require both thermal and nonthermal electrons to be present. This could be produced in a single acceleration region, where the initial acceleration process makes a nonthermal distribution but where the resulting electrons have a hybrid distribution due to lower energy electrons predominantly cooling through Coulomb collisions (which thermalize) while the higher-energy electrons maintain a power law shape by cooling via Compton scattering (Coppi, 1999). Such hybrid thermal/nonthermal spectra can be modeled in XSPEC using either COMPPS or EQPAIR. Alternatively, this could indicate that there are two separate acceleration regions: one with thermal electrons, perhaps the remnant of the hot inner flow, and one with nonthermal, perhaps magnetic reconnection regions above the disk or the jet (DGK07). This could be modeled by two separate COMPPS or EQPAIR components, one of which is set to be thermal, and the other set to be nonthermal.

Whatever the electrons are doing, they are optically thick and cover the inner disk. Hence, it is very difficult to reconstruct the intrinsic disk spectrum in these states. The

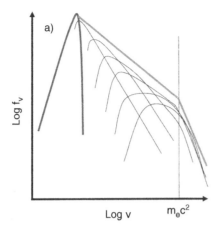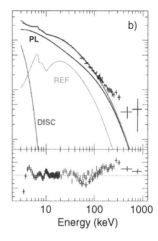

FIGURE 6.15. *a)* Spectrum resulting from a steep power law electron distribution that is optically thick ($\tau \gtrsim 1$ rather than $\tau \gg 1$). The steep power law means that the mean electron energy is low, so the spectrum is built up from multiple Compton scatterings in a manner similar to thermal Comptonization. Thus, $\epsilon\gamma \sim 1$ at $\epsilon \sim 1$, that is, at 511 keV, so the spectrum breaks because of the reduction in cross section. *b)* shows how this means that such a steep power law electron distribution cannot fit the high-energy tail seen in the very high state.

derived temperature and luminosity of the disk depend on how the tail is modeled. A simple power-law model for the tail means that it extends below the putative seed photons from the disk. Instead, in Compton upscattering, the continuum rolls over at the seed photon energy, so there are fewer photons at low energies from the tail. Therefore, the disk has to be more luminous and/or hotter in order to match the data. Compton scattering conserves photon number, so all the photons in the tail were initially part of the disk emission; thus, the intrinsic disk emission is brighter than that observed by this (geometry-dependent) factor. This means that the intrinsic disk luminosity and temperature cannot be unambiguously recovered from the data when $\tau \gtrsim 1$, where the majority of photons from the disk are scattered into the tail (Kubota et al., 2001; Kubota and Done, 2004; Done and Kubota, 2006; Steiner et al., 2009).

For more observational details of spectral states, see Remillard and McClintock (2006) and Belloni (2010).

6.3 Atomic absorption

The intrinsic continuum is modified by absorption by material along the line of sight. This can be the interstellar medium in our galaxy or the host galaxy or the X-ray source, or material associated with the source. For AGN, this can be the molecular torus, the NLR or BLR clouds, or an accretion-disk wind. For BHB, this is just an accretion-disk wind and any wind from the companion star. For magnetically truncated accretion disks such as seen in the intermediate polars (white dwarfs) and accretion powered millisecond pulsars (neutron stars), there is an accretion curtain that can be in the line of sight (de Martino et al., 2004). For extreme magnetic fields however, the disk is completely truncated, and there is only an accretion column (polars in white dwarfs), but this overlays the X-ray hot shock, giving complex absorption (Done and Magdziarz, 1998).

Again the key concept is optical depth, $\tau = \sigma(E)nR$, only now the cross section, $\sigma(E)$, has a complex dependence on energy (rather than the constant electron scattering cross section, σ_T, for energies below 511 keV). We can combine $nR = N_H$ as the number of hydrogen atoms along a line-of-sight volume with cross-sectional area of 1 cm^{-2}, so the optical depth is simply related to column density by $\tau(E) = \sigma(E)N_H$.

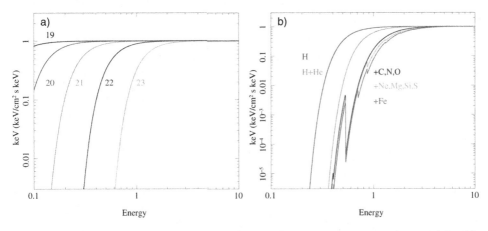

FIGURE 6.16. a) Photoelectric absorption from hydrogen alone, in a column of $\log N_H = 19, 20, 21, 22, 23$ (from left to right). b) shows how the absorption for a column of $\log N_H = 22$ changes as progressively heavier atomic numbers are added.

6.3.1 Neutral absorption

The photoelectric absorption cross section of neutral hydrogen is zero below the threshold energy of 13.6 eV, below which the photons do not have enough energy to eject the electron from the atom. It peaks at this threshold edge energy at a value of 6×10^{-18} cm^{-2} and then declines as $\approx (E/E_{\rm edge})^{-3}$. Thus, the optical depth is unity for a hydrogen column of 1.6×10^{17} cm^{-2} at 13.6 eV, whereas a typical column through our galaxy is $> 10^{20}$ cm^{-2}, showing how effectively the UV emission is attenuated.

The drop in cross section with energy means that an H column of 10^{20} cm^{-2} has $\tau = 1$ at an energy of ~ 0.2 keV, allowing the soft X-rays to be observed. Figure 6.16a shows how a hydrogen column of $\log N_H = 19, 20, 21, 22, 23$ progressively absorbs higher-energy X-rays. However, the column is not made up solely of H. There are other, heavier elements as well. These have more bound electrons, but the highest edge energy will be from the inner $n = 1$ shell (also termed the K shell) electrons, as these are the closest to the nuclear charge. Since this charge is higher, the $n = 1$ electrons are more tightly bound than those of H; for example, for He it is 0.024 keV, and C, N, O is 0.28, 0.40, and 0.53 keV. These elements are less abundant than H, so they form small increases in the total cross section. Fe is the last astrophysically abundant element, and this has a K edge energy of 7.1 keV. Figure 6.16b shows this for a column of 10^{22} cm^{-2} for progressively adding higher atomic number elements assuming solar abundances. Helium has an impact on the total cross section, but additional edges from heavier elements are important contributions to the total X-ray absorption, especially oxygen.

In XSPEC, this can be modeled using TBABS/ZTHABS (Wilms et al., 2000, where the latter has redshift as a free parameter) or PHABS/ZPHABS (Balucinska-Church and McCammon, 1992) if the abundances are assumed to be solar, or TBVARABS/ZVPHABS if the data are good enough for the individual element abundances to be constrained via their edges such as in GRS 1915+104 (Lee et al., 2002). However, with excellent spectral resolution data from gratings, the line absorption (especially from neutral oxygen) becomes important (see Section 6.3.3), and TBNEW (see J. Wilms web page at http://pulsar.sternwarte.uni-erlangen.de/wilms/research/tbabs) should be used (Juett et al., 2004).

6.3.2 Ionized absorption

Photoelectric absorption leaves an ion – that is, the nuclear charge is not balanced. Thus, all the remaining electrons are slightly more tightly bound, so all the energy levels

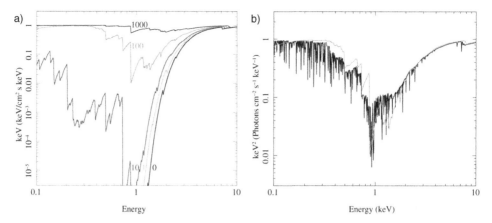

FIGURE 6.17. a) The photoelectric (bound-free) absorption of a column of $N_H = 10^{23}$ cm^{-2} and ionization state $\xi = 0, 1, 10, 100$, and $1,000$ for photoionization by a power law of $\Gamma = 2$ as calculated by ABSORI. b) shows how the opacity of partially ionized material ($\xi = 100$) can instead be dominated by bound-bound (line) transitions using a proper photoionization code such as XSTAR.

increase. The ion can recombine with any free electrons, but if the X-ray irradiation is intense, then the ion can meet an X-ray photon before it recombines, so that the absorption is dominated by photoionized ions. For H, this means there is no photoelectric absorption, since it has no bound electrons after an ionization event: thus, some fraction of the total cross section disappears. Helium may then have one electron left, so its edge moves to 0.052 keV. At higher ionizations, helium is completely ionized, so its contribution to the total cross section is lost and there are only edges from (ionized) C, N, O, and higher–atomic-number elements. Thus, the effect of going to higher ionization states is to reduce the overall cross section as the numbers of bound electrons are lower. In the limit where all the elements are completely ionized, there is no photoelectric absorption at all.

The ion populations are determined by the balance between photoionization and electron recombination. For a given element X, the ratio between X^{+i} and the next ion stage up, X^{i+1}, is given by the equilibrium reaction

$$X^{+i} + h\nu \rightleftharpoons X^{+(i+1)} + e^-. \quad (6.3)$$

The photoionization rate depends on the number density of the ion N_X^{+i} and of the number density of photons, n_γ, above the threshold energy ν_{edge} for ionization for that species. This number density can be approximated by $n_\gamma \sim L/(h\nu 4\pi r^2 c)$, where L is the source luminosity, $h\nu$ is the typical photon energy, and $4\pi r^2 c$ is the volume swept out by the photons in 1 second. Actually, what really matters is the number of photons past the threshold energy that photoionize the ion, so this depends on spectral shape as well.

Equilibrium is where the photoionization rate, $N_X^{+i} n_\gamma \sigma(X^{+i}) c$, balances the recombination rate, $N_X^{+(i+1)} n_e \alpha(X^{+(i+1)}, T)$, where $\sigma(X^{+i})$ is the photoelectric absorption cross section for X^{+i} and $\alpha(X^{+(i+1)}, T)$ is the recombination coefficient for ion $X^{+(i+1)}$ at temperature T. Hence, the ratio of the abundance of the ion to the next stage down is given by

$$\frac{N_X^{+(i+1)}}{N_X^{+i}} = \frac{n_\gamma \sigma(X^{+i}) c}{n_e \alpha(X^{+(i+1)}, T)} \propto \frac{n_\gamma}{n_e}. \quad (6.4)$$

Thus the ratio of photon density to electron density determines the ion state. If the photon density is higher, the ion meets a photon first, and so is ionized to the next stage. Conversely, if the electron density is higher, then the ion meets an electron first and

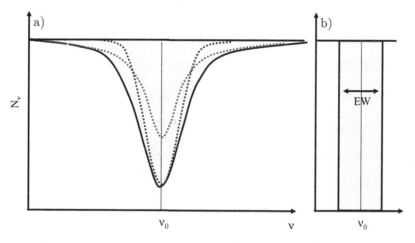

FIGURE 6.18. *a)* The Voigt profile of an absorption line (solid line), with a Doppler core (dark dotted line) and Lorentzian wings (light dotted line). *b)* The equivalent width of an absorption line is the width of a rectangular notch in the spectrum at the rest wavelength that contains the same number of photons as in the line.

recombines to the lower ion stage. The ratio of photon to electron number density can be written as $n_\gamma/n_e = \xi/(4\pi h\nu c)$, where $\xi = L/nr^2$ is the photoionization parameter. There are other ways to define this, such as $\Xi = P_{\rm rad}/P_{\rm gas} = L/(4\pi r^2 c n_e kT) = \xi/(4\pi ckT)$, but whichever description is used, the higher the ionization parameter, the higher the typical ionization state of each element.

In general, the equilibrium reaction means that there are at least two fairly abundant ionization stages for each element, so as long as the higher ionization stage is not completely ionized, then there are multiple edges from each element as each higher ion stage has a higher edge energy as the electrons are more tightly bound. This can be clearly seen in Fig. 6.17a, where the H-like oxygen (O^{+7}) edge at 0.87 keV is accompanied by the He-like (O^{+6}) edge at 0.76 keV for $\xi = 100$ and 10.

The ABSORI (Done et al., 1992) model in XSPEC calculates the ion balance for a given (rather than self-consistently computed) temperature and hence gives the photoelectric absorption opacity from the edges. However, this can be very misleading, as it neglects line opacities (see later discussion and Fig. 6.17b).

6.3.3 Absorption lines

There are also line (bound-bound) transitions as well as edges (bound-free transitions). These can occur whenever the higher shells are not completely full. Hence, oxygen can show absorption at the $n = 1$ to $n = 2$ ($1s$ to $2p$) shell transition even in neutral material, whereas elements higher than Ne need to be ionized before this transition can occur. The cross section in the line depends on the line width. This is described by a Voigt profile. The "natural" line width is set by the Heisenberg uncertainty relation between the lifetime of the transition $\Delta t \Delta \nu \lesssim \hbar$. This forms a Lorentzian profile, with broad wings. However, the ions also have some velocity due to the temperature of the material $v_{\rm thermal}^2 \sim kT_{\rm ion}/m_{\rm ion}$. Any additional velocity structure such as turbulence adds in quadrature so $v^2 = v_{\rm thermal}^2 + v_{\rm turb}^2$. These velocities Doppler shift the transition, giving a Gaussian core to the line. This combination of Gaussian core, with Lorentzian wings, is termed a Voigt profile (see Fig. 6.18a).

The line equivalent width is the width of a rectangular notch (down to zero) in the spectrum that contains the same number of photons as in the line profile, as shown in Fig. 6.18b. The equivalent width of the line grows linearly with the column density of the ion when the line is optically thin in its core, as the line gets deeper linearly with

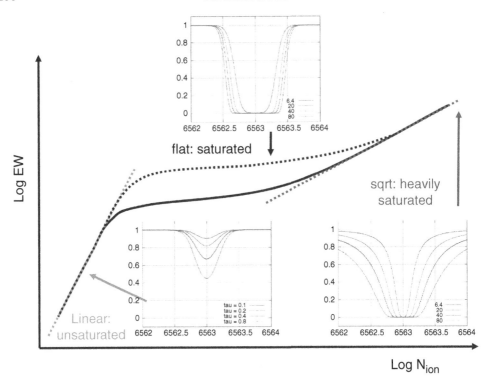

FIGURE 6.19. A curve of growth. On the unsaturated section, the line equivalent width increases linearly with increasing column. However, once the line becomes optically thick in the core, increasing the column does not lead to an increase in absorption, as the line center is black. The Doppler wings are very steep, so there is very little change in equivalent width with column (saturated) until the line becomes so optically thick that the Lorentzian wings dominate, leading to the heavily saturated increase. The point at which the line becomes saturated depends on the velocity structure. Higher turbulent velocity means that the line is broader, so it remains optically thin in the core (unsaturated) up to a higher column density.

column (called unsaturated). However, this linear behavior stops when the core of the line becomes optically thick that is, when none of the photons at the line core can escape at the rest energy of the transition. Increasing the column cannot lead to much more absorption, as there are no more photons at the line center to remove. The wings of the line can become optically thick, but Doppler wings are very steep, so the line equivalent width does not increase much as the column increases (called saturated). Eventually, the Lorentzian wings start to become important, and then the line equivalent width increases as the square root of the column density (called heavily saturated). This relation between column and line equivalent width is termed a "curve of growth." Whereas the linear section of this is unique, the point at which the line becomes saturated depends on the Doppler width of the line, that is, on the velocity structure of the material as shown in Figure 6.19.

For "reasonable" velocities, the line absorption equivalent width can be larger than the equivalent width of the edges, so that line transitions dominate the absorption spectrum. Figure 6.17b shows the total absorption (line plus edges computed using WARMABS, a model based on the XSTAR photoionization code (Kallman and Bautista, 2001; Kallman et al., 2004) compared to the more approximate ionization code that just uses edge opacities (ABSORI in XSPEC). The differences are obvious. There are multiple ionized absorption lines that dominate the spectrum as well as photoelectric edges.

Soft X-ray absorption features, especially from H and He, like oxygen, are seen in half of type 1 Seyferts (Reynolds, 1997). High-resolution grating spectra from Chandra and

XMM-Newton observations have shown these in some detail. Typically, the best datasets require several different column densities, each with different ionization parameters, in order to fit the data. By contrast, the BHB show little in the way of soft X-ray absorption (GRO J1655-40 is an extreme exception: Miller et al., 2006a), but highly inclined systems do show H- and He–like Fe Kα absorption lines in bright states (Ueda et al., 1998; Kotani et al., 2000; Yamaoka et al., 2001; Lee et al., 2002; Kubota et al., 2007).

In each case, the origin of this ionized material can be constrained by measuring the velocity width (or an upper limit on the velocity width) of a line to get the column density of that ion via a curve of growth. This can be done separately for each transition, but a better way is to take a photoionization code and use this to calculate the (range of) column density and ionization state required to produce all the observed transitions assuming solar abundance ratios. The spectrum gives an estimate for L, so the equation for the ionization parameter $\xi = L/nR^2$ can be inverted by writing $N_H = n\Delta R$ to give $R = L/(N_H \xi) \times (\Delta R/R)$. Since $\Delta R/R \lesssim 1$, the distance of the material from the source is $R \lesssim L/(N_H \xi)$. A small radius means that the material is most probably launched from the accretion disk itself, probably as a wind.

6.3.4 Winds

Wherever X-rays illuminate material, they can photoionize it. They also interact with the electrons by Compton scattering. Electrons are heated by Compton upscattering when they interact with photons of energy $\epsilon \gtrsim \Theta$ but are cooled by Compton scattering by photons with energy $\epsilon \lesssim \Theta$. Since the illuminating spectrum is a broadband continuum, the spectrum both heats and cools the electrons. The Compton temperature is the equilibrium temperature where heating of the electrons by Compton downscattering equals cooling by Compton upscattering. Section 6.2 shows that the mean energy shift is $\Delta \epsilon \sim 4\Theta\epsilon - \epsilon^2$, so integrating this over the photon spectrum gives the net heating, which is zero at the equilibrium Compton temperature, Θ_{ic}, so $0 = \int N(\epsilon)\Delta\epsilon d\epsilon = \int N(\epsilon)(4\Theta_{ic}\epsilon - \epsilon^2)d\epsilon$. Hence,

$$\Theta_{ic} = \frac{\int N(\epsilon)\epsilon^2 d\epsilon}{4\int N(\epsilon)\epsilon d\epsilon}. \quad (6.5)$$

For a photon spectrum with $\Gamma = 2.5$ between ϵ_i to $\epsilon_{\max}$, this gives $\Theta_{ic} \approx \frac{1}{4}\sqrt{\epsilon_{\max}\epsilon_i} \sim 2.5$ keV for energies spanning 1 to 100 keV. Alternatively, for a hard spectrum with $\Gamma = 1.5$, this is $\approx \epsilon_{\max}/12$ or 8 keV. The effective upper limit to $\epsilon_{\max}$ is around 100 keV, as the reduction in Klein-Nishina cross section means that higher-energy photons do not interact very efficiently with the electrons.

Thus, the irradiated face of the material can be heated up to this Compton temperature, giving typical velocities in the plasma of $v_{ic}^2 = 3kT_{ic}/m_p$. This is constant with distance from the source as the Compton temperature depends only on spectral shape (though the depth of the heated layer will decrease as illumination becomes weaker). This velocity is comparable to the escape velocity from the central object when $v_{ic}^2 \sim GM/R_{ic}$, defining a radius, R_{ic}, at which the Compton heated material will escape as a wind (Begelman et al., 1983). This is driven by the pressure gradient, and so has typical velocity at infinity of the sound speed $c_s^2 = kT_{ic}/m_p \sim v_{ic}^2 \sim GM/R_{ic}$. Thus, the typical velocity of this thermally driven wind is the escape velocity from where it was launched.

As the source approaches Eddington, the effective gravity is reduced by a factor $(1 - L/L_{\rm Edd})$, so the thermal wind can be launched from progressively smaller radii as the continuum radiation pressure enhances the outflow, forming a radiation-driven wind.

The Eddington limit assumes that the cross section for interaction between photons and electrons is only due to electron scattering. However, where the material is not strongly ionized, there are multiple UV transitions, both photoelectric absorption edges and lines. This reduces the "Eddington" limit by a factor $\sigma_{\text{abs}}/\sigma_T$, which can be as large as 4,000. This opacity is mainly in the lines, and the outflowing wind has line transitions that are progressively shifted from the rest energy to $\Delta\nu/\nu \approx v_\infty/c$. This large velocity width to the line means that its equivalent width is high, so it can absorb momentum from the line transition over a wide range in energy. This gives a UV line-driven wind.

The final way to power a wind is via magnetic driving, but this is difficult to constrain as it depends on the magnetic field configuration, so it is invoked only as a last resort.

Many of the "warm absorber" systems seen in AGN have typical velocities, columns and ionization states that imply they are launched from size scales typical of the molecular torus (e.g., Blustin et al., 2005). The much faster velocities of $\sim 0.1 - 0.2$ c implied by the broad absorption lines (BALs) seen blueward of the corresponding emission lines in the optical/UV spectra of some quasars are probably a UV line–driven wind from the accretion disk. However, the similarly fast but much more highly ionized (H and He-like Fe) absorption systems seen in the X-ray spectra of some AGN (see later in Fig. 6.26c and d) probably require either continuum driving with $L \sim L_{\text{Edd}}$ or magnetic driving (as the ionization state is so high that the line opacity is negligible with respect to σ_T).

By contrast, the BHB typically show fairly low outflow velocity of the highly ionized Fe Kα, mostly consistent with a thermally driven wind from the outer accretion disk (e.g., Kubota et al., 2007, DKG07), although the extreme absorber seen in one observation from GRO J1655-40 may require magnetic driving (Miller et al., 2006a; Kallman et al., 2009). However, this does assume that the observed luminosity, L_{obs}, measures the intrinsic luminosity, L_{int}. If there is electron scattering in optically thick, completely ionized material along the line of sight, then $L_{\text{obs}} = e^{-\tau} L_{\text{int}} \ll L_{\text{int}}$ (Done and Davis, 2008). Such scattering would strongly suppress the rapid variability power (Zdziarski et al., 2010), and indeed the variability power spectra of these data lack all high-frequency power above 0.3 to 1 Hz, rather than extending to the $\sim$10 Hz seen normally (see Fig. 6.5b). Thus, thermal winds potentially explain all of what we see in terms of absorption from BHB.

6.4 Reflection

Wherever X-rays illuminate optically thick material such as the accretion disk, the photons have some probability to scatter off an electron and so bounce back into the line of sight. This reflection probability is set by the relative importance of scattering versus photoelectric absorption. For neutral material, photoelectric absorption dominates at low energies so the reflected fraction is very small. However, the photoelectric cross section decreases with energy, so the reflected fraction increases. Iron is the last astrophysically abundant element (due to element synthesis in stars as released in supernovae; see P. Podsiadlowski, Chapter 2, this volume), so after 7.1 keV, there are no more significant additional sources of opacity. The cross section decreases as E^{-3}, becoming equal to σ_T at around 10 keV for solar abundance material (Fig. 6.20a). Above this, scattering dominates, leading to a more constant reflected fraction, but at higher energies, the photon energy is such that Compton downscattering is important, so the reflection is no longer elastic. Photons at high energy do reflect, but do not emerge at the same energy as they are incident. This gives a break at high energies as Compton scattering conserves photon *number*, and the number of photons is much less at higher energies. Thus, neutral reflection gives rise to a very characteristic peak between 20 and 50 keV, termed the reflection hump, where lower energy photons are photoelectrically absorbed and higher photons are (predominantly) downscattered (Fig. 6.20b: George and Fabian, 1991; Matt et al., 1991).

The dependence on photoelectric absorption at low energies means that the spectrum is also accompanied by the associated emission lines as the excited ion with a gap in

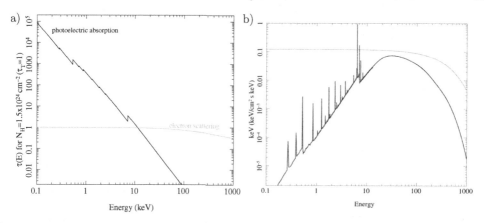

FIGURE 6.20. a) The absorption cross section for neutral material with solar abundance for a column of $\tau_T = 1$, that is, $N_H = 1.5 \times 10^{24}$ cm^{-2}, together with the full (Klein-Nishina) electron scattering cross section. b) The corresponding reflection spectrum from such material. The much larger absorption cross section at low energies means that most incident photons are absorbed rather than reflected, but Compton downscattering means that high-energy photons are not reflected elastically. The combination of these two effects makes the characteristic peak between 20 and 50 keV, termed the reflection hump, with atomic features superimposed on this. Figure courtesy of M. Gilfanov.

the K ($n = 1$) shell decays to its ground state. This excess energy can be emitted as a fluorescence line (Kα if it is the $n = 2$ to $n = 1$ transition, Kβ for $n = 3$ to $n = 1$, etc.). However, at low energies, the reflected emission forms only a very small contribution to the total spectrum, so any emission lines emitted below a few keV are strongly diluted by the incident continuum. These lines are also additionally suppressed because low–atomic-number elements have a higher probability to deexcite via Auger ionization, ejecting an outer electron rather than emitting the excess energy as a fluorescence line. This means that iron is the element that has most impact on the observed emission, as this is emitted where the fraction of reflected to incident spectrum is large and has the highest fluorescence probability. All this combines to make a reflection spectrum that contains the imprint of the iron K edge and line features, as well as the characteristic continuum peak between 20 and 50 keV (Fig. 6.20b: George and Fabian, 1991; Matt et al., 1991).

6.4.1 Ionized reflection

The dependence on photoelectric absorption at low energies means that reflection is sensitive to the ionization state of the reflecting material. Figure 6.21a shows how the absorption cross section changes as a function of ionization state using a very simple model for photoelectric absorption that considers only the photoelectric edge opacity. The progressive decrease in opacity at low energies for increased ionization state means an increase in reflectivity at these energies, as shown in Figure 6.21b for simple models of the reflected continuum (PEXRIV model in XSPEC).

However, this continuum is accompanied by emission lines and bound-free (recombination) continua, and the lines can be more important for ionized material. This is due to both an increase in emissivity (He-like lines especially have a high oscillator strength) and to the fact that the increased reflected fraction at low energies mean that these lines are not so diluted by the incident continuum. Better models of ionized reflection are shown in Figure 6.22. These include both the self-consistent emission lines and recombination continua, and the effects of Compton scattering within the disk. By definition, we only see down to a depth of $\tau(E) = 1$. Thus, the reflected continuum only escapes

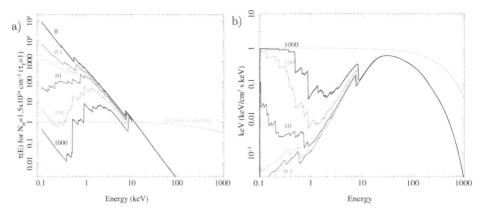

FIGURE 6.21. *a)* The ionized absorption cross section for material with solar abundance for a column of $\tau_T = 1$, that is, $N_H = 1.5 \times 10^{24}$ cm^{-2}. The curves are labeled with the value of ξ and shown together with the full (Klein-Nishina) electron scattering cross section as in Fig. 6.20. *b)* The corresponding reflection spectrum from such material. Increased ionization gives decreased opacity at low energies and hence more reflection. The reflected spectrum does not depend on photoelectric absorption at high energies and so is unchanged by ionization.

from above a depth of $\tau(E) \sim 1$. Figure 6.20a shows that the iron line is produced in a region with $\tau_T \sim 0.5$, so a fraction $e^{-\tau_T} \sim 1/3$ of the line is scattered. For neutral material, this forms a Compton downscattering shoulder to the line, but for ionized material, the disk is heated by the strong irradiation up to the Compton temperature. Compton upscattering can be important as well as downscattering, so the line and edge features are broadened (Young *et al.*, 2001; see Fig. 6.22). Models including these effects are publicly available as the XSPEC ATABLE model, REFLIONX.MOD, and this should be used rather than PEXRIV for ionized reflection. However, the incident continuum for this model is an exponential power law, so this can have problems with the continuum form at high energies (see Fig. 6.9a).

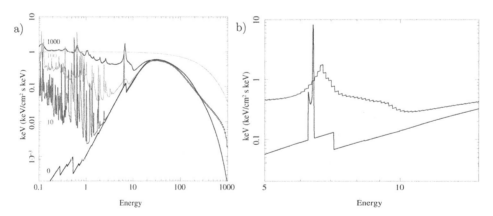

FIGURE 6.22. *a)* Ionized reflection models from a constant-density slab that include the self-consistent line and recombination continuum emission. The curves are labeled with the value of ξ and are clearly much more complex than the simple reflected continua models in Fig. 6.21. *b)* Detailed view of the iron line region. For neutral material (lower graph), around 1/3 of the line photons are scattered in the cool upper layers of the disk before escaping, forming a Compton downscattered shoulder to the line. Conversely, for highly ionized reflection (upper graph), the upper layers of the disk are heated to the Compton temperature, and Compton upscattering as well as downscattering broadens the spectral features.

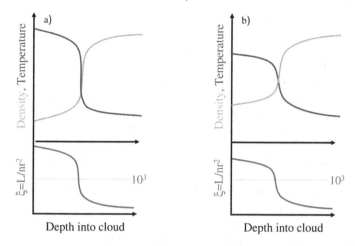

FIGURE 6.23. *a)* The ionization instability for material in hydrostatic equilibrium (or any general pressure balance) from hard spectral illumination. The high Compton temperature means that the material expands to very low density, and hence high ionization. *b)* The corresponding vertical structure from soft spectral illumination, where the lower Compton temperature gives higher surface density and hence lower ionization. Both show the rapid drop in ionization that comes from the dramatic increase in cooling from lines when partially ionized ions can exist.

6.4.2 Ionization instability: vertical structure of the disk

The slab models described previously assume that the irradiated material has constant density. Yet, if this material is a disk, it should be in hydrostatic equilibrium, so the density responds to the irradiation heating. The top of the disk is heated to the Compton temperature and thus expands, so its density drops and its ionization state is very high.

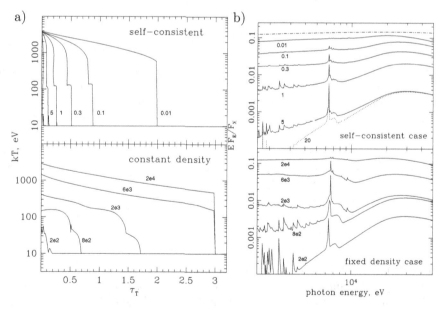

FIGURE 6.24. *a)* The vertical temperature structure from material in hydrostatic equilibrium (upper panel) shows a sharp drop in temperature due to the ionization instability. This is rather different from the much smoother drop in temperature seen for assuming the material has constant density (lower panel). *b)* The resulting differences in ionization structure of the disk photosphere give rise to different reflected spectra. In particular, the deep edge seen in the constant-density reflected spectrum at $\xi = 2{,}000$ is never present in the hydrostatic models (Nayakshin et al., 2000). Reproduced by permission of the AAS.

Farther into the disk, hydrostatic equilibrium means that the pressure has to increase in order to hold up the weight of the layers above. The Compton temperature remains the same, so the density has to increase, but an increase in density means an increase in importance of bremsstrahlung cooling. This pulls the temperature down further, but the pressure must increase so the density has to increase further still, so the cooling increases. Eventually the temperature/ionization state drops to low enough levels that not all the material is completely ionized. Bound electrons mean that line cooling can contribute, pulling the temperature down even faster, with a corresponding increase in density. The material thus makes a very rapid transition from almost completely ionized to almost completely neutral. Thus, the reflection spectrum is a composite of many different ionization parameters, with some contribution from the almost completely ionized skin, and some from the almost completely neutral material underlying the instability point, but with very little reflection at intermediate ionization states (Nayakshin et al., 2000; Done and Nayakshin, 2007).

The difference in Compton temperature for hard spectral illumination and soft spectral illumination only changes the ionization state of the skin. For hard illumination, the high Compton temperature gives a very low density and high ionization state, so the skin is almost completely ionized as described earlier. For softer illumination, the Compton temperature is lower, so the density is higher and the ionization state of the skin is lower, making it highly ionized rather than completely ionized. However, there is still the very rapid transition from highly ionized to almost neutral due to the extremely rapid increase in cooling from partially ionized material (Fig. 6.23; see also Done and Nayakshin, 2007). Figure 6.24a and b shows how this very different vertical structure for temperature-density-ionization affects the expected reflection signature.

The same underlying ionization instability, but for X-ray illumination of dense, cool clouds in pressure balance with a hotter, more diffuse medium (Krolik et al., 1981), may well be the origin of the multiple phases of ionization state seen in the AGN "warm absorbers" (e.g., Netzer et al., 2003; Chevallier et al., 2006).

6.4.3 Radial structure

This vertical structure of a disk should change with radius, giving rise to a different characteristic depth of the transition and hence a different balance between highly ionized reflection from the skin and neutral reflection from the underlying material. This does depend on the unknown source geometry, as well as the initial density structure of the disk, and how it depends on radius, which in turn depends on the (poorly understood) energy release in the disk (see, e.g., Nayakshin and Kallman, 2001). The XION model (Nayakshin) incorporates both the vertical structure from the ionization instability and its radial dependence for some assumed X-ray geometry and underlying disk properties.

However, if the material is mostly neutral, then neither vertical nor radial structure gives rise to a change in the reflected spectrum with radius. Neutral material remains neutral as the illumination gets weaker, so these models are very robust.

6.4.4 Relativistic broadening

The reflected emission from each radius has to propagate to the observer but it is emitted from material that is rapidly rotating in a strong gravitational field. There is a combination of effects that broaden the spectrum. First, the line-of-sight velocity gives a different Doppler shift from each azimuth, with maximum blueshift from the tangent point of the disk coming toward the observer, and maximum redshift from the tangent point on the receding side. Length contraction along the direction of motion means that the emission is beamed forward, so the blueshifted material is also brightened while the redshifted side of the disk is suppressed. These effects are determined only by the component of the velocity in the line of sight and so are not important for face-on disks. However, the material is intrinsically moving fast, in Keplerian rotation, so there is always time dilation (fast clocks run slow, also sometimes termed transverse redshift, as it occurs even if the velocity is completely transverse) and gravitational redshift.

All these effects decrease with increasing radius. The smaller Keplerian velocity means smaller Doppler shifts and lower boosting, giving less difference between the red and blue sides of the line. The lower velocity also means less time dilation while the larger radius means less gravitational redshift. Thus, larger radii give narrower lines, so the line profile is the inverse of the radial profile, with material furthest out giving the core of the line and material at the innermost orbit giving the outermost wings of the line (Fabian et al., 1989, 2000). The relative weighting between the inner and outer parts of the line are given by the radial emissivity, where the line strength $\propto r^{-\beta}$. This gives $\beta = 3$ for either an emissivity which follows the illumination pattern from a gravitationally powered corona, or "lamppost" point source illumination. This characteristic line profile is given by the DISKLINE (Fabian et al., 1989) and LAOR (Laor, 1991) models for Schwarzschild and extreme Kerr spacetimes, respectively.

However, these relativistic effects should be applied to the entire reflected continuum, not just the line. This can be modeled with the KDBLUR model (a re-coding of the LAOR model for convolution), or the newer KY models which work for any spin (Dovčiak et al., 2004). Figure 6.25a shows the REFLIONX.MOD reflection ATABLE convolved with KDBLUR for $r_{in} = 30, 6$ and $1.23 R_g$ for $i = 60°$. Blueshifts slightly predominate over redshifts, with the "edge" energy (actually predominantly set by the blue wing of the line) at 7.8 keV for r_{in} =6 and 1.23 compared to 7.1 keV for $r_{in} = 30$ and in the intrinsic slab spectrum (gray). Redshifts are more important for lower inclinations. Figure 6.25b shows a comparison of the iron line region for $i = 60°$ (dotted lines) with $i = 30°$ (solid lines). The "edge" energy is now ∼6.7 keV for both r_{in} =1.235 and 6.

Relativistic smearing is harder to disentangle for ionized material as the iron features are broadened by Compton scattering so there are no intrinsically sharp features to track the relativistic effects (see Fig. 6.25c and d). Nonetheless, the energies at which the line/edge features are seen are clearly shifted.

6.4.5 Observations of reflected emission in AGN: iron line and soft X-ray excess

Reflection is seen in AGN. There is a narrow, neutral iron line and reflection continuum from illumination of the torus, and there is also a broad component from the disk. This broad component is often consistent with neutral reflection from material within $50 R_g$ produced from r^{-3} illumination (Nandra et al., 2007). However, a small but significant fraction of objects require much more extreme line parameters, with $r_{in} < 3 R_g$, and much more centrally concentrated illumination $\propto r^{-\beta}$ where $\beta = 5$ to 6!! (MCG-6-30-15: Wilms et al., 2001; Fabian and Vaughan, 2003; 1H0707-495: Fabian et al., 2002, 2004; NGC4051: Ponti et al., 2006) These are generally high mass accretion rate objects, predominantly narrow-line Seyfert 1s.

However, the X-ray spectrum is not simply made up of the power law and its reflection. There is also a "soft X-ray excess," clearly seen in the spectra of most high mass accretion rate AGN below 1 keV. Figure 6.26a shows the changing shape of this soft excess for different intensity sorted spectra of Mkn766. This is rather smooth, so looks like a separate continuum component. However, fitting this for a large sample of AGN gives a typical "temperature" of this component that is remarkably constant irrespective of mass of the black hole (Czerny et al., 2003; Gierliński and Done, 2004b). This is very unlike the behavior of a disk or any component connected to a disk, requiring some (currently unknown) "thermostat" to maintain this temperature. Another, more subtle problem is that "normal" BHB do not show such a component in their spectra, so these AGN spectra do not exactly correspond to a scaled-up version of the BHB high/very high states. However, there *is* such a separate component in the most luminous state of

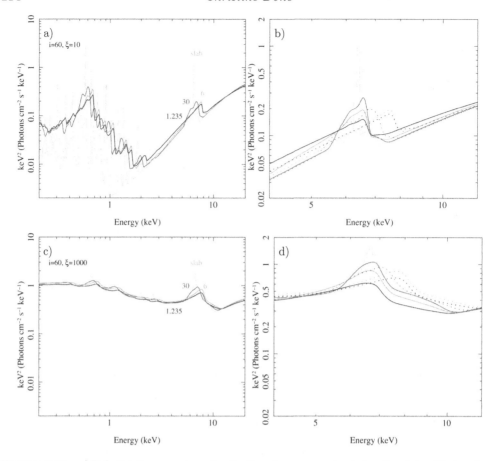

FIGURE 6.25. a) Relativistic smearing of reflection from a constant density slab with low ionization ($\xi = 10$) viewed at 60°. The intrinsic emission is convolved with the relativistic functions with $r_{\rm in} = 30$, 6 and 1.235. b) focuses in on the iron line region to show the difference in line profile for $i = 30°$ (solid lines) compared to $i = 60°$ (dotted lines). c) and d) show the same for highly ionized material ($\xi = 1{,}000$). The relativistic effects shift the shape of the line, which becomes much less marked as the spectral features are intrinsically broadened by Comptonization in the strongly irradiated material.

the brightest BHB GRS 1915+105 that may scale up to produce the soft excess in the most extreme AGN (Middleton et al., 2009; see Fig. 6.27c).

More clues to the nature of the soft excess can come from its variability. These objects are typically highly variable, and the spectrum changes as a function of intensity in a very characteristic way. At the highest X-ray luminosities, the 2- to 10-keV spectrum can often be well described by a $\Gamma \sim 2.1$ power law, with resolved iron emission line, and a soft excess that is a factor ~ 2 above the extrapolated 2 to 10 keV power-law at 0.5 keV. Conversely, at the lowest luminosities, the apparent 2- to 10-keV power-law index is much harder (and there are absorption systems from highly ionized iron), and the soft excess above this extrapolated emission is much larger. Figure 6.26a shows this spectral variability for Mkn766.

The entire spectrum at the lowest luminosity looks like moderately ionized reflection (with the iron K alpha He and H like absorption lines superimposed), but the lack of soft X-ray lines (as well as the lack of a resolved iron emission line) means it would have to be extremely strongly distorted by relativistic effects. The classic extreme iron line source, MCG-6-30-15, has similar spectral variability but with stronger "warm absorption" features around 0.7 to 1 keV and stronger ionized iron absorption lines

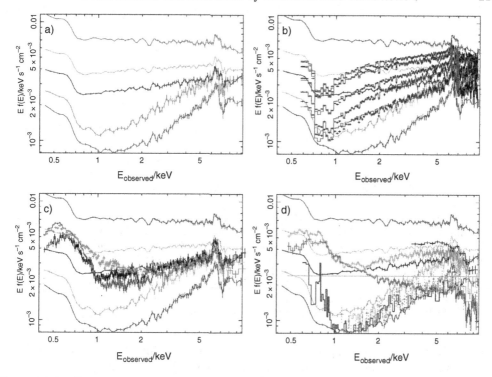

FIGURE 6.26. Intensity-sorted spectra from a selection of high mass accretion rate AGN. *a)* shows Mkn766 (Miller et al., 2007). This has a weak soft excess and soft 2- to 10-keV spectrum, together with a weak emission line at the highest luminosities, while the lowest flux states have a strong soft excess, hard 2–7 keV spectrum, with a strong drop above this, enhanced by strong He- and H-like absorption lines. *b)* shows the same for MCG-6-30-15 (Miller et al., 2008). This shows very similar variability, though it spans a smaller range. It also has a stronger "warm absorber," seen by the drop at 0.7 to 1 keV, and stronger absorption lines of He- and H-like iron on the blue wing of the line. *c)* shows PG1211+10, with even less range in variability but with ionized iron absorption that is strongly blueshifted (Reeves et al., 2008; Pounds et al., 2003) *d)* shows PDS456 (Reeves et al., 2009), where the lowest intensity spectra look like those from Mkn766, but with a stronger drop blueward of the line and even more highly blueshifted iron absorption lines. However, this can also show a rather different spectral shape, with steeper continuum. Reproduced by permission of the AAS.

(see Fig. 6.26b). In both these sources, most of the spectral variability can be modeled if there is an extremely relativistically smeared reflection component that remains constant, while the $\Gamma = 2.1$ power law varies, giving increasing dilution of the reflected component at high fluxes (Fabian and Vaughan, 2003).

The obvious way for the reflected emission to remain constant is if it is produced by far-off material, but the extreme smearing requirement conflicts with this. Instead, lightbending from a source very close to the event horizon could give both the required central concentration of the illumination pattern and apparent constancy of reflection if the variability is dominated by changes in source position giving changes in lightbending (Miniutti et al., 2003; Fabian et al., 2004; Miniutti and Fabian, 2004). As the X-ray source gets closer to the black hole, the X-ray emission is increasingly focused onto the inner disk, and so the amount of direct emission seen drops.

However, there are some physical issues with this solution. First, unlike the iron line, the soft excess requires extreme smearing (small inner radius, strongly centrally concentrated emissivity) in almost all the objects in order to smooth out the strong soft X-ray line emission predicted by reflection (Crummy et al., 2006). Second, to produce the soft excess from reflection requires that the reflecting material be moderately ionized

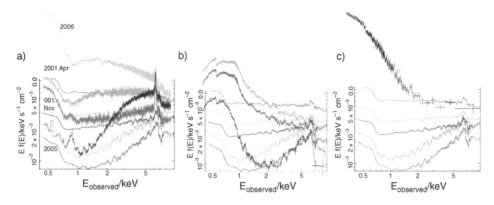

FIGURE 6.27. As in Fig. 6.26, with Mkn766 as background but now for objects showing somewhat different variability. a) shows NGC 3516 (Turner et al., 2008). Plainly there is some contribution here from complex absorption. b) shows 1H0707-495, which has a hardest spectrum that is similar to that from Mkn766, but with an enormous drop above 7 keV (Boller et al., 2002) and an extremely strong soft excess. However, the softer spectra at higher luminosity look more like the strange spectrum seen from PDS456 (Fig. 6.26d), but also have clear features from iron L emission (Fabian et al., 2009). c) shows RE J1034+396, one of the strongest soft excess known, which may be a real continuum component (Middleton et al., 2009).

over much of the disk photosphere. Yet the ionization instability for material in pressure balance means that only a very small fraction of the disk photosphere can be in such a partially ionized state (see Section 6.4.2). Either the disk is not in hydrostatic equilibrium (held up by magnetic fields?) or the soft excess is not formed from reflection (Done and Nayakshin, 2007).

Instead, it is possible that the soft excess is formed by partially ionized material seen in absorption through optically thin material rather than reflection from optically thick material. However, there is still the issue of the lack of the expected partially ionized lines – in absorption this time rather than from emission (see Fig. 6.17b). These could be smeared out in a similar way to reflection if the wind is outflowing (Gierliński and Done, 2004b; Middleton et al., 2007), but the outflow velocities required are as extreme as the rotation velocities (Schurch and Done, 2008; Schurch et al., 2009). Even including scattering in a much more sophisticated wind outflow model shows that "reasonable" outflow velocities are insufficient to blend the atomic features at low energies into a pseudo-continuum, though this can explain the broad iron line shape without requiring extreme relativistic smearing (Sim et al., 2010).

Potentially a more plausible geometry is if the absorber is clumpy and only partially covers the source. In this case there are multiple lines of sight through different columns. Any mostly neutral material gives curvature underneath the iron line, making an alternative to extreme reflection for the origin of the red wing (Miller et al., 2007, 2008; Turner et al., 2008). Such neutral material will produce an iron fluorescence line, but this line can also be absorbed, so current observations cannot yet distinguish between these two models (Yaqoob et al., 2010). This much more messy geometry perhaps gives more potential to explain the larger range of complex variability seen in some of the other high mass accretion rate AGN, as shown in Fig. 6.27a and b.

It is clearly important to find out which one of these geometries we are looking at. Either we somehow have a clean line of sight down to the very innermost regions of the disk despite it being an intense UV source which should be powering a strong wind, or we are looking through a material in strong, clumpy wind, which has important implications for AGN feedback models, giving another way to strongly suppress nuclear star formation. These questions are currently an area of intense controversy and active research.

6.4.6 Observations of reflected emission and relativistic smearing in BHB

Reflection is also seen in BHB, and here the controversy over its interpretation occurs at both low and high mass accretion rates.

The amount of reflection is determined by the solid angle subtended by the disk to the hard X-ray source, that is, the fraction of the sky that is covered by optically thick material as viewed from the hard X-ray emission region. The truncated disk models for low mass accretion rates predict that this should increase as the disk moves in toward the last stable orbit, identified with the source making a transition from the low/hard to high/soft state, while the decrease in inner disk radius means that this should also be more strongly smeared by relativistic effects. With RXTE data, the solid angle of reflection increases as expected, but the poor spectral resolution means that it is very difficult to constrain the relativistic smearing. Nonetheless, these do appear more smeared in the RXTE data (Gilfanov et al., 1999; Zdziarski et al., 1999; Ibragimov et al., 2005; Gilfanov, 2010).

These results were derived assuming neutral reflection, whereas the reflected spectrum is plainly ionized in the high/soft states (e.g., Gierliński et al., 1999). Although some of this ionization can be from photoionization by the illuminating flux, at least part of it must be due to the high disk temperature (i.e., collisional ionization: Ross and Fabian, 2007). Thus, the high/soft and very high state require fitting with complex ionized reflection models in order to disentangle the relativistic smearing and solid angle from the ionization state. Nonetheless, attempts at this using simplistic models of ionized reflection (PEXRIV) with the RXTE data gave fairly consistent answers. The high/soft data seemed to show that the solid angle subtended by the disk is of order unity, and the (poorly constrained) smearing gives $r_{in} \approx 6$ for emissivity fixed at $\beta = 3$, as expected from the potential geometries sketched in Fig. 6.12b (Gierliński et al., 1999; Zycki et al., 1998). By contrast, the very high state geometry discussed in Section 6.2.5 required that the inner disk be covered by an optically thick corona, predicting a smaller amount of reflection and smearing, again consistent with the RXTE observations (Done and Kubota, 2006).

However, the first moderate and good spectral resolution results appeared to conflict with the neat picture just described. The extent of this is best seen in Miller et al. (2009), who compiled some XMM-Newton spectra of BHB (plus a few datasets from other satellites) and fitted with the best current relativistically smeared, ionized reflection models (together with a power law and disc spectrum). Their table 3 shows that all the very high state spectra ($\Gamma > 2.4$) require a large solid angle of reflection from the very inner regions of the disk, at odds with the geometry proposed from the continuum shape where an optically thick corona completely covers the cool material (Done and Kubota, 2006).

There are even worse conflicts with the models for the low/hard state. The truncated disk/hot inner flow models clearly predict that the amount of relativistic smearing should be lower than in the high/soft state. However, XMM-Newton data from low/hard state observations also show a line that is so broad that the disc is required to extend down to the last stable orbit of a high-spin black hole. The most famous of these is GX339-4 (Miller et al., 2006b; Reis et al., 2008), though this one is probably an artifact of instrumental distortion of the data due to pileup (Fig. 6.28a; Done and Diaz Trigo, 2010). However, there are other data that also show a line in the low/hard state that is so broad that the disk is required to extend down to the last stable orbit of a rapidly spinning black hole (SAX J1711.6-3808: Miller et al. 2009; GRO J1655-40 and XTE J1650-500: Reis et al. 2009b, 2010). Although these are not so compelling as the (piled up) data from GX339-4, they still rule out the truncated disk/hot inner flow models for the low/hard state if the extreme broad line is the only interpretation of the spectral shape.

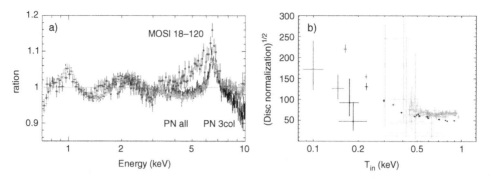

FIGURE 6.28. *a)* Residuals to a continuum model fit to the low/hard state of GX339-4. The dark gray points show the MOS data on XMM-Newton, with an obvious, extremely broad red wing to the iron line (Miller *et al.*, 2006b). However, these data suffer from pileup, and the simultaneous PN instrument on XMM-Newton, which is not piled up, show a much narrower profile (black and light gray data: Done and Diaz Trigo, 2010). *b)* Temperature versus scaled inferred inner disk radius for SWIFT (black) and RXTE (gray triangles) data from XTE J1817-330 fit to a disk plus Comptonization model. The source is in the high/soft state for temperatures above 0.4 keV, as shown by the constant radius. The two lowest-temperature points at ∼0.2 keV are low/hard states, whereas the ones in the 0.2 to 0.4-keV range are taken during the transition (intermediate state). The 3-keV lower limit to the RXTE bandpass means that it cannot follow the disk as the source goes into the transition, but the SWIFT data show that the disk radius starts to increase. A simple interpretation of the low/hard state data is that the disk returns to the last stable orbit, but the red points show how this can change by including irradiation, and the radius can be further increased to 250 to 300 in these units by going from a stress-free inner boundary to stress across the truncated disk edge. The disk radius cannot be unambiguously determined from the low/hard state data (Gierliński *et al.*, 2008). (See the figures in the original papers for color versions.)

6.5 The nature of the low/hard state in BHB and AGN

This conflict motivates us to look again at the low/hard state in particular, especially as there are other observations that also challenge the hot inner flow/truncated disk geometry. Again, this is currently an area of intense controversy and active research.

6.5.1 *Intrinsic disk emission close to the transition*

The high/soft state disk dominated spectra trace out $L \propto T^4$, giving strong evidence for a constant size scale inner radius (see Section 6.1.4). After the transition to the low/hard state, there is still a (weak) soft X-ray component that can be seen in CCD spectra (but not in RXTE because of its low-energy bandpass limit of 3 keV). This has temperature and luminosity that is consistent with the same radius as seen in the high/soft state data, implying that the disk does not truncate (Rykoff *et al.*, 2007; Reis *et al.*, 2010).

Figure 6.28b illustrates this with both RXTE (gray triangles) and SWIFT (black) data from the transient XTE J1817-330 (Rykoff *et al.*, 2007; Gierliński *et al.*, 2008). The disk goes below the RXTE bandpass just as the source makes a transition from the high/soft state, but its evolution can be followed by the lower energy data from SWIFT. Plainly the SWIFT data show that the disk does recede during (i.e., triggering!) the transition, but then in the low/hard state proper, it apparently bounces back to give the same radius as in the high/soft state. However, just after the transition, the disk is not too far recessed, so it can be strongly illuminated by the energetically dominant hard X-ray source. This changes the temperature/luminosity relation from that expected from just gravitational energy release, as shown by the light gray points in Figure 6.28. Additionally, the difference in inner boundary condition (going from the stress-free last stable orbit to a continuous stress across the truncated inner disk radius) means that the same temperature/luminosity relation implies the disk is bigger, which would move the two low/hard state points up to 250 to 300 in these units! This is even before taking into account that some of the disk photons are lost to our line of sight through Compton scattering if the disk underlies some of the hot inner flow. Putting these photons back into the disk gives higher luminosity/larger radii (Makishima et al., 2008).

These data show the difficulty in unambiguously reconstructing the inner radius of the disk when the disk component does not dominate the spectrum. They can be consistent with the disk down at the last stable orbit (Rykoff et al., 2007; Reis et al., 2010), but they can equally well be consistent with a truncated disk (Gierliński et al., 2008). However, the data during the transition make it fairly clear that the disk starts to recede, so it seems most likely to me that this continues into the low/hard state.

6.5.2 Intrinsic disk emission at very low luminosities

The transition is going to be complicated. The disk surely does not truncate in a smooth way, so there can be turbulent clumps, as well as issues with the overlap region suppressing the observed disk while also giving rise to strong illumination as discussed earlier. Instead, a much cleaner picture should emerge from a dimmer low/hard state if the disk truncates: here it should be far from the X-ray source, so irradiation and overlap effects should be negligible. Yet there is still a weak soft X-ray component, with temperature and luminosity such that the emitting area implied is very small, of order the size scale of the last stable orbit. This has now been seen in several different CCD observations of low $L/L_{\rm Edd}$ sources, so it is clearly a robust result (Reis et al., 2009a; Chiang et al., 2010; Wilkinson and Uttley, 2009).

However, putting the disk down to the last stable orbit is not a solution to these data. It is then cospatial with the hard X-ray emission, as this must be produced on small size scales. This then runs into difficulties with reprocessing. If the hard X-ray corona overlies the optically thick disk and emits isotropically, then half of the hard X-ray emission illuminates the disk. Some fraction, a (the albedo), of this is reflected, but the remainder is thermalized in the disk and adds to its luminosity. The minimum disk emission is where there is no intrinsic gravitational energy release in the disk, only this reprocessed flux. This is $L_{\rm rep} = (1-a)L_h/2 \sim L_h/3$, as the reflection albedo cannot be high for hard spectral illumination because high-energy X-rays cannot be reflected elastically. Instead they deposit their energy in the disk via Compton downscattering. Yet we see $L_{\rm soft} \sim L_h/20$. Thus the geometry must be wrong! Either the hard X-ray source is not isotropic, perhaps beamed away from the disk as part of the jet emission, or the material is a small ring rather than a disk so that its solid angle is much less than 2π for a full disk (Chiang et al., 2010; Done and Diaz Trigo, 2010).

However, for one source, XTE J1118+480, the galactic column density is so low that there are simultaneous UV and even EUVE constraints on the spectrum (Esin et al., 2001). These show that this soft X-ray component coexists with a much more luminous, cooler UV/EUV component. If the soft X-rays are the disk, what is the UV/EUV component? Alternatively, since the UV/EUV component looks like a truncated disk, what is the soft X-ray component? It must come from a much smaller emission area than the UV/EUVE emission, and one potential origin is the inner edge (rather than the top and bottom face) of the truncated disk (Chiang et al., 2010). The truncation region is probably highly turbulent, so there can be intrinsic variability produced by clumps forming and shredding, as well as reprocessing the hard X-ray irradiation. This may also explain the variability seen in this component (Wilkinson and Uttley, 2009).

Thus, in my opinion, none of the current observations require that there be a disk down to the last stable orbit in the low/hard state. More fundamentally, it is very difficult to make such a model not conflict with other observations. Reprocessing limits on the hardness of the spectrum require that the X-ray source be either patchy or beamed away from the disk (Stern et al., 1995; Beloborodov, 1999). A patchy corona would give a reflection fraction close to unity, which is not observed even considering complex ionization of the reflected emission (Barrio et al., 2003; Malzac et al., 2005). Beaming naturally associates the X-ray source with the jet but is unlikely to be able to simultaneously explain the extremely broad line parameters derived for some low/hard state sources, since the illumination pattern becomes much less centrally concentrated by the beaming. I suspect that more complex continuum modeling may make these lines less extreme, but this then removes a challenge to the truncated disk as well! Also, unlike the beaming models, the truncated disk/hot inner flow can additionally give a mechanism for the major state transitions and the variability.

6.6 Conclusions

The intrinsic radiation processes of blackbody radiation and Comptonization go a long way to explaining the underlying optical-to-X-ray continuum seen in both BHB and AGN. These, together with the atomic processes of absorption and reflection and relativistic effects in strong gravity, give us a toolkit with which to understand and interpret the spectra of black-hole accretion flows. This is currently an area of intense and exciting research, to try to understand accretion in strong gravity. If you got this far, congratulations, and come and join us: we get to play around black holes!

6.7 Acknowledgments

I thank the organizers of the IAC winter school for inviting me to give this series of lectures, finally giving me the motivation to write these things down. But I only know these things because of the many people who have given me their physical insight on radiation processes, especially Andy Fabian and Gabriele Ghisellini. I also thank ISAS and RIKEN for visits during which I developed some of these lectures. This chapter also could not have been written without the 8 hours spent at Tenerife South Airport, where their lack of a baggage check facility put paid to my plan to spend the day at a nearby surfing beach!

REFERENCES

Arévalo, P., and Uttley, P. 2006. Investigating a fluctuating-accretion model for the spectral-timing properties of accreting black hole systems. MNRAS, **367**(Apr.), 801–814.

Arnaud, K. A. 1996. XSPEC: The first ten years. Pages 17+ of: G. H. Jacoby and J. Barnes (ed), *Astronomical Data Analysis Software and Systems V*. Astronomical Society of the Pacific Conference Series, vol. 101.

Balucinska-Church, M., and McCammon, D. 1992. Photoelectric absorption cross sections with variable abundances. ApJ, **400**(Dec.), 699+.

Barrio, F. E., Done, C., and Nayakshin, S. 2003. On the accretion geometry of Cyg X-1 in the low/hard state. MNRAS, **342**(June), 557–563.

Begelman, M. C., McKee, C. F., and Shields, G. A. 1983. Compton heated winds and coronae above accretion disks. I Dynamics. ApJ, **271**(Aug.), 70–88.

Belloni, T. M. 2010 (Mar.). States and transitions in black hole binaries. Pages 53+ of: T. Belloni (ed.), *Lecture Notes in Physics*. Lecture Notes in Physics, Springer, vol. 794.

Beloborodov, A. M. 1999. Plasma ejection from magnetic flares and the X-ray spectrum of Cygnus X-1. ApJL, **510**(Jan.), L123–L126.

Blustin, A. J., Page, M. J., Fuerst, S. V., Branduardi-Raymont, G., and Ashton, C. E. 2005. The nature and origin of Seyfert warm absorbers. A&A, **431**(Feb.), 111–125.

Boller, T., Fabian, A. C., Sunyaev, R., Trümper, J., Vaughan, S., Ballantyne, D. R., Brandt, W. N., Keil, R., and Iwasawa, K. 2002. XMM-Newton discovery of a sharp spectral feature at ~ 7 keV in the narrow-line Seyfert 1 galaxy 1H 0707-49. MNRAS, **329**(Jan.), L1–L5.

Chevallier, L., Collin, S., Dumont, A.-M., Czerny, B., Mouchet, M., Gonçalves, A. C., and Goosmann, R. 2006. The role of absorption and reflection in the soft X-ray excess of active galactic nuclei. I. Preliminary results. A&A, **449**(Apr.), 493–508.

Chiang, C. Y., Done, C., Still, M., and Godet, O. 2010. An additional soft X-ray component in the dim low/hard state of black hole binaries. MNRAS, **403**(Apr.), 1102–1112.

Churazov, E., Gilfanov, M., and Revnivtsev, M. 2001. Soft state of Cygnus X-1: stable disc and unstable corona. MNRAS, **321**(Mar.), 759–766.

Coppi, P. S. 1999. The physics of hybrid thermal/non-thermal plasmas. Pages 375+ of: J. Poutanen and R. Svensson (eds.), *High Energy Processes in Accreting Black Holes*. Astronomical Society of the Pacific Conference Series, vol. 161.

Crummy, J., Fabian, A. C., Gallo, L., and Ross, R. R. 2006. An explanation for the soft X-ray excess in active galactic nuclei. MNRAS, **365**(Feb.), 1067–1081.

Cunningham, C. T. 1975. The effects of redshifts and focusing on the spectrum of an accretion disk around a Kerr black hole. ApJ, **202**(Dec.), 788–802.

Czerny, B., Nikołajuk, M., Różańska, A., Dumont, A.-M., Loska, Z., and Zycki, P. T. 2003. Universal spectral shape of high accretion rate AGN. A&A, **412**(Dec.), 317–329.

Davis, S. W., Blaes, O. M., Hubeny, I., and Turner, N. J. 2005. Relativistic accretion disk models of high-state black hole X-ray binary spectra. ApJ, **621**(Mar.), 372–387.

Davis, S. W., Done, C., and Blaes, O. M. 2006. Testing accretion disk theory in black hole X-ray binaries. ApJ, **647**(Aug.), 525–538.

de Martino, D., Matt, G., Belloni, T., Haberl, F., and Mukai, K. 2004. BeppoSAX observations of soft X-ray intermediate polars. A&A, **415**(Mar.), 1009–1019.

Di Matteo, T., Celotti, A., and Fabian, A. C. 1997. Cyclo-synchrotron emission from magnetically dominated active regions above accretion discs. MNRAS, **291**(Nov.), 805+.

Done, C., and Davis, S. W. 2008. Angular momentum transport in accretion disks and its implications for spin estimates in black hole binaries. ApJ, **683**(Aug.), 389–399.

Done, C., and Diaz Trigo, M. 2010. A re-analysis of the iron line in the XMM-Newton data from the low/hard state in GX339-4. MNRAS, **407**(July), 2287–2296.

Done, C., and Kubota, A. 2006. Disc-corona energetics in the very high state of galactic black holes. MNRAS, **371**(Sept.), 1216–1230.

Done, C., and Magdziarz, P. 1998. Complex absorption and reflection of a multitemperature cyclotron-bremsstrahlung X-ray cooling shock in BY Cam. MNRAS, **298**(Aug.), 737–746.

Done, C., and Nayakshin, S. 2007. Can the soft excess in AGN originate from disc reflection? MNRAS, **377**(May), L59–L63.

Done, C., Gierliński, M., and Kubota, A. 2007. Modelling the behaviour of accretion flows in X-ray binaries. Everything you always wanted to know about accretion but were afraid to ask. A&A Rev., **15**(Dec.), 1–66.

Done, C., Mulchaey, J. S., Mushotzky, R. F., and Arnaud, K. A. 1992. An ionized accretion disk in Cygnus X-1. ApJ, **395**(Aug.), 275–288.

Dovčiak, M., Karas, V., and Yaqoob, T. 2004. An extended scheme for fitting X-ray data with accretion disk spectra in the strong gravity regime. ApJS, **153**(July), 205–221.

Dunlop, J. S., McLure, R. J., Kukula, M. J., Baum, S. A., O'Dea, C. P., and Hughes, D. H. 2003. Quasars, their host galaxies and their central black holes. MNRAS, **340**(Apr.), 1095–1135.

Durant, M., Gandhi, P., Shahbaz, T., Fabian, A. P., Miller, J., Dhillon, V. S., and Marsh, T. R. 2008. SWIFT J1753.5-0127: A Surprising Optical/X-Ray Cross-Correlation Function. ApJL, **682**(July), L45–L48.

Esin, A. A., McClintock, J. E., and Narayan, R. 1997. Advection-dominated Accretion and the Spectral States of Black Hole X-Ray Binaries: Application to Nova MUSCAE 1991. ApJ, **489**(Nov.), 865+.

Esin, A. A., McClintock, J. E., Drake, J. J., Garcia, M. R., Haswell, C. A., Hynes, R. I., and Muno, M. P. 2001. Modeling the Low-State Spectrum of the X-Ray Nova XTE J1118+480. ApJ, **555**(July), 483–488.

Fabian, A. C., and Vaughan, S. 2003. The iron line in MCG-6-30-15 from XMM-Newton: evidence for gravitational light bending? MNRAS, **340**(Apr.), L28–L32.

Fabian, A. C., Ballantyne, D. R., Merloni, A., Vaughan, S., Iwasawa, K., and Boller, T. 2002. How the X-ray spectrum of a narrow-line Seyfert 1 galaxy may be reflection-dominated. MNRAS, **331**(Apr.), L35–L39.

Fabian, A. C., Iwasawa, K., Reynolds, C. S., and Young, A. J. 2000. Broad Iron Lines in Active Galactic Nuclei. PASP, **112**(Sept.), 1145–1161.

Fabian, A. C., Miniutti, G., Gallo, L., Boller, T., Tanaka, Y., Vaughan, S., and Ross, R. R. 2004. X-ray reflection in the narrow-line Seyfert 1 galaxy 1H 0707-495. MNRAS, **353**(Oct.), 1071–1077.

Fabian, A. C., Rees, M. J., Stella, L., and White, N. E. 1989. X-ray fluorescence from the inner disc in Cygnus X-1. MNRAS, **238**(May), 729–736.

Fabian, A. C., Zoghbi, A., Ross, R. R., Uttley, P., Gallo, L. C., Brandt, W. N., Blustin, A. J., Boller, T., Caballero-Garcia, M. D., Larsson, J., Miller, J. M., Miniutti, G., Ponti, G., Reis, R. C., Reynolds, C. S., Tanaka, Y., and Young, A. J. 2009. Broad line emission from iron K- and L-shell transitions in the active galaxy 1H0707-495. Nature, **459**(May), 540–542.

Fanidakis, N., Baugh, C. M., Benson, A. J., Bower, R. G., Cole, S., Done, C., and Frenk, C. S. 2009. Grand unification of AGN activity in the LambdaCDM cosmology. ArXiv e-prints, Nov.

Gammie, C. F., Shapiro, S. L., and McKinney, J. C. 2004. Black hole spin evolution. ApJ, **602**(Feb.), 312–319.

Gandhi, P., Makishima, K., Durant, M., Fabian, A. C., Dhillon, V. S., Marsh, T. R., Miller, J. M., Shahbaz, T., and Spruit, H. C. 2008. Rapid optical and X-ray timing observations of GX 339-4: flux correlations at the onset of a low/hard state. MNRAS, **390**(Oct.), L29–L33.

George, I. M., and Fabian, A. C. 1991. X-ray reflection from cold matter in active galactic nuclei and X-ray binaries. MNRAS, **249**(Mar.), 352–367.

Ghisellini, G. 1989. Synchrotron self Compton models for compact sources – the case of a steep power-law particle distribution. MNRAS, **236**(Jan.), 341–351.

Gierliński, M., and Done, C. 2003. The X-ray/γ-ray spectrum of XTE J1550-564 in the very high state. MNRAS, **342**(July), 1083–1092.

Gierliński, M., and Done, C. 2004a. Black hole accretion discs: reality confronts theory. MNRAS, **347**(Jan.), 885–894.

Gierliński, M., and Done, C. 2004b. Is the soft excess in active galactic nuclei real? MNRAS, **349**(Mar.), L7–L11.

Gierliński, M., and Zdziarski, A. A. 2005. Patterns of energy-dependent variability from Comptonization. MNRAS, **363**(Nov.), 1349–1360.

Gierliński, M., Done, C., and Page, K. 2008. X-ray irradiation in XTE J1817-330 and the inner radius of the truncated disc in the hard state. MNRAS, **388**(Aug.), 753–760.

Gierliński, M., Zdziarski, A. A., Poutanen, J., Coppi, P. S., Ebisawa, K., and Johnson, W. N. 1999. Radiation mechanisms and geometry of Cygnus X-1 in the soft state. MNRAS, **309**(Oct.), 496–512.

Gilfanov, M. 2010 (Mar.). X-ray emission from black-hole binaries. Pages 17+ of: T. Belloni (ed.), *Lecture Notes in Physics*. Lecture Notes in Physics, Springer, vol. 794.

Gilfanov, M., Churazov, E., and Revnivtsev, M. 1999. Reflection and noise in Cygnus X-1. A&A, **352**(Dec.), 182–188.

Gladstone, J., Done, C., and Gierliński, M. 2007. Analysing the atolls: X-ray spectral transitions of accreting neutron stars. MNRAS, **378**(June), 13–22.

Gou, L., McClintock, J. E., Liu, J., Narayan, R., Steiner, J. F., Remillard, R. A., Orosz, J. A., Davis, S. W., Ebisawa, K., and Schlegel, E. M. 2009. A determination of the spin of the black hole primary in LMC X-1. ApJ, **701**(Aug.), 1076–1090.

Gou, L., McClintock, J. E., Steiner, J. F., Narayan, R., Cantrell, A. G., Bailyn, C. D., and Orosz, J. A. 2010. The spin of the black hole in the soft X-ray transient A0620-00. ApJL, **718**(Aug.), L122–L126.

Haardt, F., and Maraschi, L. 1993. X-ray spectra from two-phase accretion disks. ApJ, **413**(Aug.), 507–517.

Hynes, R. I., Brien, K. O., Mullally, F., and Ashcraft, T. 2009. Echo mapping of Swift J1753.5-0127. MNRAS, **399**(Oct.), 281–286.

Ibragimov, A., Poutanen, J., Gilfanov, M., Zdziarski, A. A., and Shrader, C. R. 2005. Broadband spectra of Cygnus X-1 and correlations between spectral characteristics. MNRAS, **362**(Oct.), 1435–1450.

Ingram, A., and Done, C. 2010. A physical interpretation of the variability power spectral components in accreting neutron stars. MNRAS, **405**(July), 2447–2452.

Ingram, A., Done, C., and Fragile, P. C. 2009. Low-frequency quasi-periodic oscillations spectra and Lense-Thirring precession. MNRAS, **397**(July), L101–L105.

Juett, A. M., Schulz, N. S., and Chakrabarty, D. 2004. High-resolution X-ray spectroscopy of the interstellar medium: structure at the oxygen absorption edge. ApJ, **612**(Sept.), 308–318.

Kallman, T., and Bautista, M. 2001. Photoionization and high-density gas. ApJS, **133**(Mar.), 221–253.

Kallman, T. R., Bautista, M. A., Goriely, S., Mendoza, C., Miller, J. M., Palmeri, P., Quinet, P., and Raymond, J. 2009. Spectrum synthesis modeling of the X-ray spectrum of GRO J1655-40 taken during the 2005 outburst. ApJ, **701**(Aug.), 865–884.

Kallman, T. R., Palmeri, P., Bautista, M. A., Mendoza, C., and Krolik, J. H. 2004. Photoionization modeling and the K lines of iron. ApJS, **155**(Dec.), 675–701.

Kanbach, G., Straubmeier, C., Spruit, H. C., and Belloni, T. 2001. Correlated fast X-ray and optical variability in the black-hole candidate XTE J1118+480. Nature, **414**(Nov.), 180–182.

King, A. R., and Kolb, U. 1999. The evolution of black hole mass and angular momentum. MNRAS, **305**(May), 654–660.

King, A. R., Pringle, J. E., and Livio, M. 2007. Accretion disc viscosity: how big is alpha? MNRAS, **376**(Apr.), 1740–1746.

Kolehmainen, M., and Done, C. 2010. Limits on spin determination from disc spectral fitting in GX 339-4. MNRAS, **406**(Aug.), 2206–2212.

Körding, E. G., Jester, S., and Fender, R. 2006. Accretion states and radio loudness in active galactic nuclei: analogies with X-ray binaries. MNRAS, **372**(Nov.), 1366–1378.

Kotani, T., Ebisawa, K., Dotani, T., Inoue, H., Nagase, F., Tanaka, Y., and Ueda, Y. 2000. ASCA observations of the absorption line features from the superluminal jet source GRS 1915+105. ApJ, **539**(Aug.), 413–423.

Kotov, O., Churazov, E., and Gilfanov, M. 2001. On the X-ray time-lags in the black hole candidates. MNRAS, **327**(Nov.), 799–807.

Krolik, J. H., McKee, C. F., and Tarter, C. B. 1981. Two-phase models of quasar emission line regions. ApJ, **249**(Oct.), 422–442.

Kubota, A., and Done, C. 2004. The very high state accretion disc structure from the galactic black hole transient XTE J1550-564. MNRAS, **353**(Sept.), 980–990.

Kubota, A., Done, C., Davis, S. W., Dotani, T., Mizuno, T., and Ueda, Y. 2010. Testing accretion disk structure with Suzaku data of LMC X-3. ApJ, **714**(May), 860–867.

Kubota, A., Dotani, T., Cottam, J., Kotani, T., Done, C., Ueda, Y., Fabian, A. C., Yasuda, T., Takahashi, H., Fukazawa, Y., Yamaoka, K., Makishima, K., Yamada, S., Kohmura, T., and Angelini, L. 2007. Suzaku discovery of iron absorption lines in outburst spectra of the X-ray transient 4U 1630-472. PASJ, **59**(Jan.), 185–198.

Kubota, A., Makishima, K., and Ebisawa, K. 2001. Observational evidence for strong disk Comptonization in GRO J1655-40. ApJL, **560**(Oct.), L147–L150.

Laor, A. 1991. Line profiles from a disk around a rotating black hole. ApJ, **376**(July), 90–94.

Lee, J. C., Reynolds, C. S., Remillard, R., Schulz, N. S., Blackman, E. G., and Fabian, A. C. 2002. High-resolution Chandra HETGS and Rossi X-Ray Timing Explorer observations of GRS 1915+105: a hot disk atmosphere and cold gas enriched in iron and silicon. ApJ, **567**(Mar.), 1102–1111.

Li, L.-X., Zimmerman, E. R., Narayan, R., and McClintock, J. E. 2005. Multitemperature blackbody spectrum of a thin accretion disk around a Kerr black hole: model computations and comparison with observations. ApJS, **157**(Apr.), 335–370.

Liu, B. F., Yuan, W., Meyer, F., Meyer-Hofmeister, E., and Xie, G. Z. 1999. Evaporation of accretion disks around black holes: the disk-corona transition and the connection to the advection-dominated accretion flow. ApJL, **527**(Dec.), L17–L20.

Lyubarskii, Y. E. 1997. Flicker noise in accretion discs. MNRAS, **292**(Dec.), 679+.

Maccarone, T. J. 2003. Do X-ray binary spectral state transition luminosities vary? A&A, **409**(Oct.), 697–706.

Makishima, K., Takahashi, H., Yamada, S., Done, C., Kubota, A., Dotani, T., Ebisawa, K., Itoh, T., Kitamoto, S., Negoro, H., Ueda, Y., and Yamaoka, K. 2008. Suzaku results on Cygnus X-1 in the low/hard State. PASJ, **60**(June), 585–604.

Malzac, J., and Belmont, R. 2009. The synchrotron boiler and the spectral states of black hole binaries. MNRAS, **392**(Jan.), 570–589.

Malzac, J., Dumont, A. M., and Mouchet, M. 2005. Full radiative coupling in two-phase models for accreting black holes. A&A, **430**(Feb.), 761–769.

Matt, G., Perola, G. C., and Piro, L. 1991. The iron line and high energy bump as X-ray signatures of cold matter in Seyfert 1 galaxies. A&A, **247**(July), 25–34.

Mayer, M., and Pringle, J. E. 2007. Time-dependent models of two-phase accretion discs around black holes. MNRAS, **376**(Mar.), 435–456.

McClintock, J. E., Shafee, R., Narayan, R., Remillard, R. A., Davis, S. W., and Li, L.-X. 2006. The spin of the near-extreme Kerr black hole GRS 1915+105. ApJ, **652**(Nov.), 518–539.

McConnell, M. L., Zdziarski, A. A., Bennett, K., Bloemen, H., Collmar, W., Hermsen, W., Kuiper, L., Paciesas, W., Phlips, B. F., Poutanen, J., Ryan, J. M., Schönfelder, V., Steinle, H., and Strong, A. W. 2002. The soft gamma-ray spectral variability of Cygnus X-1. ApJ, **572**(June), 984–995.

Middleton, M., Done, C., and Gierliński, M. 2007. An absorption origin for the soft excess in Seyfert 1 active galactic nuclei. MNRAS, **381**(Nov.), 1426–1436.

Middleton, M., Done, C., Gierliński, M., and Davis, S. W. 2006. Black hole spin in GRS 1915+105. MNRAS, **373**(Dec.), 1004–1012.

Middleton, M., Done, C., Ward, M., Gierliński, M., and Schurch, N. 2009. RE J1034+396: the origin of the soft X-ray excess and quasi-periodic oscillation. MNRAS, **394**(Mar.), 250–260.

Miller, J. M., Homan, J., Steeghs, D., Rupen, M., Hunstead, R. W., Wijnands, R., Charles, P. A., and Fabian, A. C. 2006a. A long, hard look at the low/hard state in accreting Black Holes. ApJ, **653**(Dec.), 525–535.

Miller, J. M., Raymond, J., Fabian, A., Steeghs, D., Homan, J., Reynolds, C., van der Klis, M., and Wijnands, R. 2006b. The magnetic nature of disk accretion onto black holes. Nature, **441**(Jun.), 953–955.

Miller, J. M., Reynolds, C. S., Fabian, A. C., Miniutti, G., and Gallo, L. C. 2009. Stellar-mass black hole spin constraints from disk reflection and continuum modeling. ApJ, **697**(May), 900–912.

Miller, L., Turner, T. J., and Reeves, J. N. 2008. An absorption origin for the X-ray spectral variability of MCG-6-30-15. A&A, **483**(May), 437–452.

Miller, L., Turner, T. J., Reeves, J. N., George, I. M., Kraemer, S. B., and Wingert, B. 2007. The variable X-ray spectrum of Markarian 766. I. Principal components analysis. A&A, **463**(Feb.), 131–143.

Miniutti, G., and Fabian, A. C. 2004. A light bending model for the X-ray temporal and spectral properties of accreting black holes. MNRAS, **349**(Apr.), 1435–1448.

Miniutti, G., Fabian, A. C., Goyder, R., and Lasenby, A. N. 2003. The lack of variability of the iron line in MCG-6-30-15: general relativistic effects. MNRAS, **344**(Sept.), L22–L26.

Miyamoto, S., Kitamoto, S., Hayashida, K., and Egoshi, W. 1995. Large hysteretic behavior of stellar black hole candidate X-ray binaries. ApJL, **442**(Mar.), L13–L16.

Motch, C., Ilovaisky, S. A., Chevalier, C., and Angebault, P. 1985. An IR, optical and X-ray study of the two state behaviour of GX 339-4. Space Sci. Rev., **40**(Feb.), 219–224.

Nandra, K., O'Neill, P. M., George, I. M., and Reeves, J. N. 2007. An XMM-Newton survey of broad iron lines in Seyfert galaxies. MNRAS, **382**(Nov.), 194–228.

Narayan, R., and Yi, I. 1995. Advection-dominated accretion: Underfed black holes and neutron stars. ApJ, **452**(Oct.), 710+.

Nayakshin, S., and Kallman, T. R. 2001. Accretion disk models and their X-ray reflection signatures. I. Local spectra. ApJ, **546**(Jan.), 406–418.

Nayakshin, S., Kazanas, D., and Kallman, T. R. 2000. Thermal instability and photoionized X-ray reflection in accretion disks. ApJ, **537**(July), 833–852.

Netzer, H., Kaspi, S., Behar, E., Brandt, W. N., Chelouche, D., George, I. M., Crenshaw, D. M., Gabel, J. R., Hamann, F. W., Kraemer, S. B., Kriss, G. A., Nandra, K., Peterson, B. M., Shields, J. C., and Turner, T. J. 2003. The ionized gas and nuclear environment in NGC 3783. IV. Variability and modeling of the 900 kilosecond Chandra spectrum. ApJ, **599**(Dec.), 933–948.

Nowak, M. A. 1995. Toward a unified view of black-hole high-energy states. PASP, **107**(Dec.), 1207+.

Ponti, G., Miniutti, G., Cappi, M., Maraschi, L., Fabian, A. C., and Iwasawa, K. 2006. XMM-Newton study of the complex and variable spectrum of NGC 4051. MNRAS, **368**(May), 903–916.

Pounds, K. A., Reeves, J. N., King, A. R., Page, K. L., O'Brien, P. T., and Turner, M. J. L. 2003. A high-velocity ionized outflow and XUV photosphere in the narrow emission line quasar PG1211+143. MNRAS, **345**(Nov.), 705–713.

Poutanen, J., and Svensson, R. 1996. The two-phase pair corona model for active galactic nuclei and X-ray binaries: how to obtain exact solutions. ApJ, **470**(Oct.), 249+.

Pszota, G., Zhang, H., Yuan, F., and Cui, W. 2008. Origin of X-ray emission from transient black hole candidates in quiescence. MNRAS, **389**(Sept.), 423–428.

Reeves, J., Done, C., Pounds, K., Terashima, Y., Hayashida, K., Anabuki, N., Uchino, M., and Turner, M. 2008. On why the iron K-shell absorption in AGN is not a signature of the local warm/hot intergalactic medium. MNRAS, **385**(Mar.), L108–L112.

Reeves, J. N., O'Brien, P. T., Braito, V., Behar, E., Miller, L., Turner, T. J., Fabian, A. C., Kaspi, S., Mushotzky, R., and Ward, M. 2009. A Compton-thick wind in the high-luminosity quasar, PDS 456. ApJ, **701**(Aug.), 493–507.

Reis, R. C., Fabian, A. C., and Miller, J. M. 2010. Black hole accretion discs in the canonical low-hard state. MNRAS, **402**(Feb.), 836–854.

Reis, R. C., Fabian, A. C., Ross, R. R., and Miller, J. M. 2009b. Determining the spin of two stellar-mass black holes from disc reflection signatures. MNRAS, **395**(May), 1257–1264.

Reis, R. C., Fabian, A. C., Ross, R. R., Miniutti, G., Miller, J. M., and Reynolds, C. 2008. A systematic look at the very high and low/hard state of GX339-4: constraining the black hole spin with a new reflection model. MNRAS, **387**(July), 1489–1498.

Reis, R. C., Miller, J. M., and Fabian, A. C. 2009a. Thermal emission from the stellar-mass black hole binary XTE J1118+480 in the low/hard state. MNRAS, **395**(May), L52–L56.

Remillard, R. A., and McClintock, J. E. 2006. X-ray properties of black-hole binaries. ARA&A, **44**(Sept.), 49–92.

Reynolds, C. S. 1997. An X-ray spectral study of 24 type 1 active galactic nuclei. MNRAS, **286**(Apr.), 513–537.

Ross, R. R., and Fabian, A. C. 2007. X-ray reflection in accreting stellar-mass black hole systems. MNRAS, **381**(Nov.), 1697–1701.

Różańska, A., and Czerny, B. 2000. Vertical structure of the accreting two-temperature corona and the transition to an ADAF. A&A, **360**(Aug.), 1170–1186.

Rykoff, E. S., Miller, J. M., Steeghs, D., and Torres, M. A. P. 2007. Swift observations of the cooling accretion disk of XTE J1817-330. ApJ, **666**(Sept.), 1129–1139.

Schnittman, J. D., Homan, J., and Miller, J. M. 2006. A precessing ring model for low-frequency quasi-periodic oscillations. ApJ, **642**(May), 420–426.

Schurch, N. J., and Done, C. 2008. Funnel wall jets and the nature of the soft X-ray excess. MNRAS, **386**(May), L1–L4.

Schurch, N. J., Done, C., and Proga, D. 2009. The impact of accretion disk winds on the X-ray spectra of active galactic nuclei. II. Xscort + hydrodynamic simulations. ApJ, **694**(Mar.), 1–11.

Shafee, R., McClintock, J. E., Narayan, R., Davis, S. W., Li, L.-X., and Remillard, R. A. 2006. Estimating the spin of stellar-mass black holes by spectral fitting of the X-ray continuum. ApJL, **636**(Jan.), L113–L116.

Shakura, N. I., and Sunyaev, R. A. 1973. Black holes in binary systems. Observational appearance. A&A, **24**, 337–355.

Shimura, T., and Takahara, F. 1995. On the spectral hardening factor of the X-ray emission from accretion disks in black hole candidates. ApJ, **445**(June), 780–788.

Sim, S. A., Miller, L., Long, K. S., Turner, T. J., and Reeves, J. N. 2010. Multidimensional modelling of X-ray spectra for AGN accretion disc outflows – II. MNRAS, **404**(May), 1369–1384.

Steiner, J. F., Narayan, R., McClintock, J. E., and Ebisawa, K. 2009. A simple Comptonization model. PASP, **121**(Nov.), 1279–1290.

Stern, B. E., Poutanen, J., Svensson, R., Sikora, M., and Begelman, M. C. 1995. On the geometry of the X-ray–emitting region in Seyfert galaxies. ApJL, **449**(Aug.), L13+.

Titarchuk, L. 1994. Generalized Comptonization models and application to the recent high-energy observations. ApJ, **434**(Oct.), 570–586.

Turner, T. J., Reeves, J. N., Kraemer, S. B., and Miller, L. 2008. Tracing a disk wind in NGC 3516. A&A, **483**(May), 161–169.

Ueda, Y., Inoue, H., Tanaka, Y., Ebisawa, K., Nagase, F., Kotani, T., and Gehrels, N. 1998. Detection of absorption-line features in the X-ray spectra of the galactic superluminal source GRO J1655-40. ApJ, **492**(Jan.), 782+.

Uttley, P., and McHardy, I. M. 2001. The flux-dependent amplitude of broadband noise variability in X-ray binaries and active galaxies. MNRAS, **323**(May), L26–L30.

Uttley, P., McHardy, I. M., and Vaughan, S. 2005. Non-linear X-ray variability in X-ray binaries and active galaxies. MNRAS, **359**(May), 345–362.

van Paradijs, J. 1996. On the Accretion Instability in Soft X-Ray Transients. ApJL, **464**(June), L139+.

Vasudevan, R. V., and Fabian, A. C. 2007. Piecing together the X-ray background: bolometric corrections for active galactic nuclei. MNRAS, **381**(Nov.), 1235–1251.

Wardziński, G., and Zdziarski, A. A. 2000. Thermal synchrotron radiation and its Comptonization in compact X-ray sources. MNRAS, **314**(May), 183–198.

Wilkinson, T., and Uttley, P. 2009. Accretion disc variability in the hard state of black hole X-ray binaries. MNRAS, **397**(Aug.), 666–676.

Wilms, J., Allen, A., and McCray, R. 2000. On the absorption of X-rays in the interstellar medium. ApJ, **542**(Oct.), 914–924.

Wilms, J., Reynolds, C. S., Begelman, M. C., Reeves, J., Molendi, S., Staubert, R., and Kendziorra, E. 2001. XMM-EPIC observation of MCG-6-30-15: direct evidence for the extraction of energy from a spinning black hole? MNRAS, **328**(Dec.), L27–L31.

Yamaoka, K., Ueda, Y., Inoue, H., Nagase, F., Ebisawa, K., Kotani, T., Tanaka, Y., and Zhang, S. N. 2001. ASCA Observation of the superluminal jet source GRO J1655-40 in the 1997 outburst. PASJ, **53**(Apr.), 179–188.

Yaqoob, T., Murphy, K. D., Miller, L., and Turner, T. J. 2010. On the efficiency of production of the Fe Kα emission line in neutral matter. MNRAS, **401**(Jan.), 411–417.

Young, A. J., Fabian, A. C., Ross, R. R., and Tanaka, Y. 2001. A complete relativistic ionized accretion disc in Cygnus X-1. MNRAS, **325**(Aug.), 1045–1052.

Yu, W., and Yan, Z. 2009. State transitions in bright galactic X-ray binaries: luminosities span by two orders of magnitude. ApJ, **701**(Aug.), 1940–1957.

Zdziarski, A. A., Johnson, W. N., and Magdziarz, P. 1996. Broad-band γ-ray and X-ray spectra of NGC 4151 and their implications for physical processes and geometry. MNRAS, **283**(Nov.), 193–206.

Zdziarski, A. A., Lubinski, P., and Smith, D. A. 1999. Correlation between Compton reflection and X-ray slope in Seyferts and X-ray binaries. MNRAS, **303**(Feb.), L11–L15.

Zdziarski, A. A., Misra, R., and Gierliński, M. 2010. Compton scattering as the explanation of the peculiar X-ray properties of Cyg X-3. MNRAS, **402**(Feb.), 767–775.

Zhang, S. N., Cui, W., and Chen, W. 1997. Black hole spin in X-ray binaries: observational consequences. ApJL, **482**(June), L155+.

Życki, P. T., Done, C., and Smith, D. A. 1998. Evolution of the accretion flow in Nova MUSCAE 1991. ApJL, **496**(Mar.), L25+.

Życki, P. T., Done, C., and Smith, D. A. 1999. X-ray spectral evolution of GS 2023+338 (V404 Cyg) during decline after outburst. MNRAS, **305**(May), 231–240.

7. Observational characteristics of accretion onto black holes II: environment and feedback

ROB FENDER

7.1 Introduction

I'll begin this chapter with one of my favorite quotes about black holes:

> Of all the conceptions of the human mind, from unicorns to gargoyles to the hydrogen bomb, the most fantastic, perhaps, is the black hole; a hole in space with a definite edge into which anything can fall and out of which nothing can escape; a hole with a gravitational force so strong that even light is caught and held in its grip; a hole that curves space and warps time. Like unicorns and gargoyles, black holes seem more at home in the realms of science fiction and ancient myth than in the real Universe. Nonetheless, well-tested laws of physics predict firmly that black holes exist. (Thorne, 1994)

... and add that not only do the laws of physics predict that black holes exist, but a vast array of observational evidence points to their existence in a range of masses and environments, throughout the universe. Whether or not they really exist is debated by some and is hard to prove to the satisfaction of others, but there are clearly a vast number of objects ($>10^{16}$) in the observable universe that conform very closely to our concept of a black hole (i.e., high accretion efficiency but otherwise "dark," no apparent surface, simple scaling of properties over a range $\geq 10^8$ in mass). In this contribution to the XXI Canary Islands Winter School of Astrophysics, I focus on these objects, specifically on the observational consequences of accretion onto them, and the associated feedback to the surrounding environment.

This review is meant to be read together with that of Chris Done, "Observational Characteristics of Accretion onto Black Holes I," Chapter 6 in these proceedings.

7.1.1 The role of black-hole feedback over cosmological time

By *black-hole feedback* I mean the radiation and kinetic energy that are released to the surrounding medium as a result of the accretion process (and, in some as yet unproven scenarios, also powered by the black-hole spin). What the potential accretion power is, and how it divides between these radiative and kinetic feedbacks (and also in some cases possible advection of energy across black-hole event horizons) will become clear as this article progresses.

Figure 7.1 summarizes very simplistically some key epochs for black hole accretion and the related feedback over cosmological time.

As far as our current understanding holds, the first major role for black hole accretion was in playing a role in reionizing the universe following the "dark ages." The exact epoch and distribution of this reionization remains unclear and is a key goal of several new astronomical facilities (e.g., LOFAR, MWA, in the low-frequency radio band), but it seems clear that the universe has been in its current highly ionized state since a redshift $z \sim 8$.

Over the next 5 billion years there were plentiful galaxy mergers and the central black holes of galaxies – active galactic nuclei (AGN) – were kept well fed and grew rapidly (e.g., Yu and Tremaine, 2002, and references therein). If, as we currently assume, this growth was limited by the Eddington limit, then it would occur exponentially on the "Salpeter time scale" ($\sim 5 \times 10^7$ yr).

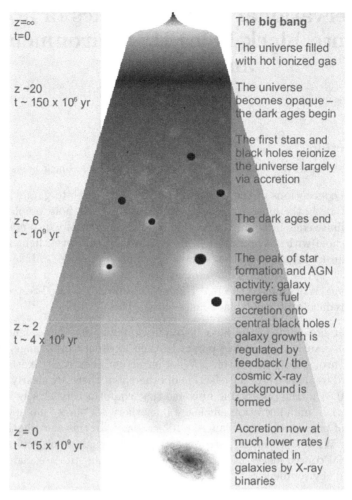

FIGURE 7.1. A simple illustration of the different phases of black hole accretion, the related feedback, and its consequences, over cosmological time.

During this phase kinetic feedback from these central black holes – in the form of jets and winds – probably acted to stall the growth of galaxies, reheating cooling flows and preventing surrounding gas from cooling and condensing into stars. Hints of this kind of feedback have come from observations of a fairly tight correlation between the mass of the galaxy bulges and the central black hole (e.g., Ferrarese and Merritt, 2000), where the gravitational influence of the black hole alone should not be enough to regulate the growth of the bulge. In addition to this, combined X-ray and radio images of the centers of galaxy clusters have revealed strong evidence that the powerful jets of the central AGN are acting to stall and heat the inward flow of the surrounding cluster gas (Fabian et al., 2006; Fig. 7.2).

During this phase, the cosmological X-ray background (CXB) was formed from the combined accretion luminosity of all of these black holes. The presence of this CXB is beautifully illustrated by ROSAT observations of the moon (Fig. 7.3). The spectrum of the CXB long suggested that it originated in a large population of accreting sources, something that was confirmed in the Chandra and XMM-Newton eras by its spatial resolution into discrete sources (e.g., Moretti et al., 2003, and references therein).

Since a redshift of ≥ 1, the Eddington ratio of accretion onto supermassive black holes has dropped fairly rapidly, and in the current (and probably future) universe, the accretion luminosity of galaxies is dominated by X-ray binaries (XRBs). In our own galaxy this is roughly split between accretion onto neutron stars and black holes, accreting

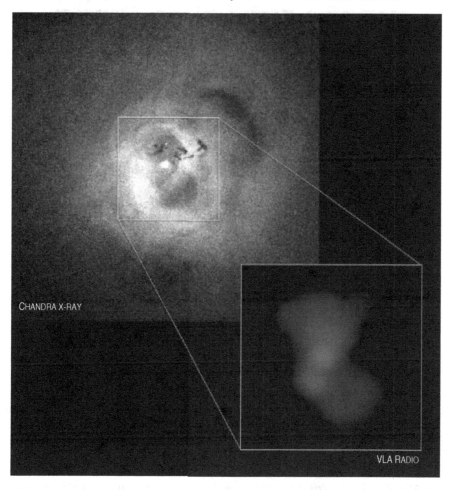

FIGURE 7.2. X-ray and radio images of the environment at the center of the Perseus cluster of galaxies. Jets from the central AGN, clearly visible in the radio band, correspond to cavities in the hot X-ray–emitting gas, demonstrating that the jets are pushing back strongly on the gas. In turn, ripples further out in the hot gas (X-rays) suggest that previous episodes of jet activity, and/or sound waves propagating away from the center, carry the effect of this feedback throughout the central cluster gas. From Fabian et al. (2006).

at Eddington ratios ≥ 0.1, while the central black hole (Sgr A*) is accreting at an (X-ray) Eddington ratio more than a million times lower (as are, presumably, a sizeable population of intermediate-mass black holes). At redshift $z \sim 2$, it would have taken $\geq 10^7$ XRBs in outburst simultaneously to challenge a central supermassive black hole of the most luminous AGN; within our own galaxy, nowadays, the central black hole (albeit a relatively low-mass one) is completely outshone by any single XRB in outburst (and many that are quasi-persistent). The giants are sleeping, but may occasionally wake for a stretch; Ponti et al. (2010) have demonstrated evidence that Sgr A* made a small but significant outburst in the past ~ 100 years.

7.2 Scaling of black hole accretion/accretion efficiency

7.2.1 Simple scalings

Uniquely among objects in the universe, black holes span a range of $>10^8$ in mass while being described by only three numbers:
- Mass
- Spin
- Charge

FIGURE 7.3. ROSAT X-ray image of the moon. The moon shows a bright crescent, reflected X-rays from the sun. However, the dark side of the moon is clearly much darker than the background, beautifully illustrating the existence of a noninstrumental cosmic X-ray background (CXB). In recent years most of the CXB has been resolved into individual AGN – that is, the CXB is the integrated signature of black-hole accretion and growth. From Schmitt et al. (1991).

This has led to black holes being described as "giant elementary particles" (compare with neutron stars, which can be described as "giant atomic nuclei"). This means of course that information loss (and entropy increase) to the universe when something collapses to a black hole or crosses the event horizon of a black hole is enormous.[1]

The single most important scaling to recall is that of the length scale:

$$R_G = \frac{GM}{c^2}. \tag{7.1}$$

This means that size of a black hole scales linearly with its mass, from which simple scaling many other relations can be derived (note that the event horizon of a black hole is at 1–2 R_G, depending on spin [but perhaps not in the way you'd expect – nonrotating black holes are "larger"]).

As an (important) example, how does the temperature of an accretion disk around a black hole scale with black-hole mass (assume that both black holes are accreting at their Eddington luminosities)?

In answering this, recall that Eddington luminosity is linearly proportional to mass, and that length scale is also proportional to mass (see earlier discussion), so area (of a 2-D object such as an accretion disk) scales with mass^2. Assuming blackbody emission (i.e., radiatively efficient – probably a good approximation for accretion at or near Eddington rates, see later discussion),

$$L = \sigma A T^4, \tag{7.2}$$

---

[1] Or maybe not. See, for example, *The Black Hole War* by Susskind (2009).

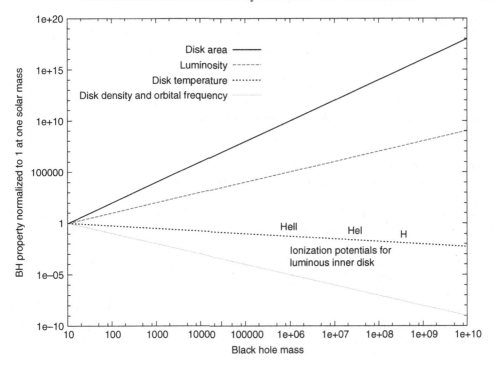

FIGURE 7.4. Simple scalings of key accretion parameters with black-hole mass (for black holes at comparable Eddington ratios).

which leads, putting in the foregoing mass scalings, to

$$M \propto M^2 T^4 \tag{7.3}$$

$$T \propto M^{-1/4}. \tag{7.4}$$

In other words, the temperature of an optically thick accretion flow decreases slowly with increasing black-hole mass, meaning that the accretion disks around the most massive black holes in luminous AGN are around two orders of magnitude cooler than those around XRBs at the same Eddington ratio (and consequently will peak in the UV rather than in X-rays). This and other scalings are illustrated in Figure 7.4.

7.2.2 Accretion efficiency

The efficiency of the accretion process is easy to describe theoretically for most objects but has some nuances, for black holes in particular, that it is well worth being aware of.

Some key points are:

- The potential efficiency of accretion is determined by the ratio of mass to radius – the larger the ratio, the greater the efficiency. For a range of different objects, this simply means that denser objects accrete more efficiently, although bear in mind that for a black hole, the mean "density" (of the volume within the event horizon) decreases with mass, whereas the accretion efficiency stays the same (a remarkable fact on its own).
- The accretion efficiency of a black hole *may* be determined by the innermost stable circular orbit (ISCO) which could limit the inner edge of the accretion disk. This radius varies from 6 R_G for a nonrotating ("Schwarzschild") black hole to 1 R_G for a maximally rotating ("maximal Kerr") black hole.

The accretion efficiency of neutron stars is comparable to that of black holes (the surface of a neutron star lies at only 2 to 3 R_G, comparable to the ISCO of a moderately rotating black hole). The accretion efficiency of white dwarfs (in cataclysmic variables – see lectures by Brian Warner) is much less as they are ~1,000 times larger (and as a result, nuclear fusion is a more efficient release of energy for them).

7.2.3 Radiative efficiency

For a standard accretion flow, the observed radiative luminosity (i.e., not including kinetic power in outflows) should be linearly proportional to the mass accretion rate $\dot{m}$,

$$L = \eta \dot{m} c^2, \tag{7.5}$$

where η is referred to as the radiative efficiency. For black holes, theory (i.e., the accretion efficiency arguments outlined in the previous section) gives us $0.05 \leq \eta \leq 0.4$ (depending on the spin of the BH).

In the context of black holes, this is radiatively efficient accretion (in other words, getting out about as much radiation as we'd expect for accretion onto something of this mass and radius).

However, there are solutions (usually at low accretion rates = low densities, sometimes called advection-dominated accretion flows [ADAFs] – see, e.g., Narayan and Yi, 1994) which have a dependence of η on accretion rate, e.g., $\eta \propto \dot{m}$. This would result in a quadratic dependence of luminosity on accretion rate $L \propto \dot{m}^2$ (e.g., Mahadevan, 1997).

This is (an example of) radiatively inefficient accretion (in other words, getting nowhere near as much radiation out as we'd expect for accretion onto something of this mass and radius).

The evidence for radiatively inefficient accretion in black holes comes in several forms. These include the spectral energy distributions of hard-state and quiescent black-hole binaries, and the observation that black-hole binaries show lower quiescent luminosities than neutron star systems with similar binary parameters.

Of course the change in gravitational potential remains the same, so where does this missing energy go? In many cases, the evidence for radiatively inefficient accretion has been taken as direct evidence for the advection of this energy across black-hole event horizons (e.g., Garcia et al., 2001) and has sometimes been pushed as the strongest evidence to date for event horizons. However, it has also been demonstrated that in essentially all cases where ADAF models provide a good fit to the spectral energy distribution, a powerful jet is present (e.g., Fender et al., 2003). It may be that the jet can carry away all of the "missing" energy (the jury is still out on this one – see also Körding et al., 2006b).

The Soltan argument

A clear and simple argument, first put forward by Soltan (1982), links the observed CXB with the current mass distribution of black holes via the radiative efficiency of the accretion process. The argument has been revisited several times over the years (Fabian and Iwasawa, 1999; Yu and Tremaine, 2002).

Put in very simple terms, if black holes have grown over cosmological time by accretion, and that accretion process has left its signature in the form of the CXB, then the flux from the CXB should act as a measure of the growth of black holes. The scaling between the CXB flux and the amount of mass accreted relates to the accretion efficiency. Working through a simplified case, we start by reminding ourselves that the energy released by the accretion of mass M onto a black hole is

$$E = \eta M c^2. \tag{7.6}$$

We can use this to think about how the energy density from these accretion processes would appear to us now:

$$\epsilon(1+z) = \eta \rho c^2, \tag{7.7}$$

where ϵ is the observed mean energy density in the CXB, z is the mean redshift at which the CXB was generated, and ρ is the mean mass density added to the black holes during the accretion process (in other words, mass density $\to$ radiation density via η, and then redshifted to the present where we observe it).

We can measure ϵ directly from the CXB (applying both a bolometric correction and a correction for a large (30% to 50%) fraction of heavily obscured sources), and we can measure ρ directly from the local mass function of black holes (not trivial!). Note that there is a clear assumption here, which is that most of the mass observed today in the local BH mass function was added during the phase at which the CXB was created.

So, let's have a go ourselves.

Assuming that a large fraction (30% to 50%) of AGN were obscured, Fabian and Iwasawa (1999) estimated the local energy density due to the CXB as

$$\epsilon \sim 10^{-15} \quad \text{erg cm}^{-3} \text{ s}^{-1}. \tag{7.8}$$

Assuming this arises at redshift $z = 2$, we can estimate $\eta\rho$:

$$\eta\rho \sim 3 \times 10^{-36} \quad \text{g cm}^{-3}. \tag{7.9}$$

We can convert this to solar masses per cubic megaparsec ($M_\odot = 2 \times 10^{33}$ g, 1 pc = 3×10^{18} cm^3):

$$\eta\rho \sim 5 \times 10^4 \quad M_\odot \text{ Mpc}^{-3}. \tag{7.10}$$

Typical recent "direct" (usually based on stellar velocity dispersions) measurements of the local black-hole space density are in the range 3 to 5×10^5 $M_\odot$ Mpc^{-3}.

So what does this imply for the average accretion efficiency η during AGN growth? Putting in the numbers, it implies

$$\eta \sim 0.1, \tag{7.11}$$

i.e., AGN growth was radiatively efficient. This is a pretty significant conclusion but, as we shall see later, it does not tell us the whole story. This is because we find that at such radiative efficiencies, black holes can switch between accretion states that have very different kinetic outputs.

7.2.4 *(Mass) unification schemes*

The very simple scalings discussed earlier suggest that it should in some way be possible to quantitatively scale the observable properties of black-hole accretion between XRBs and AGN. An early discussion of how this might look was presented in Sams *et al.* (1996) and during the 2000s has become a reality. These scalings, described later, have served to link the fields of black-hole XRB (BHXRB) and AGN research in a very strong way and have illustrated how each field can shed light on the other.

Luminosity unification

In the early 2000s two groups independently realized that the strong correlations between X-ray and radio luminosity observed in XRBs may be extendable to all black holes by the inclusion of a mass term. Merloni *et al.* (2003) and Falcke *et al.* (2004) showed that a "fundamental plane of black hole activity" exists linking these three measurable parameters across the whole mass spectrum of accreting black holes, from 10 to 10^9 $M_\odot$. The Merloni *et al.* (2003) version of the plane is presented in Figure 7.5; the Falcke *et al.* (2004) result has been more recently updated and refined by Körding *et al.* (2006a).

The results are a key step in unification of black-hole accretion and kinetic feedback. The Merloni *et al.* (2003) sample includes AGN of (probably) all accretion states, and as such, the correlation implies that radio-quiet "soft" states, which would cause deviation

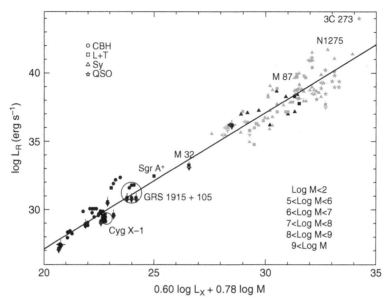

FIGURE 7.5. A relation between the radio luminosity, mass and X-ray luminosity of accreting black holes from X-ray binaries to the most massive AGN (from Merloni *et al.*, 2003).

from the plane, do not occur at much lower luminosities than in XRBs (where, so far, they are only observed at X-ray Eddington ratios $\geq 1\%$). The Falcke *et al.* (2004) and Körding *et al.* (2006a) samples use only objects at lower Eddington ratios, in the so-called hard accretion state, and this selection reveals an even tighter correlation than that of Merloni *et al.* (2003), implying a very clear scaling of accretion and feedback in BH at relatively low Eddington ratios, with little requirement for any other parameter. Note that both samples use only core, and not total (i.e., including extended lobes), radio emission, which is important for the discussion of the possible spin-powering of jets later on.

Timing unification

As noted earlier (Fig. 7.4), the expected scaling of frequencies (e.g., orbital frequency) with black-hole mass is simply $\nu \propto 1/M$. These frequencies are typically measured by using X-ray power density spectra. However, the picture is not so simple – it was also known that in X-ray binaries (both black hole and neutron star), these frequencies increased with luminosity as the luminosity from a given system changed with time (i.e., mass was not changing). Therefore, a unification of timing properties, as with the luminosity scalings we just touched on, would require at least three parameters: frequency ν, mass M, and mass accretion rate $\dot{m}$.

Such a plane connecting black-hole X-ray binaries and AGN was first reported in McHardy *et al.* (2006). The form of the relation was

$$\nu \propto \frac{\dot{m}}{M^2} \qquad (7.12)$$

(within uncertainties of order $\pm$ 0.1 in the exponents). Although this is not the only interpretation, this result can be expressed in terms of Eddington ratios, which leads to

$$\nu \propto \frac{\dot{m}_{\text{Edd}}}{M}, \qquad (7.13)$$

producing the expected simple scaling with mass. The interpretation of the result $\nu \propto \dot{m}$ is not trivial; *if* ν corresponds to an orbital frequency, it would imply $r \propto \dot{m}^{-2/3}$ (but

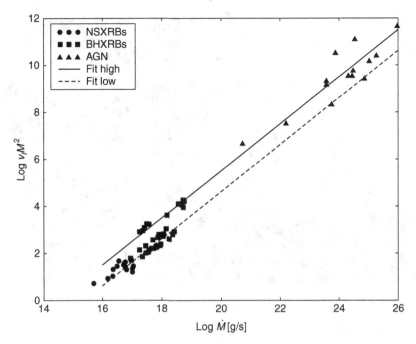

FIGURE 7.6. Scalings of timing features (breaks and QPOs in power density spectra), with mass accretion rate and black-hole mass. The observed relation, $\nu M^2 \propto \dot{m}$, can be refigured as $\nu \propto \dot{m}_{\rm Edd}/M$, where $\dot{m}_{\rm Edd}$ is the mass accretion rate in Eddington units; in this approach, the expected $\nu \propto 1/M$ result is obtained. From Körding et al. (2007); see also McHardy et al. (2006).

note that the observed frequencies are much lower than those corresponding to the ISCO, even at the highest Eddington ratios, and so if they are orbital, they come from some considerable distance away from the black hole).

Körding et al. (2007) refined and extended this result, including many more X-ray binaries and further showing that the relation may extend to lower-luminosity XRBs and even neutron stars (Fig. 7.6). There may also be an "adjustment" to the timing plane depending on the accretion "state" of the source (see Chapter 6 by C. Done in this contribution).

7.3 An introduction to jets

Tightly collimated outflows, or *jets*, are a common and yet poorly understood aspect of astrophysics. They are observed in many forms of accreting system, and also, apparently, in some other environments (such as spin-powered radio pulsars or young massive stars). In many cases they can be demonstrated to carry a large amount of kinetic power (sometimes comparable to, or even exceeding, the total bolometric radiative luminosity of a system) and probably also help to extract angular momentum from accretion flows (although apparently accretion can proceed well without them – see the "soft" states of black holes). They are the clearest example of strong kinetic feedback from accreting black holes (although strong disk winds may also have a large impact in softer states).

7.3.1 Jets from accreting systems

As noted previously, jets are observed from a very wide range of accreting objects. Figure 7.7 presents images of jets from a wide variety of objects.

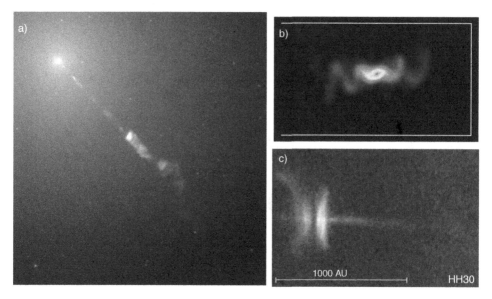

FIGURE 7.7. A montage of images of jets from accreting objects. *a)* shows the long straight jet – the first observed in astrophysics – from the core of the giant elliptical galaxy M87 at the core of the Virgo cluster (Biretta *et al.*, 2005). *b)* shows the precessing jet from the binary system SS 433, in which the accretor is either a neutron star or stellar-mass black hole (Blundell and Bowler, 2004). *c)* shows a jet from a young stellar object, in which a protostellar system is producing the jet, which is beautifully shown to be clearly perpendicular to the midplane of the accreting torus (Burrows *et al.*, 1996). Reproduced by permission of the AAS.

Synchrotron emission

The most important emission mechanism associated with relativistic jets (and indeed many high-energy phenomena) is synchrotron radiation. Synchrotron radiation results from the acceleration of electrons (and possibly in some cases positrons) as they spiral around magnetic field lines. Astrophysical shocks, often believed to be associated with jets, which are highly supersonic, both compress and amplify the ambient magnetic field and accelerate the electrons, leading almost ubiquitously to synchrotron radiation. There are many texts describing in detail the process of synchrotron radiation (e.g., Longair, *High Energy Astrophysics, Volumes I and II*); key aspects of synchrotron radiation that provide simple astrophysical diagnostics are:

- When optically thin, the spectral index of the observed emission (α, where the flux density varies with frequency as $S_\nu \propto \nu^\alpha$) is a direct measure of the underlying (nonthermal) power-law distribution of electrons ($p = 1 - 2\alpha$, where $N(E)dE \propto E^{-p}$).
- The emission can be highly linearly polarized, up to $\sim 70\%$, which in the optically thin case has an electric vector perpendicular to the magnetic field in the emitting region.
- For a given size and luminosity of a synchrotron emitting source, a minimum total internal energy can be calculated, an argument that dates back to Burbidge (1958).
- The maximum *brightness temperature* for a synchrotron emitting region, in a steady state, is $T_{B,\max} \sim 10^{12}$ K. Above this value, inverse Compton emission (effectively cooling) rapidly reduces the energy of the electrons. In fact, the ratio of synchroton to inverse Compton emission from a region of magnetic field, photons, and electrons is simply the ratio of the magnetic to photon energy densities.

7.3.2 *The speed of jets*

Jets from black holes and neutron stars have been observed to move very fast – in some cases with speeds exceeding $0.99c$. The most important term is the bulk Lorentz factor

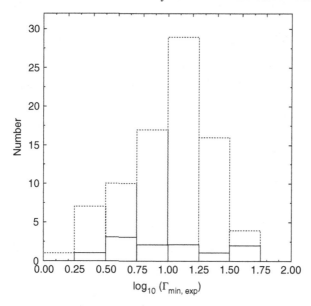

FIGURE 7.8. A comparison of measured Lorentz factors for AGN (dotted line, mostly blazars) with the much smaller sample of X-ray binaries (solid line), under the assumption that the apparent tight collimation of X-ray binary jets is due to their large Lorentz factors (which retard the apparent lateral expansion). See Miller-Jones et al. (2006) for more details.

of the flow $\Gamma = (1 - \beta^2)^{-1/2}$, where $\beta = v/c$, the ratio of the jet speed to the speed of light.

In most cases we can only actually measure a lower limit to the jet velocity from radio (and occasionally, X-ray) images. Using this, we know that several jets from black-hole X-ray binaries have speeds in excess of $0.9c$, as do many tens of AGN. Placing an upper limit on the speeds is not, however, easy – Figure 7.8 shows the results of one attempt to do this, based on the very narrow (less than a few degrees) observed opening angles of X-ray binary jets; if this method is appropriate, then X-ray binary jets may have Lorentz factors as high as the fastest blazars ($\Gamma \geq 20$).

Apparent superluminal motion

In several cases involving both AGN and XRBs, simple division of the jet projected motion by the travel time leads to the conclusion that the jets are moving faster than the speed of light. Figure 7.9 illustrates how a combination of simple geometry and nonadditive speed of light can result in this effect (Rees, 1966).

Although apparent superluminal motion does not really mean objects are moving with $v > c$, it does require high velocities (in fact, $v \geq \frac{c}{\sqrt{2}}$) in order to work, so it is an indicator of high velocities. Note that you can have cases where both jets are redshifted and deboosted but one still shows apparent superluminal motion (as is the case for GRS 1915+105). Also, it is a simple and useful result that if you measure an apparent velocity β_{app}, that is actually a lower limit on the true bulk Lorentz factor of the flow.

Aberration and one-sided jets

The very high velocities reached by relativistic jets from black holes (and neutron stars) mean that some interesting special relativistic effects come into play.

A key factor in understanding these effects is the relativistic Doppler factor

$$\delta_{\text{app,rec}} = \gamma^{-1}(1 \mp \beta \cos\theta)^{-1}, \qquad (7.14)$$

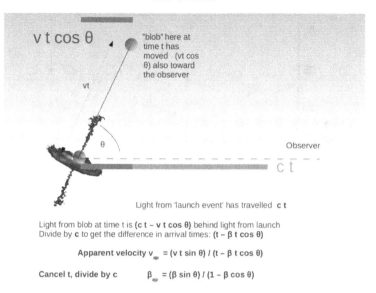

FIGURE 7.9. An illustration of the simple geometrical effects that result in apparent superluminal motion for rapidly moving objects.

where the shift from rest frame to observed frequencies is

$$\nu_{\text{obs}} = \delta \nu_{\text{rest-frame}}. \tag{7.15}$$

Note that both δ_{app} and δ_{rec} can be less than 1 (as is the case for the famous "microquasar" GRS 1915+105), meaning that both approaching and receding jets are redshifted. This can occur when the angle to the line of sight is large (for GRS 1915+105 it is $>60°$), in which case special relativistic time dilation wins out over the line-of-sight velocity component (this is the same reason that the line-emitting jets of the famous jet system SS 433 show a significant redshift even when in the plane of the sky with no line-of-sight velocity component).

Relativistic aberration also affects radiating objects, including jets, that are moving at significant fractions of the speed of light. The flux observed by an observer from a source with relativistic Doppler factor δ will be modified by a factor δ^k, where $2 \leq k \leq 3$. The result of this is that most of the flux from relativistically-moving jets is "beamed" into the forward direction (within an opening angle $\sim 1/\Gamma$, where Γ is the bulk Lorentz factor of the jet; see Fig. 7.10). This is the origin of the phenomena sometimes seen in FR II radio galaxies whereby only one highly collimated jet (the approaching one) is seen but two lobes (which are more or less at rest, and therefore not beamed) are observed. In the case of the receding jet, decelerating makes it more visible.

Many of the points discussed in this chapter are explored in more detail in Fender (2006) and references therein (especially Hughes, 1991).

7.3.3 The power of jets

Several different methods exist for estimating the power from jets, some of which are summarized here.

Synchrotron minimum energy

On of the most common methods for estimating the power in jets is by the approach of minimum energy, first discussed in the context of the lobes of radio galaxies by Burbidge (1958). The observation of synchrotron radiation from a source of measurable volume allows the calculation of this minimum energy, because there are at least two components

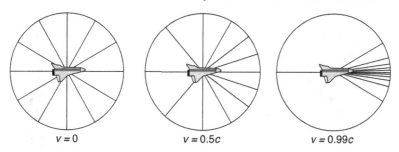

FIGURE 7.10. An illustration of relativistic aberration. A source of radiation that is isotropic in its rest frame (left panel, in this imaginary case the Space Shuttle), will appear to have its radiation field concentrated in the forward direction as it approaches relativistic speeds (see middle and right-hand panels). Astrophysical jets can often reach speeds in excess of $0.9c$ and are as a result very strongly "beamed," as in the right-hand panel. From The Physics and Relativity FAQ, by Scott Chase, Michael Weiss, Philip Gibbs, Chris Hillman, and Nathan Urban, part of the Usenet Physics FAQ.

required for synchrotron emission (electrons and magnetic energy), and they have very different dependencies on the magnetic field:
- The energy in the emitting electrons varies as $B^{-1.5}$.
- The energy in the magnetic field varies as B^2.

This is illustrated in Figure 7.11 and is the most common approach to estimating the power of jets. The size of the emitting region is typically estimated either directly from radio images (for AGN), or from variability time scales (for X-ray binaries).

Other methods

It is also possible to estimate the power of a jet not directly from the emission of the jet itself, but rather from the impact it has had on the external medium. This, again, is a method that was pioneered for AGN (e.g., Kaiser and Alexander, 1997) but which in

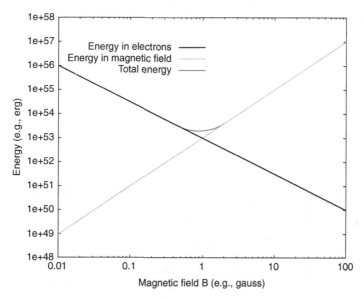

FIGURE 7.11. The minimum energy of a synchrotron emitting volume can be found from the different dependencies of the energy in electrons and in the magnetic field on the magnetic-field strength. Both higher and lower magnetic fields are possible but push up the energy requirements.

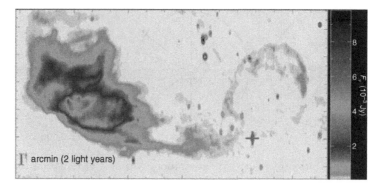

FIGURE 7.12. A large-scale bubble that we believe has been created by the action of the jet from Cygnus X-1 (marked with the cross) on the ambient medium, which on one side appears to be overdense at the edge of a H II region (to the left). Fitting models to the bubble allow us to make estimates of the jet power independent of emission from the jet itself, and we find a power comparable to the total X-ray luminosity (Gallo et al., 2005).

recent years has found application for XRBs. One of the best examples is the large-scale bubble around the black-hole binary Cygnus X-1 (Fig. 7.12).

Similar methods have been applied to the large-scale radio and X-ray nebula around the binary jet source SS 433 and also find very large power requirements.

Our current best view of the power of jets as a function of X-ray luminosity and/or accretion rate, with XRBs as a starting point, is summarized in Körding et al. (2006b). In brief, it seems that at relatively high X-ray Eddington ratios ($\geq 1\%$), the radiative and kinetic (jet) powers are comparable. At lower Eddington ratios, we currently favor a scenario in which the jet power scales linearly with mass accretion rate, whereas the X-ray luminosity falls off as something like $\dot{m}^2$ (see previous sections on radiative efficiency) and as a result systems become "jet dominated." The "missing" accretion power may be advected across the event horizon or carried away in the jet, and current systematic uncertainties in measurements of both $\dot{m}$ and jet power do not at this time allow us to draw strong conclusions either way. We note that for *neutron stars* we appear to see the expected linear scaling of both radiative and jet luminosities with accretion rate, as nothing can be advected.

In the next part of this chapter I discuss coupling of accretion states and type of jet, but in advance of this I'd note that the major transient ejections we observe during state transitions are not that much more powerful, if at all, than the bright quasi-steady hard-state jets that precede them.

7.4 Disk-jet coupling in X-ray binaries

One of the major advances in our understanding of the relation between accretion and the various flavors of feedback from accreting black holes has come from combined radio, near-infrared, and X-ray monitoring of BHXRBs.

Most of these systems are highly variable, undergoing large changes (orders of magnitude) in the mass accretion rate onto the central black hole on time scales observable by humans (in fact, on time scales suitable for PhDs) – see Figure 7.13. By observing simultaneously the X-ray flux from the hot inner accretion flow, and near-infrared and radio flux from the jet, we find repeating empirical connections between the two. These in turn have relevance for our understanding of feedback from AGN, for which (with a few exceptions, e.g., Marscher et al., 2002) the time scales for such variability are much longer than human time scales and we have to rely instead on population statistics (and their associated selection effects).

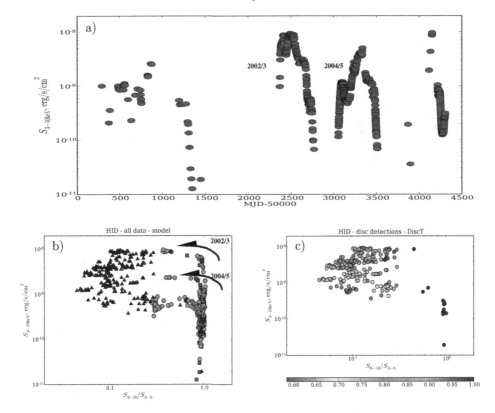

FIGURE 7.13. a) Outbursting behavior of the BHXRB GX 339-4 in both light curve and b) hardness–intensity diagram (HID) form. The source is "semitransient," undergoing an outburst every ∼2 years; however, the patterns of behavior are very similar to those observed from other lower duty-cycle transients. Typically BHXRBs in outburst will rise and fall from quiescence in hard states but at the highest Eddington ratios (≥ 0.01) will describe an anticlockwise pattern in the HID (but also note that the pattern is not identical between two outbursts of the same source). c) shows the measured temperature of the accretion disk (color scale, in keV) as a function of location in the HID – the disk temperature falls with flux in the softer states in a way that is consistent with a blackbody emission from a fixed emitting area (the apparent detections of hot disks in the hard state are probably spurious). From Dunn et al. (2008).

7.4.1 The hard state

BHXRBs at luminosities below ∼1% of the Eddington ratio (in X-ray luminosity, L_X), seem to be always observed in the "hard" X-ray state. This state is characterized by a relatively hard photon index ($\Gamma \sim 1.5$, where the relation between photon index and spectral index is $\alpha = (1 - \Gamma)$), which is observed, at high luminosities at least, to have a high-energy cutoff typically in the range 50 to 200 keV (note that this photon/spectral index is more or less the same as that produced by optically thin synchrotron emission, leading to some models in which the X-rays are actually produced by the jet – e.g., Markoff et al., 2001; Russell et al., 2010). This spectral shape is inconsistent with, for example, blackbody emission from an accretion disk and is generally modelled as Comptonization of lower-energy seed photons in a hot "corona" (see Chapter 6 by Chris Done in these proceedings). Short time scale (e.g., 0.01 to 10 Hz) variability is also stronger in this state than in the "soft" X-ray state (see later discussion).

In terms of outflow, the hard state seems to be always associated with a more or less steady (on short time scales at least), more or less flat-spectrum (spectral index $-0.5 \leq \alpha \leq +0.5$) radio component. This radio emission has been spatially resolved in two cases (Cyg X-1 and the pseudo-hard "plateau" state of GRS 1915+105) into a compact jet

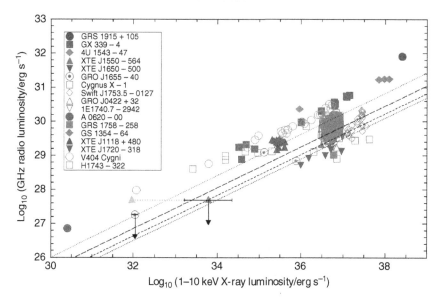

FIGURE 7.14. The relation between radio and X-ray luminosities for hard-state black holes, including all data available as of 2010. The overall slope of the relation is $\sim +0.6$, but clearly there is significant spread in the normalizations (probably beyond that attributable to uncertainties in distance measurements). From Calvelo et al. (2010).

(note also that in AGN a flat-spectrum core radio component has in several cases also been resolved into a jetlike structure). A crucial step forward in our understanding of the coupling of accretion and outflow came when the correlation between the X-ray and radio fluxes in the hard state was established for the BHXRB GX 339-4 (the same source as plotted in Fig. 7.13). Hannikainen et al. (1998) and Corbel et al. (2002) found that the radio and X-ray luminosities from this system were correlated as $L_R \propto L_X^{0.7}$. This is significant in a number of ways, not least because it implies that as the luminosity decreases, the ratio L_R/L_X will increase. Furthermore, theoretical models for the scaling of jet power with the emitted radio luminosity, of the form $L_R \propto P_J^{1.4}$, meant that the observed relation implied that $L_X \propto P_J^2$. This meant that as the luminosity declined, the ratio of jet power to X-ray luminosity would rise rapidly. Even six years ago, combined with the best estimates of jet power, this meant that at the very lowest luminosities ("quiescence"), the kinetic power output in the jet would exceed the X-ray luminosity, resulting in "jet-dominated states" (Fender et al., 2003). Furthermore, if the jet power scaled linearly with the accretion rate, as it did in some theoretical models, the observed scaling implied $L_X \propto \dot{m}^2$, which, the reader will recall, is the scaling expected for ADAFs.

This scaling is therefore very important for our understanding of black-hole accretion, and Gallo et al. (2003) argued that it was a universal one for all BHXRBs by plotting all measured (quasi-)simultaneous observations of radio and X-ray emission from hard state sources. Since this time, a couple of important developments have taken place – first, Gallo et al. (2006) detected the transient BHXRBs A0620-00 in quiescence at precisely the level predicted from an extrapolation of the hard-state relation at higher luminosities. Second, it became increasingly apparent that a number of BHXRBs fell significantly below the GX 339-4 relation, sometimes by an order of magnitude or more, although still showing the approximately flat-spectrum emission. The reason that some systems are "radio quiet" is not currently well understood. Note also that the refined mean slope of the correlation is $\sim +0.6$, slightly flatter than the measurement for GX 339-4 alone. Figure 7.14 presents the most up-to-date compilation of the radio and X-ray relation for hard-state BHXRBs.

7.4.2 Intermediate states and major ejections

Above $\sim 1\%$ of the Eddington ratio, BHXRBs enter a phase in which they can be in either hard or soft states at the same overall X-ray luminosity. As can be seen from Figure 7.13, the typical pattern of behavior is for sources to rise well above this 1% Eddington level in the hard state, make a transition to a soft(er) state, decline in luminosity in this soft state, and finally return to the hard state at a lower level (often, but not always, close to the 1% level).

The approximately steady jet observed in the hard state seems to persist in the initial phases of the first (hard → soft) state transition (Corbel et al., 2004) but has a tendency to become more erratic (in flux and spectral index). Somewhere around the middle of the state transition, however, things change dramatically. One or more radio flares are seen, which have in many cases been directly resolved by radio interferometers in seemingly discrete, certainly powerful, ejection events (e.g., Mirabel and Rodríguez, 1994; Fender et al., 1999; Miller-Jones et al., 2005, for the spectacular jets of GRS 1915+105; Corbel et al., 2002, for the radio and X-ray jets of XTE J1550-564).

In X-rays, as well as a softening spectrum, these high-luminosity "intermediate" states (sometimes called "very high states") often show very strong quasi-periodic oscillations (Casella et al., 2005).

7.4.3 The soft state

After the phase of radio flaring, the core radio flux is much diminished, often dropping to levels well below the flux expected for the hard state at that L_X. The X-ray spectrum is now dominated by a relatively low temperature (0.1 to 1 keV) blackbody-like component that is usually identified with an optically thick accretion disk, and the X-ray variability, like the jet, is also much less than in the hard state.

Many more details of observations and models of the X-ray states can be found in the accompanying contribution by Chris Done.

7.4.4 Toward a unified model

By 2003 it was realized that the patterns of behavior seen in GX 339-4 were, broadly speaking, common to all BHXRB transients. We therefore argued that there was a more or less common pattern of X-ray to radio (and hence accretion to jet) coupling in *all* BHXRBs, based on a detailed study of four systems (Fender et al., 2004). In addition, we attempted to provide some simple theoretical intepretation of the results. This model is presented in Figure 7.15. More recently, we have tested this model with a much larger sample of systems and outbursts, and have found no serious exceptions to the proposed empirical couplings (Fender et al., 2009). The theoretical interpretations remain in some cases poorly tested (e.g., the increasing jet velocity during the state transition) and in others rather contentious (e.g., the variation of the accretion disk radius, also during the state transition). We futhermore attempted to add a description of X-ray timing properties, specifically changes in the total X-ray variability, and the appearance of strong QPOs, to the jet coupling, but the results were inconclusive, due in large part to poor radio sampling (compared to X-rays; this is something that will change dramatically with the new generation of wide-field high-sensitivity radio facilities currently under construction).

Insights in the near-infrared

A further major change in our concept of the "radio jets" from BHXRBs, and one that is very relevant for estimates of their power, came when it was realized that the jet spectrum in some cases extended to (at least) the near-infrared (1–3 μm) band. This near-infrared emission is very prominent in most sources in bright hard states and arises from a region of the jet approximately 10^5 times closer to the central black hole

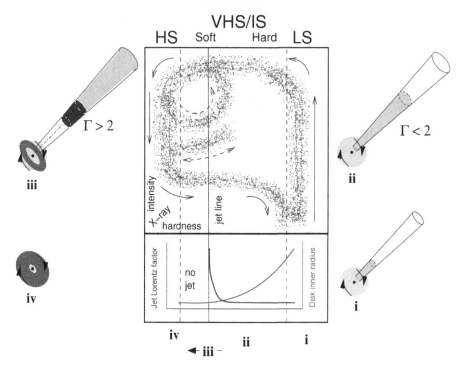

FIGURE 7.15. A schematic of the simplified model for the jet-disk coupling in black-hole binaries presented by Fender et al. (2004). The central box panel represents an X-ray hardness-intensity diagram (HID); "HS" indicates the "high/soft state," "VHS/IS" the "very high/intermediate state," and "LS" the "low/hard state." In this diagram, X-ray hardness increases to the right and intensity upward. The lower panel indicates the variation of the bulk Lorentz factor of the outflow with hardness – in the LS and hard-VHS/IS, the jet is steady with an almost constant bulk Lorentz factor $\Gamma < 2$, progressing from state i to state ii as the luminosity increases. At some point – usually corresponding to the peak of the VHS/IS – Γ increases rapidly, producing an internal shock in the outflow (iii) followed in general by cessation of jet production in a disk-dominated HS (iv). At this stage, fading optically thin radio emission is only associated with a jet/shock that is now physically decoupled from the central engine. As a result, the solid arrows indicate the track of a simple X-ray transient outburst with a single optically thin jet production episode. The dashed loop and dotted track indicate the paths that GRS 1915+105 and some other transients take in repeatedly hardening and then crossing zone iii – the "jet line" – from left to right, producing further optically thin radio outbursts. Sketches around the outside illustrate our concept of the relative contributions of jet, "corona," and accretion disk at these different stages. From Fender et al. (2004).

than the radio emission. Global correlations of the near-infrared jet flux (and optical flux, which probably arises from the irradiated outer accretion disk) with X-ray emission show similar patterns to those observed in radio (Russell et al., 2006). This in turn means that it should be a better place to look for short–time-scale correlations between the jet and the accretion flow, and indeed that is what has been observed in some cases (e.g., Casella et al., 2010; Fig. 7.16).

7.4.5 Broader relevance

So we have learned a lot about the coupling of accretion and ejection in black-hole X-ray binaries – so what? Is there a broader relevance?

We can hope that there is, because the mass-unification schemes (Section 2.4) strongly indicate, as theoretically expected, that black-hole accretion scales in a very simple way with black-hole mass and should be more or less scale free. But this does not tell us whether or not the *patterns* of behavior observed in BHXRBs are also observed in AGN.

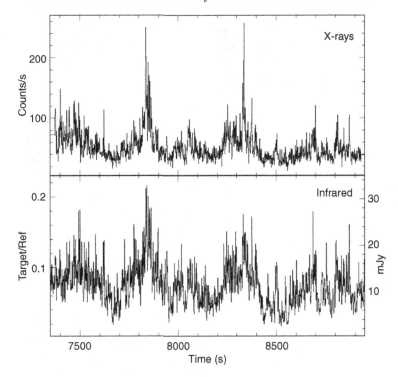

FIGURE 7.16. High–time-resolution simultaneous X-ray and near-infrared observations of GX 339-4. The near-infrared (2 μm band) flux is almost certainly from the jet and lags the X-rays by $\sim$100 ms, demonstrating the very clear and fast coupling between the accretion flow and outflow. From Casella et al. (2010).

In order to tackle this question, Körding et al. (2006c) took the HID used for X-ray binaries, in which the "hardness" is effectively a measure of the ratio of "coronal" to "disk" emission, and generalized it to physical measures that could also be used for AGN. This tool, the disk fraction luminosity diagram, was populated using a large sample of AGN with measured X-ray (i.e., coronal) and optical (disk) fluxes and compared to radio detections (or upper limits) of these sources. The result (Fig. 7.17) is a figure that emphatically supports the idea that AGN above a comparable Eddington ratio ($\sim$1%) have a range in states, but below this are only "hard," just like BHXRBs, and furthermore that the harder states are more "radio loud" (again just like BHXRBs). In addition, as already mentioned, Marscher et al. (2002) report evidence for observations of similar disk-jet coupling on a time scale of years from the AGN 3C120.

If it is true that we can take our understanding of the disk-jet coupling in BHXRBs and apply it to AGN, then this is a major step forward. It means that at high Eddington ratios (such as at the peak of starburst and AGN activity, $2 \leq z \leq 4$), supermassive black holes will always be more or less radiatively efficient, but may switch between modes with powerful jets and those without. This is very relevant, as it is observed that only $\sim$5% of optically selected luminous AGN are actually "radio loud" (actually, "radio loud" is a poorly defined term for AGN, but still...).

It further implies that when AGN drop below $\sim$1% Eddington, they probably switch to modes in which the output of released accretion power is primarily in the form of kinetic energy in the jet. Because in the current universe, at $z = 0$, this is most certainly the case, this is very relevant for our (and future) cosmic epoch.

Combining these findings in more detail with models of galaxy formation, mergers, clusters and feeding of AGN should provide a clearer picture of the process of galaxy formation.

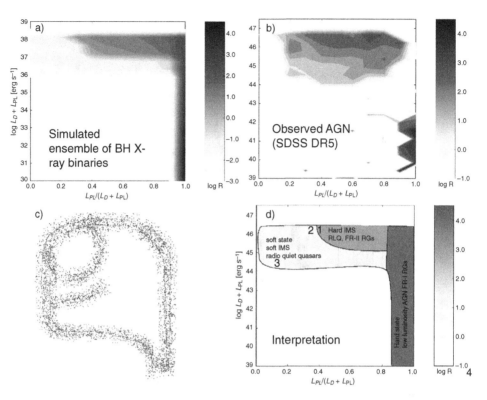

FIGURE 7.17. Does the same disk-jet coupling happen in supermassive black holes? The work of Körding et al. (2006c) suggests it may. In these plots the color scale is "radio loudness" (in this case a measure of the ratio of flux to (optical+X-ray) luminosity. a) shows a simulation of BHXRB behavior, based on the empirical patterns discussed earlier in this chapter; b) is a sample of (SDSS AGN + LLAGN) with X-ray measurements and either radio detections or radio upper limits – it looks strikingly similar to the pattern for BHXRB; c) looks unnervingly like a turtle's head; d) is a simplistic interpretation of the a) and b) results, comparing classes of AGN with BHXRB states.

7.5 Open issues

In this section I briefly summarize some of the (many) open issues in understanding the feedback from black-hole accretion.

7.5.1 Patterns of disk-jet coupling

We have made great progress in the past 15 years or so in quantifying the power of jets from accreting black holes (especially BHXRBs) and understanding empirically how they connect to accretion rates and states. This is a big step forward – we can take this understanding and, given population models of binaries, AGN, and so forth, model in a much clearer way how the feedback from accretion has been split between radiation and kinetic power as the population evolve.

Naturally, however, there remain many open issues, not least with the empirical models, which suffer primarily from poor radio coverage (a typical bright X-ray outburst may have hundreds of RXTE observations, each one of which can be analyzed spectrally and temporally, compared to typically 10 radio observations, each of which is a flux measurement, nothing more). This situation should improve with the next generation of radio facilities, currently under construction. Some issues include:

- What exactly is the relation between timing properties and the "moment" of jet production?

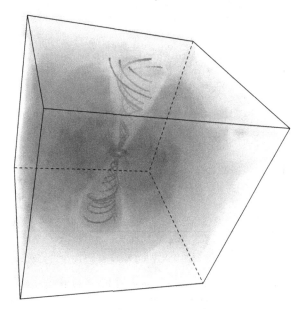

FIGURE 7.18. A still from a state-of-the-art simulation of jet formation around a rotating black hole. The simulations support the production of collimated relativistic jets powered by the black-hole spin. See McKinney and Blandford (2009) for details.

- When does the jet reactivate in the return to the hard state?
- What is the origin of the hysteresis observed between luminosity and accretion "state"?

Several other open issues remain, which have been with us since the start. These include the production mechanism of the jets, the origin of the flat-spectrum radio cores, and whether or not black hole jets could be powered by black hole spin. In the following sections I briefly discuss each of these in turn.

7.5.2 How are jets produced? Black-hole spin?

This is the big theoretical question, and one toward which much progress is being made, but so far no strong conclusion can be reached. In some subrelativistic cases, it is possible that "simple" hydrodynamics is enough to produce the jetlike outflow, but for most systems, and especially black holes, a magnetic field is almost certainly required to produce the jet.

When observations of BHXRBs provided the clue that hard states produced jets and soft states did not, this seemed like a big clue as to what is required to produce a jet. Furthermore, it seemed to confirm the need for a large-scale magnetic field out of the plane of the disk, something which (maybe) could only be supported in the hard state with its large-scale height accretion flow. Note that although this interpretation is alive and well, another possibility has always existed, namely that for some reason we just don't see the jet in the soft state (beamed out of line of sight, cooled?).

Some of the most recent work on simulations of jets (e.g., Koide et al., 2002; De Villiers et al., 2005; McKinney and Blandford, 2009 – see Fig. 7.18) does indicate that a large-scale height accretion flow coupled with a rotating central black hole produces the fastest and most powerful jets.

However, the observational evidence supporting the spin-powering of black hole jets remains uncertain. Within the field of AGN, there has long been a recognition of a radio loud–radio quiet "dichotomy," the origin of which is uncertain but has been shown to correlate at various times, and in various definitions, with things such as black hole mass

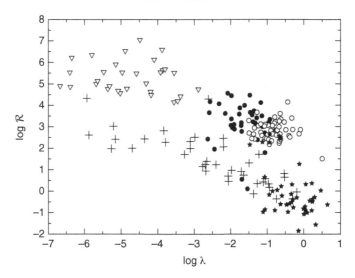

FIGURE 7.19. The apparent radio loud–radio quiet dichotomy in AGN. Radio loudness $\mathcal{R}$ vs. Eddington ratio λ. BLRGs are marked by the filled circles, radio-loud quasars by the open circles, Seyfert galaxies and LINERs by the crosses, FR I radio galaxies by the open triangles, and PG quasars by the filled stars. From Sikora et al. (2007), whose interpretation is that to be present in the upper, more radio-loud track requires a high black-hole spin. Reproduced by permission of the AAS.

and host galaxy type. The most recent major works in this area are those by Sikora et al. (2007) (Fig. 7.19) and Tchekhovskoy et al. (2010), both of which argue the case for an influence of black-hole spin (although not exclusively – Sikora et al., 2007, clearly also note that at high Eddington ratios the AGN analogue of XRB accretion states is likely to complicate the issue).

In parallel with these latest works on AGN populations, however, an increasing number of rather precise measurements of black-hole spin have been reported from X-ray spectroscopy, usually, but not exclusively, of BH XRBs. In Fender et al. (2010), we compare these reported spin measurements with our best estimates of the jet power and speed and find no correlation. This does not disprove the spin-powering conjecture, of course – the radio or X-ray measurements could be in some way wrong, or maybe spin-powering is just not effective in XRBs compared to AGN – but it does keep the question alive.

7.5.3 Other issues

Although the two areas just discussed are key areas that, subjectively, attract my interest, there are many others that are important. Some of these are listed next (and many others are no doubt missed...).

How does AGN feedback work?

The works on modeling and BH spin relate, via AGN populations, directly to the question of AGN feedback, and how it may serve to regulate galaxy growth, the action of which is not itself well understood. I think trying to take more clearly what we've understood from X-ray binaries, and merging it with the understanding of the more diverse environments encountered for AGN, is the way to go.

What is the role of slow dense winds?

There is a lot of evidence in AGN, and some in XRBs, that at high Eddington ratios the majority of the kinetic output is dominated by a slower, broader wind and not a fast, collimated jet. In fact it has recently been suggested that it is the activation of this wind, possibly associated to the onset of a radiation-driven instability, that is responsible for suppression of the jet in the soft state (Miller et al., 2006; Neilsen and Lee, 2009). Quantifying this area is of great importance – for example, is it possible that although the jet is suppressed in the soft state, the kinetic power (and maybe angular momentum) output remain approximately constant across state transitions?

How is the flat spectrum produced?

Since Blandford and Konigl (1979), it has been more or less accepted that the flat (spectral index ~ 0) radio–mm–infrared cores seen in some AGN and hard-state BHXRBs arise from an approximately conical jet. That this jet requires "reheating" in order to produce broadband flat spectra has been recognized since Blandford and Konigl (1979), but exactly how this is done has remained unclear. Some recents steps toward a solution of this problem have been explored in Pe'er and Casella (2009) and Jamil et al. (2010), but there remain many assumptions about the nature of this component that need to be tested.

Can we image the event horizon?

The largest angular diameter event horizons that we know of are those associated with the galactic center BH, Sgr A*, and the relatively nearby AGN, M87. It is straightforward to show that global mm VLBI should be able to achieve angular resolutions very close to the predicted size of the event horizon (for latest prospects, see Doeleman et al., 2009), and it is clear that sometime in the next decade or two such a test will be made. It is not clear that all the accreting matter around these black holes (and there undoubtedly is some) will allow a simple interpretation of the results, but I mention it here, as my last point, as possibly the best chance we have of actually "seeing" a black hole.

REFERENCES

Biretta, J. A., Sparks, W. B., and Macchetto, F. 2005. Hubble Space Telescope observations of superluminal motion in the M87 jet. 1999. ApJ, **520**(Aug.), 621–626.

Blandford, R. D., and Konigl, A. 1979. Relativistic jets as compact radio sources. ApJ, **232**(Aug.), 34–48.

Blundell, K. M., and Bowler, M. G. 2004. Symmetry in the changing jets of SS 433 and its true distance from us. ApJ, **616**(Dec.), L159–L162.

Burbidge, G. R. 1958. Possible sources of radio emission in clusters of galaxies. ApJ, **128**(July), 1–8.

Burrows, C. J., Stapelfeldt, K. R., Watson, A. M., Krist, J. E., Ballester, G. E., Clarke, J. T., Crisp, D., Gallagher, J. S. III, Griffiths, R. E., Hester, J. J., Hoessel, J. G., Holtzman, J. A., Mould, J. R., Scowen, P. A., Trauger, J. T., Westphal, J. A. 1996. Hubble Space Telescope observations of the disk and jet of HH 30. ApJ, **473**(Dec.), 437–L162.

Calvelo, D. E., Fender, R. P., Russell, D. M., Gallo, E., Corbel, S., Tzioumis, A. K., Bell, M. E., Lewis, F., and Maccarone, T. J. 2010. Limits on the quiescent radio emission from the black hole binaries GRO J1655-40 and XTE J1550-564. MNRAS, **409**(Dec.), 839–845.

Casella, P., Belloni, T., and Stella, L. 2005. The ABC of low-frequency quasi-periodic oscillations in black hole candidates: analogies with Z sources. ApJ, **629**(Aug.), 403–407.

Casella, P., Maccarone, T. J., O'Brien, K., Fender, R. P., Russell, D. M., van der Klis, M., Pe'Er, A., Maitra, D., Altamirano, D., Belloni, T., Kanbach, G., Klein-Wolt, M., Mason, E., Soleri, P., Stefanescu, A., Wiersema, K., and Wijnands, R. 2010. Fast infrared variability from a relativistic jet in GX 339-4. MNRAS, **404**(May), L21–L25.

Corbel, S., Fender, R. P., Tomsick, J. A., Tzioumis, A. K., and Tingay, S. 2004. On the origin of radio emission in the X-ray states of XTE J1650-500 during the 2001-2002 outburst. ApJ, **617**(Dec.), 1272–1283.

Corbel, S., Fender, R. P., Tzioumis, A. K., Tomsick, J. A., Orosz, J. A., Miller, J. M., Wijnands, R., and Kaaret, P. 2002. Large-scale, decelerating, relativistic X-ray jets from the microquasar XTE J1550-564. Science, **298**(Oct.), 196–199.

De Villiers, J.-P., Hawley, J. F., Krolik, J. H., and Hirose, S. 2005. Magnetically driven accretion in the Kerr metric. III. Unbound outflows. ApJ, **620**(Feb.), 878–888.

Doeleman, S., Agol, E., Backer, D., Baganoff, F., Bower, G. C., Broderick, A., Fabian, A., Fish, V., Gammie, C., Ho, P., Honman, M., Krichbaum, T., Loeb, A., Marrone, D., Reid, M., Rogers, A., Shapiro, I., Strittmatter, P., Tilanus, R., Weintroub, J., Whitney, A., Wright, M., and Ziurys, L. 2009. Imaging an event horizon: sub-mm-VLBI of a super massive black hole. *Astro2010: The Astronomy and Astrophysics Decadal Survey*. Science White Papers, no. 68.

Dunn, R. J. H., Fender, R. P., Körding, E. G., Cabanac, C., and Belloni, T. 2008. Studying the X-ray hysteresis in GX 339-4: the disc and iron line over one decade. MNRAS, **387**(June), 545–563.

Fabian, A. C., and Iwasawa, K. 1999. The mass density in black holes inferred from the X-ray background. MNRAS, **303**(Feb.), L34–L36.

Fabian, A. C., Sanders, J. S., Taylor, G. B., Allen, S. W., Crawford, C. S., Johnstone, R. M., and Iwasawa, K. 2006. A very deep Chandra observation of the Perseus cluster: shocks, ripples and conduction. MNRAS, **366**(Feb.), 417–428.

Falcke, H., Körding, E., and Markoff, S. 2004. A scheme to unify low-power accreting black holes. Jet-dominated accretion flows and the radio/X-ray correlation. A&A, **414**(Feb.), 895–903.

Fender, R. 2006. *Jets from X-Ray Binaries*. Compact stellar X-ray sources. Walter Lewin & Michiel van der Klis (eds.), Cambridge Astrophysics Series, No. 39. Cambridge, UK: Cambridge University Press, p. 381–419.

Fender, R. P., Belloni, T. M., and Gallo, E. 2004. Towards a unified model for black hole X-ray binary jets. MNRAS, **355**(Dec.), 1105–1118.

Fender, R. P., Gallo, E., and Jonker, P. G. 2003. Jet-dominated states: an alternative to advection across black hole event horizons in "quiescent" X-ray binaries. MNRAS, **343**(Aug.), L99–L103.

Fender, R. P., Gallo, E., and Russell, D. 2010. No evidence for black hole spin powering of jets in X-ray binaries. MNRAS, **406**(Aug.), 1425–1434.

Fender, R. P., Garrington, S. T., McKay, D. J., Muxlow, T. W. B., Pooley, G. G., Spencer, R. E., Stirling, A. M., and Waltman, E. B. 1999. MERLIN observations of relativistic ejections from GRS 1915+105. MNRAS, **304**(Apr.), 865–876.

Fender, R. P., Homan, J., and Belloni, T. M. 2009. Jets from black hole X-ray binaries: testing, refining and extending empirical models for the coupling to X-rays. MNRAS, **396**(July), 1370–1382.

Ferrarese, L., and Merritt, D. 2000. A fundamental relation between supermassive black holes and their host galaxies. ApJ, **539**(Aug.), L9–L12.

Gallo, E., Fender, R. P., Kaiser, C., Russell, D., Morganti, R., Oosterloo, T., and Heinz, S. 2005. A dark jet dominates the power output of the stellar black hole Cygnus X-1. Nature, **436**(Aug.), 819–821.

Gallo, E., Fender, R. P., Miller-Jones, J. C. A., Merloni, A., Jonker, P. G., Heinz, S., Maccarone, T. J., and van der Klis, M. 2006. A radio-emitting outflow in the quiescent state of A0620-00: implications for modelling low-luminosity black hole binaries. MNRAS, **370**(Aug.), 1351–1360.

Gallo, E., Fender, R. P., and Pooley, G. G. 2003. A universal radio-X-ray correlation in low/hard state black hole binaries. MNRAS, **344**(Sept.), 60–72.

Garcia, M. R., McClintock, J. E., Narayan, R., Callanan, P., Barret, D., and Murray, S. S. 2001. New evidence for black hole event horizons from Chandra. ApJ, **553**(May), L47–L50.

Hannikainen, D. C., Hunstead, R. W., Campbell-Wilson, D., and Sood, R. K. 1998. MOST radio monitoring of GX 339-4. A&A, **337**(Sept.), 460–464.

Hughes, P. A. 1991. *Beams and Jets in Astrophysics*. Cambridge University Press (Cambridge Astrophysics Series, No. 19), 593 p.

Jamil, O., Fender, R. P., and Kaiser, C. R. 2010. iShocks: X-ray binary jets with an internal shocks model. MNRAS, **401**(Jan.), 394–404.

Kaiser, C. R., and Alexander, P. 1997. A self-similar model for extragalactic radio sources. MNRAS, **286**(Mar.), 215–222.

Koide, S., Shibata, K., Kudoh, T., and Meier, D. L. 2002. Extraction of black hole rotational energy by a magnetic field and the formation of relativistic jets. Science, **295**(Mar.), 1688–1691.

Körding, E., Falcke, H., and Corbel, S. 2006a. Refining the fundamental plane of accreting black holes. A&A, **456**(Sept.), 439–450.

Körding, E. G., Fender, R. P., and Migliari, S. 2006b. Jet-dominated advective systems: radio and X-ray luminosity dependence on the accretion rate. MNRAS, **369**(July), 1451–1458.

Körding, E. G., Jester, S., and Fender, R. 2006c. Accretion states and radio loudness in active galactic nuclei: analogies with X-ray binaries. MNRAS, **372**(Nov.), 1366–1378.

Körding, E. G., Migliari, S., Fender, R., Belloni, T., Knigge, C., and McHardy, I. 2007. The variability plane of accreting compact objects. MNRAS, **380**(Sept.), 301–310.

Mahadevan, R. 1997. Scaling laws for advection-dominated flows: applications to low-luminosity galactic nuclei. ApJ, **477**(Mar.), 585+.

Markoff, S., Falcke, H., and Fender, R. 2001. A jet model for the broadband spectrum of XTE J1118+480. Synchrotron emission from radio to X-rays in the low/hard spectral state. A&A, **372**(June), L25–L28.

Marscher, A. P., Jorstad, S. G., Gómez, J.-L., Aller, M. F., Teräsranta, H., Lister, M. L., and Stirling, A. M. 2002. Observational evidence for the accretion-disk origin for a radio jet in an active galaxy. Nature, **417**(June), 625–627.

McHardy, I. M., Koerding, E., Knigge, C., Uttley, P., and Fender, R. P. 2006. Active galactic nuclei as scaled-up Galactic black holes. Nature, **444**(Dec.), 730–732.

McKinney, J. C., and Blandford, R. D. 2009. Stability of relativistic jets from rotating, accreting black holes via fully three-dimensional magnetohydrodynamic simulations. MNRAS, **394**(Mar.), L126–L130.

Merloni, A., Heinz, S., and di Matteo, T. 2003. A fundamental plane of black hole activity. MNRAS, **345**(Nov.), 1057–1076.

Miller, J. M., Raymond, J., Fabian, A., Steeghs, D., Homan, J., Reynolds, C., van der Klis, M., and Wijnands, R. 2006. The magnetic nature of disk accretion onto black holes. Nature, **441**(June), 953–955.

Miller-Jones, J. C. A., Fender, R. P., and Nakar, E. 2006. Opening angles, Lorentz factors and confinement of X-ray binary jets. MNRAS, **367**(Apr.), 1432–1440.

Miller-Jones, J. C. A., McCormick, D. G., Fender, R. P., Spencer, R. E., Muxlow, T. W. B., and Pooley, G. G. 2005. Multiple relativistic outbursts of GRS1915+105: radio emission and internal shocks. MNRAS, **363**(Nov.), 867–881.

Mirabel, I. F., and Rodríguez, L. F. 1994. A superluminal source in the Galaxy. Nature, **371**(Sept.), 46–48.

Moretti, A., Campana, S., Lazzati, D., and Tagliaferri, G. 2003. The resolved fraction of the cosmic X-ray background. ApJ, **588**(May), 696–703.

Narayan, R., and Yi, I. 1994. Advection-dominated accretion: A self-similar solution. ApJ, **428**(June), L13–L16.

Neilsen, J., and Lee, J. C. 2009. Accretion disk winds as the jet suppression mechanism in the microquasar GRS 1915+105. Nature, **458**(Mar.), 481–484.

Pe'er, A., and Casella, P. 2009. A model for emission from jets in X-ray binaries: consequences of a single acceleration episode. ApJ, **699**(July), 1919–1937.

Ponti, G., Terrier, R., Goldwurm, A., Belanger, G., and Trap, G. 2010. Discovery of a superluminal Fe K Echo at the galactic center: the glorious past of Sgr A* preserved by molecular clouds. ApJ, **714**(May), 732–747.

Rees, M. J. 1966. Appearance of relativistically expanding radio sources. Nature, **211**(July), 468–470.

Russell, D. M., Fender, R. P., Hynes, R. I., Brocksopp, C., Homan, J., Jonker, P. G., and Buxton, M. M. 2006. Global optical/infrared-X-ray correlations in X-ray binaries: quantifying disc and jet contributions. MNRAS, **371**(Sept.), 1334–1350.

Russell, D. M., Maitra, D., Dunn, R. J. H., and Markoff, S. 2010. Evidence for a compact jet dominating the broad-band spectrum of the black hole accretor XTE J1550-564. MNRAS, **405**(July), 1759–1769.

Sams, B. J., Eckart, A., and Sunyaev, R. 1996. Near-infrared jets in the Galactic microquasar GRS1915+105. Nature, **382**(July), 47–49.

Schmitt J. H. M. M., Snowden S. L., Aschenbach B., Hasinger G., Pfeffermann E., Predehl P., Trumper J., 1991, A soft X-ray image of the moon. Nature, **349**(February), 583–587.

Sikora, M., Stawarz, Ł., and Lasota, J.-P. 2007. Radio loudness of active galactic nuclei: observational facts and theoretical implications. ApJ, **658**(Apr.), 815–828.

Soltan, A. 1982. Masses of quasars. MNRAS, **200**(July), 115–122.

Susskind, L. 2009. *The Black Hole War: My Battle with Stephen Hawking to Make the World Safe for Quantum Mechanics*. Published by Little, Brown and Company, Hachette Book Group, 237 Park Avenue, New York, NY 10017 (www.HachetteBookGroup.com)

Tchekhovskoy, A., Narayan, R., and McKinney, J. C. 2010. Black hole spin and the radio loud/quiet dichotomy of active galactic nuclei. ApJ, **711**(Mar.), 50–63.

Thorne, K. 1994. *Black Holes and Time Warps: Einstein's Outrageous Legacy*. Published by W. W. Norton & Company, Inc, 500 Fifth Avenue, New York, NY 10110.

Yu, Q., and Tremaine, S. 2002. Observational constraints on growth of massive black holes. MNRAS, **335**(Oct.), 965–976.

8. Computing black-hole accretion
JOHN F. HAWLEY

8.1 Introduction

Astronomers have a remarkably successful theory for stars and stellar evolution. This success is due in part to the simplicity of spherical symmetry and steady-state equilibrium. Stars can be modeled using a series of time-independent equations that depend on only one spatial coordinate, namely the radius of the star. But the universe is a much more dynamic and active place than is implied by the stars alone. Some of the most energetic photons that astronomers observe originate not within stars but in orbiting disks of gas. This realization has brought the study of *accretion disks* to the forefront of high-energy astrophysics.

The idea of an orbiting disk of gas in a context other than that of a nascent solar system or spiral galaxy can be traced at least as far back as the work of astronomer Gerard Kuiper on mass transfer in close binary stellar systems. He noted that in such systems, gas can flow through a stream from one star to the other. Kuiper realized that the gas would possess sufficient angular momentum that it must go into orbit around the attracting star, forming a ring.

In 1955, John Crawford and Robert Kraft published a paper (Crawford and Kraft, 1956) that proposed an orbiting ring model for AE Aquarii, a short-period binary star system that showed significant episodic variability. The masses of the stars and the sizes of their orbits were such that mass transfer from one star to the other was likely. Recognizing that the high angular momentum of the gas ejected from the one star would prevent direct infall onto the second, they proposed that the gas must form a ring, as Kuiper had suggested. The issue then becomes how the gas moves from the ring down to the central star, a process known as accretion.

The theory of accretion disks found its most significant application with the opening of new regions of the electromagnetic spectrum to observational scrutiny. Almost immediately, new classes of objects were discovered, such as compact galactic X-ray sources. The temperature of ordinary stars is not sufficient to produce radiation with energies much above the ultraviolet. The discovery of radio-loud quasars was equally surprising. Although these powerful radio sources were identified with starlike point sources, their spectral properties were inconsistent with emission from stars. Astronomers soon discovered that quasars are among the most remote objects in the universe. Their apparent brightness combined with their great distance implies that quasars have enormous luminosity. A typical quasar's luminosity greatly exceeds that of typical spiral galaxies. Many quasars also exhibit rapid variability, implying that their light originates in a very compact region. Collimated beams of energetic plasma emerge from the centers of radio galaxies and many quasars. These *jets* can be observed at very high resolution using long-baseline radio interferometry, leaving no doubt that they are energized, launched, and collimated from a region less than a parsec in size.

To explain the extreme properties of quasars, Donald Lynden-Bell proposed that their source of energy is gravitational, released in the formation and growth of supermassive black holes through accretion (Lynden-Bell, 1969). The brightest of the quasars must consume yearly a quantity of gas equivalent to more than 1,000 solar masses. Such extreme fueling rates cannot last indefinitely; eventually the quasar must fade. Long after the quasars have disappeared, however, the supermassive black holes remain to power active galactic nuclei (AGN), such as radio galaxies and Seyfert galaxies, with a more modest rate of star consumption and a correspondingly lower luminosity.

How can a phenomenon as apparently simple as accretion account for such diverse systems as X-ray binaries, quasars, and active galaxies? Accretion disks are powered by the release of gravitational energy; the deeper the gravitational potential, the greater the energy available. The most basic issue governing the structure and evolution of accretion disks then is, how does gas in an accretion disk spiral down toward the central mass? The trajectory of gas orbiting a compact body is determined by its angular momentum, which in turn will be a conserved quantity of the motion, assuming no additional forces are applied. Unless there is some way for the gas to lose some of its angular momentum, it cannot drop down to a lower orbit. This means there must be some physical process by which angular momentum is removed from the gas and transferred outward through the disk. How could this be accomplished?

Orbital dynamics (Kepler's laws) dictate that particles in orbits closer to the central body must move with higher velocity. Hence the velocities in the disk constitute a shear flow, or, more precisely, a differentially rotating flow. In terrestrial fluids, shear flows are smoothed out by the exchange of momentum between neighboring fluid elements lying transverse to the direction of the shear flow. This transfer occurs due to the action of viscosity, ν. A sufficiently viscous disk will transfer angular momentum outward, and the gas will spiral down onto the central star. The difficulty with this straightforward explanation is that the gas in the astrophysical disk is not sufficiently viscous. The importance of viscosity is measured by the *Reynolds number*, $Re = LV/\nu$, a dimensionless quantity equal to the ratio of the characteristic length L and velocity V of the system to the viscosity. Low Reynolds numbers describe smooth, viscous flow, as is seen in many common terrestrial systems. In accretion disks the viscosity is very small, and the Reynolds number is correspondingly huge. If ordinary viscosity were the only available mechanism to transport angular momentum, the disk would simply orbit with no energy release.

Although this might seem to be a problem for the accretion disk model, the high Reynolds number has often been viewed in quite a different light, beginning with the first detailed model. Because "the Reynolds number [of the gas] is extremely high," Crawford and Kraft argued in their 1956 paper, "it is quite certain that the gas is turbulent." Turbulence, it was assumed, would be enough to ensure outward transport of angular momentum. Turbulence is characterized by complex, disorganized fluid motions on all length scales, in contrast to viscous, laminar flow in which the fluid moves smoothly in layers characterized by a single velocity. Viscosity makes flows laminar; turbulence can appear only when there is insufficient viscosity to smooth out all small velocity fluctuations that would otherwise increase in magnitude. Experiments with terrestrial fluids have shown that the Reynolds number must be greater than some critical value before a fluid will become turbulent. Once turbulence develops, the velocity fluctuations rapidly mix fluid elements. In an accretion disk, turbulence would presumably redistribute the fluid's angular momentum.

Turbulent mixing is not the only possible solution to the angular momentum problem in accretion disks, however. Lynden-Bell's 1969 disk model was based on angular momentum transport by magnetic fields embedded in the ionized gas in the accretion disk. The presumption was that magnetic transport would dominate over turbulent transport, with the presence of turbulence taken for granted. In any case, any detailed analysis was impeded by the complexity of the time-dependent, multidimensional equations that describe the physics of disks.

One of the most important and influential of the early disk papers is Shakura and Sunyaev (1973), which develops a detailed mathematical model for accretion disks in black-hole binary systems. Angular momentum transport was assumed to be due to a combination of turbulent and magnetic stresses whose magnitude should be proportional to the principal physical scales within the disk, such as the rotation speed, disk thickness, and gas temperature. Shakura and Sunyaev introduced α as a new parameter for the (unknown) constant of proportionality between the transport and the disk properties, specifically $t_{r\phi} = \alpha P$, where $t_{r\phi}$ is a relevant component of the stress tensor and P is the pressure in the disk. The use of this α parameter has allowed substantial progress to be made in modeling accretion disk systems despite the ignorance of the underlying physics of angular momentum transport. But that progress has been limited to certain specific disk behaviors for which the α parameterization is appropriate. Many disk phenomena don't seem to fit comfortably into that framework.

Over the years the central problems of accretion disk theory have remained: how do they accrete, and what radiation does that accretion produce? In these sections I outline some of the theoretical progress that has been made in understanding basic accretion physics and describe efforts to use computational techniques to model black-hole accretion in ever greater detail and realism.

8.2 Basic accretion disk properties

Broadly speaking, accretion disks can be separated into three categories: protostellar disks, disks formed by mass transfer in binary star systems, and disks in AGN and quasars. This review focuses on black-hole binary and AGN disks that are hot enough to be significantly or fully ionized, and their high conductivity means that currents flow freely. Magnetic fields strongly influence the structure and evolution of these disks, and high conductivity ensures that the field will be well coupled to the gas.

Far less is understood about astrophysical disks than is understood about stars; disks are in many ways dynamically much more complicated. Unlike stars, whose energy derives from relatively well-understood nuclear reactions in their cores, accretion disks generate their luminosity as a byproduct of the accretion process itself. Another difficulty lies with the geometry: stars are spheres and vary in only one dimension, whereas disks are multidimensional. This is intrinsic to the presence of orbital angular momentum, the defining disk characteristic. Many critical aspects of disks vary over all three dimensions. Further, whereas stars are relatively stable and slow either to expand or to contract, accretion disks are governed by fluid moving at supersonic orbital speeds. Their behavior is much more dynamic and variable than is that of most stars. Thus, the equations that describe accretion disks are much more complex than the standard equations of stellar structure.

As the origin of accretion disk luminosity is gravity, for an accretion rate of $\dot{M}$ falling toward a star of mass M, the luminosity will be on order $GM\dot{M}/R$, where R is some fiducial radius such as the star's surface. In the case of free fall, this energy would be released as matter hit the stellar surface, but in a disk system the gravitational energy is released more slowly as material spirals in. In this case, half of the energy at radius R is available to be radiated locally and half is retained as orbital kinetic energy. Of course, if the central star has a solid surface, the gas will eventually arrive there and release the remaining energy in a boundary layer. If, on the other hand, the central star is a black hole, the energy may be permanently lost. This is one of the ways a black hole differs from ordinary stars: the black hole has no surface. For a black hole, the total accretion luminosity will depend on how much gravitational energy is radiated by the accreting gas before it is lost down the hole. This could be essentially zero if the gas is simply free-falling straight in! This is very unlikely however, as the gas will have some angular momentum.

The luminosity of an accretion disk is characterized by the *Eddington luminosity* L_E. This quantity is the luminosity at which the radiative momentum flux is balanced by the gravitational force from a central object. Because both the radiation flux and gravity diminish as the inverse square of the distance, L_E depends only on the central mass M,

$$L_E = \frac{4\pi G m_p M c}{\sigma_T} = 1.3 \times 10^{38} \frac{M}{M_\odot} \text{ergs s}^{-1}, \tag{8.1}$$

where G is the gravitational constant, m_p the proton mass, c the speed of light, and σ_T the Thomson scattering cross section. The ratio $M/M_\odot$ normalizes the central mass to one solar mass ($M_\odot = 1.989 \times 10^{33}$ g). If the luminosity in a spherical source exceeded L_E, accretion would not be possible. Disks are not spherical, and under some circumstances luminosities in excess of L_E can be produced. Generally, however, most source luminosities tend to be below their Eddington value, and L_E is a useful fiducial benchmark. For example, a binary X-ray source luminosity of 10^{37} ergs s^{-1} ~ 0.1 L_E is fairly typical.

We may define an Eddington accretion rate $\dot{M}_E$ in a form most suitable for black-hole accretion as $\dot{M}_E \equiv L_E/\eta c^2$, where η is the efficiency of accretion defined as the fraction of the rest-mass energy converted into photons. The Eddington accretion rate corresponds to $1.4 \times 10^{17} \eta^{-1}(M/M_\odot)$ g s^{-1}, which is about a solar mass per year for a supermassive 10^9 $M_\odot$ black hole. For black-hole accretion, $\eta = 0.1$ is the usual assumption; the actual value depends on factors such as how much orbital energy is extracted prior to the accreted mass entering the hole, how efficiently that energy is converted to radiation, and whether that radiation escapes or is itself captured by the hole.

At this point, a brief review of steady-state thin disk theory is in order. For simplicity we will assume Newtonian gravity in this discussion. The relativistic version of steady thin disk theory is given by Novikov and Thorne (1973). Relativity changes some details of the model, but not the underlying principles. In this model the gas orbits with the local Keplerian frequency $\Omega \propto R^{-3/2}$ and local thickness of the disk H (the pressure scale height) is much less than its orbital radius R, for example, $H/R \ll 1$. This *thin disk* approximation implies that the sound speed c_s is much less than the orbital velocity, with $c_s \approx \Omega H$. All quantities can be vertically averaged and assumed to be time independent. In this limit the mass accretion rate follows from simple mass conservation and is given by

$$\dot{M} = -2\pi R \Sigma \langle u_R \rangle_\rho, \tag{8.2}$$

where Σ is vertically integrated surface density and $\langle u_R \rangle_\rho$ is a density-weighted, vertically averaged radial inflow velocity. Conservation of angular momentum implies that the angular momentum flux through the disk will consist of inward transport of orbiting material balanced by an outward transport by stress. The net value carried in will be determined by the boundary value at the inner edge of the disk, as in

$$-\frac{\dot{M}}{2\pi}R\Omega + \Sigma R W_{R\phi} = \text{const.}, \tag{8.3}$$

where $W_{R\phi}$ is the height-averaged, density-weighted r, ϕ stress tensor component. Conservation of energy gives us a radial energy flux

$$\mathcal{F}_E = \frac{\dot{M}R\Omega^2}{4\pi} + \Sigma R \Omega W_{R\phi} \tag{8.4}$$

whose divergence is related to the local disk emissivity $\mathcal{Q}$,

$$\mathcal{F}_E = \frac{3GM\dot{M}}{4\pi R^2}\left[1 - \frac{2}{3}\left(\frac{R_o}{R}\right)^{1/2}\right] = 2\mathcal{Q}, \tag{8.5}$$

where R_o is the inner radius of the disk. For a black hole, this radius will be the point where the accretion stress is no longer significant. Note that in steady-state thin disk theory, the luminosity follows from basic conservation laws and depends only on the accretion rate; the stress doesn't matter. What a specific prescription for the stress provides are all the details, such as disk density, vertical profile, and other quantities.

In the thin disk approximation, the optically thick disk emits thermally at the effective temperature needed to radiate the energy released locally, which is given by

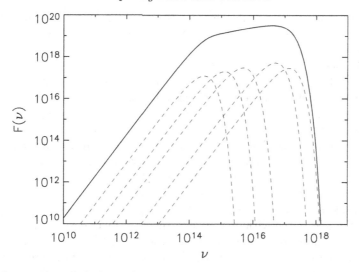

FIGURE 8.1. An accretion disk thermal spectrum is composed of the superposition of blackbody spectra at all radii of the disk.

divergence of the energy flux. Specifically,

$$2\sigma T_{eff}^4 = -\frac{1}{R}\frac{d}{dR}(R\mathcal{F}_E) = \frac{3GM\dot{M}}{4\pi R^3}. \tag{8.6}$$

This means that

$$T_{eff} \propto R^{-3/4}, \tag{8.7}$$

a standard thin disk result. Figure 8.1 illustrates the composite disk spectrum that results from the integration over radius of all the individual thermal spectra. The peak temperature in the disk spectrum comes from the location of the inner edge of the disk, and this will be determined by the point where the stress goes to zero and no further extraction of orbital energy occurs.

Although continuum disk spectra of the type depicted in Figure 8.1 are observed in many systems, many accreting black holes manifest multiple states, each of which has its own characteristic properties, including its own intrinsic time variability. Typically, one sees transitions between high, intermediate, and low states, often on regular time intervals. These distinct spectral states almost certainly correspond to distinct dynamical states of accretion, as yet poorly understood. What are the physical processes that cause a massive accretion disk to behave with such regularity and cycle between such distinctive states? The standard thin disk model cannot answer these questions.

The mass of a hole defines a standard length, $GM/c^2 = 1.5 \times 10^5$ cm, and a standard time, $GM/c^3 = 5 \times 10^{-6}$ s. The Schwarzschild radius of a nonrotating hole is $R_g = 2GM/c^2$, or $2M$ as it is usually written (by setting G and c to 1). In these units, the effective temperature of a disk in a black hole system is

$$T_{eff} = 3.5 \times 10^7 \left(\frac{R}{R_g}\right)^{-3/4} \left(\frac{M}{M_\odot}\right)^{-1/4} \left(\frac{\eta\dot{M}}{\dot{M}_E}\right)^{1/4} \text{ K}. \tag{8.8}$$

Note that the effective temperature goes *down* as the mass of the black hole goes up; the area available for radiating the energy increases more rapidly than the energy released. For a stellar-mass black hole, the characteristic temperature in the inner disk is on order 10^7 and the emission is primarily in the X-ray band, whereas a supermassive black hole will have a temperature $\sim 10^5$ and emission in the ultraviolet.

The substantial range of possible masses is one thing that distinguishes black-hole accretion from, say, accretion onto a neutron star or white dwarf. Another is that black holes themselves have only two significant properties: mass, M, and angular momentum a (in principle, black holes can also have a net electric charge, but this is not expected to be important for black holes in nature). Instead of a surface, a black hole has an *event horizon*, the point of "no return" where all future-pointing worldlines are directed inward. Put another way, the horizon is the point where the escape velocity equals c. For a nonrotating Schwarzschild black hole, the horizon is at $2M$, whereas rotating holes have smaller event horizons, down to M for a maximally rotating hole, $a/M = 1$. Again, the presence of an event horizon means that arbitrary amounts of matter and energy can be swallowed by the hole. With an ordinary star, all accreted matter must hit the surface where any remaining infall energy will be released.

Although black-hole accretion was originally proposed as a mechanism to explain the huge luminosities associated with quasars and their ilk, there is now considerable interest in *underluminous* black-hole accretion flows. Much of this interest stems from detailed observations of the galactic center Sgr A*. The unprecedented spatial resolution afforded by the Chandra observatory has revealed a remarkably underluminous source, given the density and temperature of the surrounding gas. Sources such as the galactic center are tremendously underluminous compared with AGNs despite the presumed availability of a seemingly adequate gas supply. Certainly these holes are not being fed by a standard radiative Keplerian disk.

One of the more important aspects of black holes for disk accretion is the presence of an innermost stable circular orbit (ISCO). In Newtonian gravity, the gravitational potential is $\propto 1/R$, and it is always possible to have a circular orbit at any radius around a point mass. With black holes, the effective potential (gravitational plus centrifugal) has a maximum. For a Schwarzschild hole, the location of the ISCO is $R_{\rm ISCO} = 6M$. The ISCO lies closer to the horizon for rotating holes. Once a particle reaches the ISCO with the angular momentum appropriate to a circular orbit there, it requires no further loss of angular momentum in order to fall into the hole. That is, accretion can proceed without any additional stress. Because of this, it has become a standard assumption that the stress goes to zero at the ISCO. This provides the boundary condition on the angular momentum transport that is required to close the steady-state thin disk equations. The efficiency of accretion then becomes equal to the binding energy of the ISCO orbit. For a Schwarzschild hole, $\eta = 0.057$; η increases to about 40% for a maximally spinning black hole where the ISCO moves very close to the horizon. Note, however, that photons radiated from such a location would have a very difficult time reaching a distant observer, and those that did would be significantly redshifted, so in practical terms it is likely that the net efficiency remains well below the theoretical limit.

Finally, rotating black holes can themselves be a significant source of energy. In general relativity, the geometric properties of spacetime are determined by the presence of mass, and spacetime determines what constitutes inertial motion (free-falling trajectories). A rotating black hole's influence on spacetime is so strong that a zero–angular momentum free-falling trajectory must move in the direction of the hole's rotation; this is known as the "dragging of inertial frames." Near a rotating black hole, there is a region known as the ergosphere within which it is no longer possible to be at rest with respect to an observer at infinity. A rocket within the ergosphere could accelerate in the direction of the hole's rotation and emerge from the ergosphere with positive energy at infinity that is greater than the rest mass of the rocket exhaust thrown down the hole. The excess energy comes from the rotation of the hole itself. A rocket is a somewhat artificial means of tapping into the hole's rotation energy, but Blandford and Znajek (1977) pointed out that a suitably arranged magnetic field threading the horizon could also do so. The hole's rotation creates an electromotive force (EMF) leading to an outward-directed Poynting flux (and an inward-directed Poynting flux of negative energy).

Clearly, there are a wide range of theoretical questions that are impossible to approach through the standard, analytic time-stationary thin disk model. But the prospects for rapid progress are excellent. First, we now have a much better understanding of the physical processes that underlie accretion, namely the origins and properties of angular momentum transport. This has established the notion that the accretion process can be profitably studied from a basic set of equations of compressible magnetohydrodynamics (MHD) without the need for artificial parameterizations for the stress. Second, the rapid increase in computational power and the construction of new codes using a variety of algorithms have made it possible to carry out full time-dependent and three-dimensional simulations of accretion disks.

In the following sections I review the basic physics of the magnetorotational instability (MRI), briefly describe some aspects of the simulation codes used currently in astrophysics for accretion studies, and describe some of the results from those simulations.

8.3 Turbulent Transport and the MRI

Accretion disks are powered by the gravitational potential energy released as orbiting matter slowly descends toward the central star. This infall requires a torque of some sort to remove angular momentum from the orbiting gas. The obvious candidate for such a torque is molecular viscosity, but in an accretion disk, viscosity is too small to be important (or, more precisely, the accretion disk is too big – disks have very high Reynolds numbers). Something more substantial is required, and other possibilities include large-scale torques from magnetic fields driving an MHD wind from the disk, tidal torques in a binary system, spiral waves, internal magnetic stresses, and stresses resulting from turbulence within the disk. This review focuses on the last two candidates, turbulence and magnetic fields. It transpires that these two are fundamentally linked. Students interested in a thorough review of all aspects of angular momentum transport in disks should see Balbus (2003).

We begin the study of the dynamics of black-hole accretion with the standard (non-relativistic) equations of MHD. These are the equations of mass conservation,

$$\frac{\partial \rho}{\partial t} + \nabla \cdot \rho \mathbf{v} = 0; \tag{8.9}$$

the equation of momentum conservation,

$$\frac{\partial \rho \mathbf{v}}{\partial t} + \nabla \cdot \rho \mathbf{v}\mathbf{v} + \nabla \left(P + \frac{B^2}{8\pi} \right) - \frac{1}{4\pi}(\mathbf{B} \cdot \nabla)\mathbf{B} + \rho \nabla \Phi = 0; \tag{8.10}$$

energy conservation,

$$\frac{\partial E}{\partial t} + \nabla \cdot \mathbf{v} \left[\frac{\rho v^2}{2} + \rho \Phi + \frac{\gamma}{\gamma - 1} P + \frac{\mathbf{B}}{4\pi} \times (\mathbf{v} \times \mathbf{B}) \right] = -\nabla \cdot \mathbf{F}_{\mathrm{rad}}; \tag{8.11}$$

and the induction equation

$$\frac{\partial \mathbf{B}}{\partial t} - \nabla \times (\mathbf{v} \times \mathbf{B}) = 0. \tag{8.12}$$

In these equations ρ is the mass density, $\mathbf{v}$ the velocity, P the pressure, assumed here to be given by an adiabatic equation of state $P \propto \rho^\gamma$, $\mathbf{B}$ is the magnetic field, and Φ is the external gravitational potential. The total energy E is composed of kinetic, thermal, gravitational, and magnetic components.

In an accretion disk that is in an averaged steady state, we can define a fluctuation velocity $\mathbf{u}$ that is obtained from the total velocity by subtracting the circular orbit velocity, that is, $\mathbf{u} = \mathbf{v} - \Omega R \hat{\phi}$. From the equation of angular momentum conservation, we can identify specific terms corresponding to the transport of angular momentum, namely,

$$R\left[\rho u_R(R\Omega + u_\phi) - \frac{B_R B_\phi}{4\pi}\right]. \tag{8.13}$$

The first term corresponds to the direct transport of angular momentum by the radial velocity. The second term is the Reynolds stress; net outward transport of angular momentum requires that the averaged radial and angular velocity fluctuations, $\langle u_R u_\phi\rangle$, be correlated so as to produce a net positive value. The last term is the Maxwell stress, the transport due to magnetic torques. Again, field components cannot be completely random, but must be correlated to produce a net torque. Within a differentially rotating disk, such a correlation arises naturally, since radial gradients in the angular velocity Ω in the magnetic induction equation generate toroidal fields from radial fields.

Given the right combination of velocity fluctuations and magnetic fields, substantial angular momentum transport can be produced within a turbulent flow. Because turbulence is dissipative, there must be a source of energy for it to be maintained. In an accretion system, that energy is in the differential rotation that can be tapped if the net stress is positive. Further, the onset of turbulence requires that the laminar orbital flow be destabilized in such a way as to lead naturally to self-sustaining turbulence. As it turns out, purely hydrodynamic flows cannot accomplish this, whereas magnetized flows can.

8.3.1 Hydrodynamic Stability

To explore the question of the existence of turbulence in an orbiting fluid, we begin with the relatively straightforward question of the linear stability of differential rotation. There are a variety of ways to illustrate the dynamical stability of a differentially rotating system; here we consider the simple dynamics of orbiting particles. Set up a standard cylindrical coordinate system (R, ϕ, z) centered on a central gravitating body. Ignoring pressure gradient forces, the equations of motion for a particle orbiting in a plane in the field of central potential Φ are

$$\ddot{R} - R\dot{\phi}^2 + \frac{\partial \Phi}{\partial R} = 0 \tag{8.14}$$

and

$$R\ddot{\phi} + 2\dot{R}\dot{\phi} = 0, \tag{8.15}$$

where we use dots for time derivatives. Now transform these equations to a local frame, centered on a circular orbit $R = R_0$, with angular frequency Ω_0, and introduce quasi-Cartesian x and y variables corresponding to the radial and toroidal directions,

$$R = R_0 + x \qquad \phi = \Omega_0 t + \frac{y}{R_0}. \tag{8.16}$$

Substituting this into the previous two equations, canceling leading order terms, and retaining terms linear in x and y leads to

$$\ddot{x} - 2\Omega\dot{y} = -x\frac{d\Omega^2}{d\ln R}, \tag{8.17}$$

$$\ddot{y} + 2\Omega\dot{x} = 0. \tag{8.18}$$

The angular velocity gradient $d\Omega^2/d\ln R$ refers to the circular velocity profile $\Omega(R)$; this term corresponds to the local gravitational tidal force. The remaining terms are the familiar Coriolis forces.

This system of equations has solutions for x and y of the form $e^{i\omega t}$, provided that

$$\omega^2 = 4\Omega^2 + \frac{d\Omega^2}{d\ln R} \equiv \kappa^2. \tag{8.19}$$

The quantity on the right-hand side is the square of the *epicyclic frequency*, which also can be written in terms of the specific angular momentum ($L \equiv R^2\Omega$) gradient

$$\kappa^2 \equiv \frac{1}{R^3}\frac{dL^2}{dR}. \tag{8.20}$$

If only gravitational forces are present, a particle will execute periodic retrograde epicycles around its circular orbit, as seen from within a corotating frame provided $\kappa^2 > 0$. This is the *Rayleigh stability criterion*; differential rotation is unstable only if the angular momentum decreases outward. That is most unlikely in astrophysical disks where the rotation law is determined by the gravitational field. For accretion disks with a Keplerian profile, $L \propto R^{1/2}$, the Rayleigh stability condition is more than satisfied.

8.3.2 The Magnetorotational Instability

Now we turn to the question of disk stability when a magnetic field is added. Assume that the field is weak, in the sense that the magnetic energy density is small compared with the thermal (and therefore also the rotational) energies. For the moment, let the field have some vertical component, but otherwise it may have an arbitrary geometry. Use the local system of equations and consider incompressible, small displacements (x and y) from circular motion of the form $e^{i(kz-\omega t)}$ (Balbus and Hawley, 1998),

$$\ddot{x} - 2\Omega\dot{y} = -x\frac{d\Omega^2}{d\ln R} - (\mathbf{k}\cdot\mathbf{v}_A)^2 x, \tag{8.21}$$

$$\ddot{y} + 2\Omega\dot{x} = -(\mathbf{k}\cdot\mathbf{v}_A)^2 y, \tag{8.22}$$

where the Alfvén speed $\mathbf{v}_A$ is defined by

$$\mathbf{v}_A = \frac{\mathbf{B}}{\sqrt{4\pi\rho}}. \tag{8.23}$$

These are simply the equations for Alfvén waves, modified by Coriolis forces and a tidal centrifugal force in the radial direction. This system of equations has solutions for x and y of the form $e^{i\omega t}$, with the significantly modified dispersion relation

$$\omega^4 - \omega^2(\kappa^2 + 2(\mathbf{k}\cdot\mathbf{v}_A)^2) + (\mathbf{k}\cdot\mathbf{v}_A)^2\left((\mathbf{k}\cdot\mathbf{v}_A)^2 + \frac{d\Omega^2}{d\ln R}\right) = 0. \tag{8.24}$$

This is a simple quadratic in ω^2, and one can show that there are values of $(\mathbf{k}\cdot\mathbf{v}_A)^2$ for which $\omega^2 < 0$ if and only if

$$\frac{d\Omega^2}{dR} < 0, \tag{8.25}$$

that is, when the angular *velocity* decreases outward. This is a dramatic alteration from the Rayleigh criterion, since this *is* the behavior of a Keplerian (or any other astrophysical disk) rotation law.

The magnetic term $(\mathbf{k}\cdot\mathbf{v}_A)^2$ enters equations 8.21 and 8.22 in the same manner as a spring constant. Consider a system of two orbiting masses connected by a spring (Balbus and Hawley, 1992b), with x and y representing the relative separation coordinates of the masses (Fig. 8.2). When the connecting spring is stretched so that one mass is orbiting slightly farther out than the other, the inner mass rotates more rapidly on the lower orbit. The tug of the spring pulls back on this mass, causing it to lose angular momentum. This loss compels the mass to descend to a yet lower orbit, with higher angular velocity, while the gain of angular momentum in the outer mass sends it to a higher orbit with lower angular velocity. The separation grows, the spring stretches more, and the process runs away. Note that this is an instability only if the spring is relatively weak; if the spring is too strong, the mass points will simply oscillate as they orbit. The instability is strongest

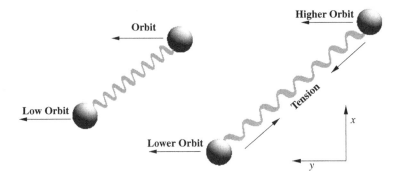

FIGURE 8.2. The action of the MRI is similar to that of a spring connecting two orbiting bodies. The tugging of the spring transfers angular momentum from the lower object to the higher, causing the relative angular velocities between the two objects to increase. This increases their separation and the force from the spring.

if the spring constant $(\mathbf{k}\cdot\mathbf{v}_A)^2$ is comparable to Ω^2. Specifically, the maximum growth rate is

$$|\omega_{max}| = -\frac{1}{2}\frac{d\Omega}{d\ln R}, \quad (8.26)$$

which, for a Keplerian rotation law $\Omega \sim R^{-3/2}$, is obtained when

$$(\mathbf{k}\cdot\mathbf{v}_A)^2 = \frac{15}{16}\Omega^2. \quad (8.27)$$

The growth rate is quite large, amounting to an amplification factor of about 100 in energy per orbit. Figure 8.3 shows the growth rate of the MRI as a function of vertical wavenumber for the case of a purely vertical field.

The MRI does not just lead to outward angular momentum transport; it arises *because* of outward angular momentum transport, created by the Maxwell stress. In the absence

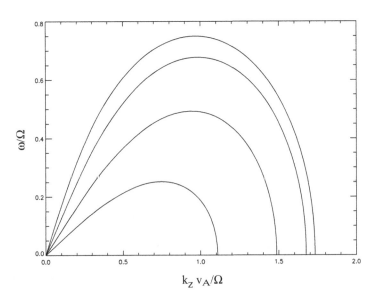

FIGURE 8.3. Growth rate of the MRI for a vertical magnetic field as a function of $(k_z \cdot v_A)/\Omega$. The different curves correspond to different values of k_r; the uppermost curve has $k_r = 0$, which produces the maximum growth rate for the vertical field case.

of a magnetic field, differentially rotating flows are linearly stable by the Rayleigh criterion, and the Coriolis force is a strongly restoring force. Hydrodynamic and magnetic differentially rotating systems are fundamentally different in their basic stability properties.

Several points bear emphasis. First, there is a qualitative difference in the behavior of the cases $B \to 0$ and $B = 0$. The maximum growth rate of the instability is independent of the strength of the magnetic field. When the magnetic field is finite, however small, there will always be *some* wavenumber, however large, that makes $\mathbf{k} \cdot \mathbf{v}_A$ comparable to Ω. What the magnetic field does is to set the scale for dynamically unstable wavelengths, a scale that is comparable to the distance an Alfvén wave travels in an orbital period. Inertial waves, which characterize the field-free response of the disk, have no associated characteristic wavenumber scale. They have instead a characteristic frequency, Ω. The interesting wavenumbers when a field is present are then of order Ω/v_A.

In practice, of course, there is a minimum field strength set by the condition that the wavelength of maximum growth cannot be so small that dissipative physics (conductivity or viscosity) is important. Even then, the instability is present at long wavelengths, but with a growth rate proportional to the magnetic field for a given wavelength. In fact, linear analyses that include resistivity (Jin, 1996), ambipolar diffusion (Blaes and Balbus, 1994), and the Hall effect (Wardle, 1999) show the instability to be present for many nonideal astrophysical plasmas. Generally speaking, resistive effects take over when the field slips through the fluid faster than the growth rate of the MRI. For a resistivity η and a wavenumber k, this occurs when $kv_A \sim k^2\eta$. The effect of viscosity ν is to damp fluid motions, and the viscous effect becomes dominant when the viscous frequency is on the order of the orbital frequency, $\Omega \sim k^2\nu$. The flows can be characterized by a Reynolds number, $\mathrm{Re} = v_A^2/\nu\Omega$ and a magnetic Reynolds number (or, more precisely, the Elsasser number) $\mathrm{Re}_m = v_A^2/\eta\Omega$. Nominally, the MRI is suppressed when either number is less than 1.

Second, the instability is even more robust than this simplified presentation indicates. The maximum growth rate (eq. 8.26) is independent not only of the strength of the magnetic field, but also of the geometry of the magnetic field. Even a purely toroidal field is unstable (Balbus and Hawley, 1992a) with the same maximum growth rate as seen for axisymmetric disturbances, although this is achieved only for very high poloidal wavenumbers (Balbus and Hawley, 1998). Indeed, equation 8.26 probably represents the maximum possible growth rate that *any* instability tapping into the free energy of differential rotation can achieve (Balbus and Hawley, 1992b).

Third, the MRI is stabilized for a given wavelength when the field becomes strong enough. The magnetic tension forces become stronger than the centrifugal acceleration. In a disk with vertical pressure scale height $H = c_s/\Omega$, MRI stabilization occurs for wavelengths shorter than H when the Alfvén and sound speeds become comparable, $v_A \sim c_s$. The toroidal field MRI becomes stabilized when the Alfvén speed becomes comparable to the *orbital* speed, a much larger value. In any case, if stabilization of the MRI occurs due to strong fields, that does not mean that the magnetic field is thus ignorable! Rather, one moves into a magnetically dominated accretion regime. There still exist *global* MRI modes where the disk exchanges angular momentum with gas at large radius (Gammie and Balbus, 1994). It is also possible that novel accretion states occur when Lorentz forces dominate. Such magnetically dominated accretion flows have been proposed by Meier (2005) to explain the "low-hard" states. The possible behaviors of such strongly magnetic accretion flows are largely unexplored to date.

Fourth, the instability has more general astrophysical implications than those associated with accretion disks. It is instructive to examine the general analysis of Balbus (1995), who considered the local axisymmetric stability of a weakly magnetized, stratified system that is differentially rotating with angular frequency $\Omega(R, z)$. For the case of an unmagnetized adiabatic fluid with an equation of state $P \propto \rho^\gamma$, the stability criteria are

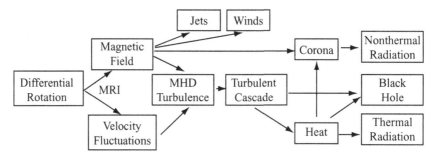

FIGURE 8.4. Schematic showing how the MRI catalyzes the transformation of orbital energy into heat, light, and outflows within accretion systems.

the well-known classical *Høiland criteria* (Tassoul, 1978):

$$\frac{-1}{\gamma \rho}(\nabla P) \cdot \nabla \ln P \rho^{-\gamma} + \frac{1}{R^3}\frac{\partial L^2}{\partial R} > 0, \tag{8.28}$$

$$\left(-\frac{\partial P}{\partial z}\right)\left(\frac{\partial L^2}{\partial R}\frac{\partial \ln P \rho^{-\gamma}}{\partial z} - \frac{\partial L^2}{\partial z}\frac{\partial \ln P \rho^{-\gamma}}{\partial R}\right) > 0. \tag{8.29}$$

Here $L^2 \equiv R^4 \Omega^2$ is the square of the specific angular momentum. Instability can arise from unfavorable entropy gradients (the Schwarzschild criterion) or from unfavorable angular momentum gradients (the Rayleigh criterion). In the presence of a magnetic field, however, Balbus showed that the Høiland criteria must be replaced by the following:

$$\frac{-1}{\gamma \rho}(\nabla P) \cdot \nabla \ln P \rho^{-\gamma} + \frac{1}{R^3}\frac{\partial \Omega^2}{\partial R} > 0, \tag{8.30}$$

$$\left(-\frac{\partial P}{\partial z}\right)\left(\frac{\partial \Omega^2}{\partial R}\frac{\partial \ln P \rho^{-\gamma}}{\partial z} - \frac{\partial \Omega^2}{\partial z}\frac{\partial \ln P \rho^{-\gamma}}{\partial R}\right) > 0. \tag{8.31}$$

Two features of the Balbus criteria are particularly striking: first, the criteria are the same as the Høiland criteria with the substitution of the angular velocity Ω for the angular momentum L. This seemingly minor change has enormous significance. In a Keplerian disk $L \propto R^{1/2}$, but $\Omega \propto R^{-3/2}$, so the disk is linearly stable by criterion 8.28, but *unstable* by 8.30. Second, the stability criteria for the magnetized plasma makes no reference to the magnetic field. It is the *presence* of the magnetic field, not its amplitude, that renders the disk unstable.

That the existence of the MRI is independent of the strength of the magnetic field (for ideal MHD) is one of its more surprising aspects. The concept of a "weak" magnetic field is generally expressed in the ratio of the total field energy to other energies of the system, such as thermal or kinetic. But such a concept misses the point: what is important is the existence of a tangential Maxwell stress force, a force that has no hydrodynamic analogue. The magnetic field enables degrees of freedom in the plasma that are unavailable with hydrodynamics alone. The magnetic field strength sets the wavelength of the most unstable mode of the MRI, but does not determine whether or not the system is unstable.

For accretion disks, then, magnetic fields make all the difference. Figure 8.4 presents a schematic diagram of the flow of energy through the accretion disk. The source of the energy to power the system comes from the differential rotation. The minimum energy state would be solid body rotation, Ω = constant. But Newton's laws do not admit that as a dynamical possibility in a non–self-gravitating disk. Instead, the disk will remain

MRI unstable and rotating with a nearly Keplerian distribution. The action of the MRI creates turbulent magnetic and velocity fluctuations that transport angular momentum outward in a bid to reach an unattainable state of minimum energy. That turbulence is dissipated into heat. Some of that heat will be radiated from the disk, although some will be swallowed by the black hole. Large-scale magnetic fields may go to power winds or jets from the system, and fields will buoyantly rise into the corona, which can be a source of significant nonthermal radiation. The MRI makes accretion disks accrete, and the challenge is to understand all the consequences.

8.4 Computational astrophysics

The steady-state thin disk model is based on conservation laws and, as such, provides a basic "zeroth-order" description of an accretion system. The details matter, however, particularly when trying to provide theoretical interpretation to the new observations that are now coming from both space- and ground-based observatories. Past efforts to develop those details were hampered by the lack of understanding of the angular momentum transport process. With the identification of a physical process that leads to turbulence and to angular momentum transport, theoretical studies of accretion have moved from dimensional analysis to genuine dynamics.

Unfortunately, there is no analytic theory of MHD turbulence; the full three-dimensional equations of MHD are too complex. Things become much more difficult if one wishes to include detailed thermodynamics and radiation transport! Since the goal is, as far as is possible, to let the equations themselves determine the properties of accreting systems, we must necessarily adopt a computational approach. Fortunately, computers are powerful enough now that it is feasible to begin to carry out a computational research program of multidimensional, time-dependent black-hole accretion.

Computational astrophysics is, in the end, not that different from traditional theory. In any theoretical inquiry, one must formulate a specific problem to solve that incorporates the essential physics and identifies the appropriate equations. One must then choose the best method of solution that, in numerical work, is the appropriate algorithm for solving a discretized version of the equations. Then one must implement that algorithm in the form of a program for the computer to execute. In the execution of that program, one chooses a specific set of initial and boundary conditions, and the result is a particular (approximate) solution to the general problem. One hopes that analysis of such specific numerical simulation (or a finite set of individual simulations) will yield general insights. But this is not guaranteed, nor can one be assured that the necessarily artificial setup of the simulated problem truly allows for the behaviors that are observed in nature.

For the present review, we focus on some of the numerical approaches that have been applied to local simulations of the MRI and global accretion disk simulations. Although the discussion will center on finite difference techniques, there are many other numerical techniques and algorithms that have been applied to astrophysics and to the accretion problem, such as spectral and particle methods. In generating a numerical solution, one must go from a continuous equation to a discrete system. In a spectral method, the solution is approximated as a finite sum of a set of basis functions. In a particle method, discrete particles that represent a much larger number of particles or even a continuous fluid carry a certain mass and move in response to imposed forces. In a finite difference technique, one discretizes space into some number of grid zones; the continuous fluid variables are likewise reduced to a discrete number of points, in both time and space. Temporal and spatial derivatives are then represented as *finite differences*. The partial differential equations under study are thus reduced to a series of coupled algebraic equations that can be solved on the computer. The timestep used to evolve the equations is limited by the well-known *Courant condition*, which states that Δt must be sufficiently

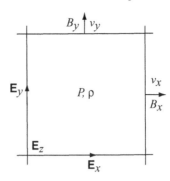

FIGURE 8.5. Location of variables on the grid for the ZEUS code. Scalars such as density and pressure are located in the grid center. Vectors are located at the appropriate zone face. The EMFs, **E**, that determine the magnetic field through the induction equation run along the zone edges.

small that $c\Delta t$ is less than the grid spacing Δx, where c is the maximum local speed of a signal, such as sound speed or Alfvén speed, plus fluid velocity.

We briefly consider and compare two finite difference approaches that are in widespread use within astrophysical research. One is epitomized by the ZEUS code, a finite difference, operator-split, staggered-mesh code. The ZEUS algorithm is described in detail by Stone and Norman (1992a, 1992b); subsequent developments for MHD are described in Hawley and Stone (1995). Another widely used approach is the flux conservative technique, most famously implemented for hydrodynamics in the form of the piecewise parabolic method (PPM) code of Colella and Woodward (1984). The extension of the technique to MHD is more recent, and up-to-date examples of flux conservative MHD codes include ATHENA (Stone et al., 2008) and PLUTO (Mignone et al., 2007). The following discussion is meant only to give the most general points of these numerical algorithms. Interested persons should consult the literature for the details.

Finite difference schemes such as ZEUS employ simple, relatively low-order Taylor expansions to represent derivatives. (The error that arises because one uses a truncated Taylor expansion is called, appropriately enough, *truncation error*.) ZEUS uses a *staggered grid* approach in which scalar quantities such as density and pressure are located in zone centers, and momenta and velocities are located at the appropriate zone face, for example, the x momentum on the x face of the grid zone (see Fig. 8.5). This technique has the advantage of locating various terms appropriately for taking derivatives. For example, the pressure gradient force acts on the momentum, and having the pressure spatially offset from the momentum naturally produces a centered, nominally second-order derivative for that term. Similarly, the divergence of the velocity accounts for Pdv work on the internal energy. Further improvements to the scheme are made by assuming that the substructure within the zone is represented by a function of some order. The basic ZEUS uses piecewise linear or parabolic representations for the variables.

ZEUS breaks down the timestep update into two steps, transport and source. If we consider the basic hydrodynamic equations, we have mass conservation

$$\frac{\partial \rho}{\partial t} + \nabla \cdot \rho \mathbf{v} = 0, \tag{8.32}$$

momentum conservation

$$\frac{\partial \rho \mathbf{v}}{\partial t} + \nabla \cdot \rho \mathbf{v}\mathbf{v} + \nabla P + \rho \nabla \Phi = 0, \tag{8.33}$$

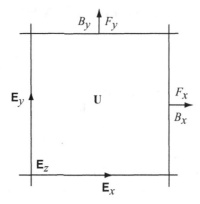

FIGURE 8.6. Location of variables on the grid for the conservative ATHENA algorithm. The conserved state vector $\mathbf{U}$ is located in the grid center. Fluxes are located at the appropriate zone face. To preserve the $\nabla \cdot \mathbf{B} = 0$ constraint, the magnetic fields and the EMFs that evolve them must be located as in the ZEUS code.

and an internal energy equation

$$\frac{\partial e}{\partial t} + \nabla \cdot e\mathbf{v} + P\left(\nabla \cdot \mathbf{v}\right) = 0. \tag{8.34}$$

The divergence term, for example, $\nabla \cdot \rho \mathbf{v}$, accounts for the motion of the gas through space; hence, it is referred to as the transport piece. The remaining terms, such as the pressure gradient, Pdv work, and gravitational accelerations in the momentum equation, are the source terms. Doing various portions of an equation in sequence, constantly updating the variables, is a technique called *operator splitting*. In a two- or three-dimensional problem, one further breaks the problem down into one-dimensional sweeps, a process called *directional splitting*. Completing all directional sweeps for both the transport and the source terms completes the update for one timestep.

The use of the internal energy equation in ZEUS deserves further comment. By solving this equation, ZEUS ensures entropy conservation. There is, however, no term in the equation as it is written to account for entropy-increasing dissipation processes such as shock waves. In ZEUS this is accomplished by adding a viscosity to the internal energy and momentum equations. This viscosity is generally referred to as an *artificial viscosity* because it is usually not equivalent in functional form to a physical viscosity term. The disadvantage of solving the internal energy equation is that total energy is not conserved and kinetic and magnetic (in MHD) energy can be lost from the system due to purely numerical factors unless viscous and resistive terms are included; it is reasonably straightforward to include these terms in ZEUS. Although ZEUS simulations generally solve the internal energy equation, there is nothing preventing one from using the ZEUS algorithm to solve the total energy equation instead. The choice is problem dependent.

The flux conservative algorithm is based around the formal conservation laws written in the form

$$\frac{\partial \mathbf{U}}{\partial t} + \nabla \cdot \mathbf{F}(\mathbf{U}) = \vec{S}, \tag{8.35}$$

where $\mathbf{U}$ is the vector of conserved quantities, and $\mathbf{F}(\mathbf{U})$ is the conserved flux. The remaining term, $\vec{S}$, is a source term that might, for example, include external forces such as gravity. In the case of simple, one-dimensional hydrodynamics, $\mathbf{U} = (\rho, \rho v, E)$ and $\mathbf{F} = (\rho v, \rho v^2, Ev + P)$. From the conserved state $\mathbf{U}$, one must construct the derived variables of velocity and pressure and internal energy. This is straightforward in Newtonian mechanics with a simple equation of state. These variables are located on the grid as shown in Figure 8.6.

Ignoring the source term, we can construct a simple finite difference form as the following:

$$\mathbf{U}_i^{n+1} - \mathbf{U}_i^n = -\frac{\Delta t}{\Delta x}\left(\mathbf{F}_{i+1/2}^{n+1/2} - \mathbf{F}_{i-1/2}^{n+1/2}\right), \qquad (8.36)$$

where the subscript refers to the ith gridzone location and the superscript refers to the nth time level. The trick, of course, is developing a procedure for calculating accurate fluxes at the half-timestep level at gridzone interfaces. Flux-conservative schemes do this by considering the problem of two distinct fluid states interacting for a time interval Δt. For a gas with a simple equation of state, one can obtain a solution to that *Riemann problem* by using conservation laws and the Riemann invariants (Colella and Woodward, 1984). Since that requires solving a nonlinear algebraic system, and the exact solution is not really needed for an adequate flux calculation, in practice most codes use an approximate linearized solution to the Riemann problem. In any case, that solution allows one to compute the values of the fluxes at the zone faces at the half-timestep interval. Because those fluxes are based on an approximate solution to the full fluid equations, these schemes can do an excellent job solving for shocks, rarefactions, and other waves within the fluid. The fluxes are obtained in a directionally split fashion. One can then solve equation 8.36 in directional sweeps, or use a sequence of flux calculations to update the system (Colella, 1990) in a nondirectionally split manner.

Equation 8.36 shows why schemes based around this formalism are called *conservative*. The fluxes are computed at the half time level for each gridzone interface, and each flux is used twice: once as a flux coming in to a zone, and once as a flux exiting the adjacent zone. If one sums up this equation over all zones, the fluxes cancel except for any boundary values, and the total integrated value of $\mathbf{U}$ remains unchanged. Since $\mathbf{U}$ is a vector of conserved quantities, mass, momentum and energy, the scheme guarantees that they will be globally conserved. Note that the internal energy is derived from the total energy by subtracting off other components such as the kinetic and magnetic energies. Different amounts of truncation error in the total energy and its components can lead to errors in the internal energy. For example, there is no guarantee that entropy will be conserved or only increase, or that the pressure should remain positive. Things can become difficult where the pressure is a very small fraction of the total energy so that the pressure is comparable to the truncation error in E. It is generally straightforward to fix such pressures where they occur by enforcing a minimum, or floor, pressure.

Hydrodynamic flux conservative schemes have been around in their present form for over 25 years. The development of robust schemes for MHD has been a more recent achievement. The thing that makes MHD more difficult is that in addition to the induction equation that describes the time evolution of the field,

$$\frac{\partial \mathbf{B}}{\partial t} - \nabla \times (\mathbf{v} \times \mathbf{B}) = 0, \qquad (8.37)$$

there is also the constraint equation $\nabla \cdot \mathbf{B} = 0$ – that is, there are no magnetic monopoles. A constraint equation is not an evolution equation. If the divergence-free condition is satisfied by the initial conditions, the induction equation guarantees that it will remain so as the field evolves. A numerical realization of the $\nabla \times \nabla \cdot$ operator need not equal zero, however, and a field that begins divergence-free may quickly evolve to contain monopoles that may lead to unphysical magnetic accelerations (Brackbill and Barnes, 1980). The problem is particularly difficult if one has adopted a directionally split approach, since the constraint equation is inherently three-dimensional.

Several techniques have been used to handle the constraint equation. One is to evolve a vector potential instead of $\mathbf{B}$. Another is to allow monopoles to develop but then correct the field to restore the constraint. This technique is usually called *flux cleaning*. The ZEUS and ATHENA codes use a technique called *constrained transport*, or CT

(Evans and Hawley, 1988). In the CT scheme, the fields are considered to be fluxes Φ through the zone face, and the EMFs that evolve them lie along the zone edges (see Fig. 8.5). The induction equation is solved in the form

$$\frac{\partial \Phi}{\partial t} = \oint_{\partial S} (\mathbf{v} \times \mathbf{B}) \cdot dl. \tag{8.38}$$

The point is that each zone edge enters the line integral twice, once with one sign for one zone face, once with the other sign for the adjacent zone face. If one then takes a numerical divergence of the sum of the EMFs, the result is zero, thus preserving the constraint. This is basically the analogue to flux conservation for the mass, energy, and momentum equations, as described earlier. Also, as above, conservation of the flux constraint occurs regardless of how one computes the EMF, which is a separate issue.

Turning now to the problem of black-hole accretion, we consider the behavior of matter moving in the gravitational field of a rotating black hole. In GR, gravity is a result of the geometry of spacetime, which is, in turn, determined by the distribution of matter. Spacetime geometry is encapsulated in the metric tensor g, which is defined in terms of the differential spacetime line element ds as $ds^2 = g_{\mu\nu} dx^\mu dx^\nu$, where dx is some arbitrary coordinate system. (Note that by convention, Greek indices vary over four integers, three for space and one for time.) For black-hole accretion, we can simplify things by assuming that the spacetime is fixed and corresponding to the Kerr metric for a rotating black hole (or the Schwarzschild metric in the limit of zero rotation). We can then specify a convenient coordinate system and carry out our computations. The use of a slightly more complicated coordinate system than (say) spherical coordinates is not really the challenge. What poses a greater difficulty is *special* relativity and the strong coupling between the four-velocity and the fluid variables, particularly for high boost factors, and the need to include the displacement current in the electromagnetic evolution equations. Velocities tend to be on order c, and that places a rather stringent limit on the timestep that can be used. On the other hand, the upper bound on the velocity means that there is a limit as to how small the timestep can become.

In recent years there have been several efforts to develop and use general relativistic MHD (GRMHD) for black-hole accretion. These include Koide et al. (2001); De Villiers and Hawley (2003); Gammie et al. (2003); Komissarov (2004); Duez et al. (2005); Anninos et al. (2005); and Antón et al. (2006). The first full GRMHD black-hole accretion simulation was performed years ago by Wilson (1975), although the available computer power then limited his simulations to 20 by 30 gridzones! It should also be noted that it is often not necessary to use the full GR approach to study black-hole accretion. It can be sufficient to use a pseudo-Newtonian potential of the form $\Phi \propto 1/(R - R_g)$, where R_g is the gravitational radius, analogous to the horizon. The pseudo-Newtonian potential captures the behavior of the ISCO while avoiding the complexities associated with the relativistic equations.

The most familiar coordinate system used for the Kerr black hole spacetime is Boyer-Lindquist coordinates. Boyer-Lindquist coordinates are defined in terms of t corresponding to a clock at infinity and a radial coordinate for which the proper surface area of a sphere at radius r is $4\pi r^2$. The angular coordinates are thus the familiar ones from standard spherical coordinates. The line element then has the form

$$ds^2 = g_{tt} dt^2 + 2 g_{t\phi} dt \, d\phi + g_{rr} dr^2 + g_{\theta\theta} d\theta^2 + g_{\phi\phi} d\phi^2. \tag{8.39}$$

The $g_{t\phi}$ term is due to the black hole's rotation and is responsible for the "dragging of inertial frames" phenomenon, which causes a free-falling, zero–angular-momentum particle to move in the ϕ direction. This can be an important effect, as it allows some of the rotational energy of the hole to be transferred to surrounding matter and fields. The spin measure for a black hole is the angular momentum per unit mass a. The upper limit for this is $a/M = 1$. Boyer-Lindquist coordinates have the advantage that the t and ϕ

stress-energy terms correspond to the familiar conserved energy and angular momentum. The disadvantage of these coordinates is that the metric terms become singular at the black hole horizon.

The equations of general relativistic MHD are the same statements of conservation principles as before, usually expressed in a covariant form as follows. The continuity equation is

$$\nabla_\mu (\rho U^\mu) = 0, \tag{8.40}$$

and the conservation of stress-energy is

$$\nabla_\mu T^{\mu\nu} = 0, \tag{8.41}$$

where the stress-energy is composed of fluid and electromagnetic components

$$T^{\mu\nu} \equiv T^{\mu\nu}_{(fluid)} + T^{\mu\nu}_{(EM)}, \tag{8.42}$$

which separately are

$$T^{\mu\nu}_{(fluid)} = \rho h U^\mu U^\nu + P g^{\mu\nu} \tag{8.43}$$

and

$$T^{\mu\nu}_{(EM)} = \frac{1}{4\pi} \left(F^\mu{}_\alpha F^{\nu\alpha} - \frac{1}{4} F_{\alpha\beta} F^{\alpha\beta} g^{\mu\nu} \right). \tag{8.44}$$

In these equations the density ρ, pressure P, and enthalpy h are the fluid frame values, and U^ν is the four-velocity. The electromagnetic terms are written in terms of the electromagnetic tensor $F^{\mu\nu}$. In the rest frame of the fluid, the magnetic induction and the electric field are, respectively, $B^\mu = {}^*F^{\mu\nu}U_\nu$ and $E^\mu = F^{\mu\nu}U_\nu$. Maxwell's equations can be expressed in the compact form

$$\nabla_\mu F^{\mu\nu} = 4\pi J^\nu, \tag{8.45}$$

along with

$$\nabla_\mu {}^*F^{\mu\nu} = 0. \tag{8.46}$$

By making certain definitions, it is possible to rewrite these equations into forms that look very much like the usual nonrelativistic MHD equations (De Villiers and Hawley, 2003). First, the equation of mass conservation can be written

$$\partial_t D + \frac{1}{\sqrt{\gamma}} \partial_j (D \sqrt{\gamma} V^j) = 0, \tag{8.47}$$

where $D = \rho W$ is an auxiliary density variable, $V^j = U^j/U^t$ is the transport velocity constructed from the four-velocity U^μ, and W is the relativistic boost- or gamma-factor. Spatial indices are indicated by roman characters $i, j = 1, 2, 3$. Similarly, the internal energy equation can be written

$$\partial_t e + \frac{1}{\sqrt{\gamma}} \partial_j (e \sqrt{\gamma} V^j) + P \partial_t W + \frac{P}{\sqrt{\gamma}} \partial_j (W \sqrt{\gamma} V^j) = 0, \tag{8.48}$$

where $e = \epsilon D$ is the auxiliary energy variable, and ϵ is the specific internal energy. The equation of momentum conservation

$$\partial_t (S_j - \alpha b_j b^t) + \frac{1}{\sqrt{\gamma}} \partial_i \sqrt{\gamma} (S_j V^i - \alpha b_j b^i)$$

$$+ \frac{1}{2} \left(\frac{S_\epsilon S_\mu}{S^t} - \alpha b_\mu b_\epsilon \right) \partial_j g^{\mu\epsilon} + \alpha \partial_j \left(P + \frac{\|b\|^2}{2} \right) = 0 \tag{8.49}$$

is written in terms of the auxiliary four-momentum, $S_\mu = (\rho h + \|b\|^2) W U_\mu$, where $h = 1 + \epsilon + P/\rho$ is the relativistic enthalpy, b^μ is the magnetic field four-vector in the rest frame of the fluid, and $\|b\|^2 = g^{\mu\nu} b_\mu b_\nu$. The momentum is subject to the normalization condition $g^{\mu\nu} S_\mu S_\nu = -(\rho h + \|b\|^2)^2 W^2$, which is algebraically equivalent to the more familiar velocity normalization $U^\mu U_\mu = -1$. The term $\partial_j g^{\mu\epsilon}$ accounts for inertial accelerations due to terms such as gravitational, Coriolis, and centrifugal forces.

The induction equation is solved in the form

$$F_{\alpha\beta,\gamma} + F_{\beta\gamma,\alpha} + F_{\gamma\alpha,\beta} = 0, \qquad (8.50)$$

which involves only ordinary derivatives regardless of coordinate system. This can be put into a form suitable for the CT scheme by defining

$$\mathcal{B}^r = F_{\phi\theta}, \ \mathcal{B}^\theta = F_{r\phi}, \ \mathcal{B}^\phi = F_{\theta r}. \qquad (8.51)$$

By assuming infinite conductivity (the flux-freezing condition), we have $F^{\mu\nu} U_\nu = 0$. The induction equation is then placed in a familiar form,

$$\partial_t \left(\mathcal{B}^i\right) - \partial_j \left(V^i \mathcal{B}^j - \mathcal{B}^i V^j\right) = 0, \qquad (8.52)$$

along with the constraint equation

$$\partial_j \left(\mathcal{B}^j\right) = 0. \qquad (8.53)$$

The point of all this manipulation is that the equations have been recast into a form that is very similar to the Newtonian equations solved in ZEUS. This is the basis of the GRMHD code developed by De Villiers and Hawley (2003).

It is also possible to use a flux-conservative scheme to solve the GRMHD equations. This is the approach taken by Gammie et al. (2003) in the development of their code called HARM (see also Noble et al., 2006). HARM evolves the conservative state vector defined as

$$\mathbf{U} = \sqrt{-g}\left(\rho U^t, T^t_t, T^i_t, B^i\right), \qquad (8.54)$$

where the terms correspond to mass, energy, momentum, and magnetic flux. The geometry is included in the state vector through $\sqrt{-g}$. One particular difficulty for the conservative approach is deriving the primitive state vector,

$$\mathbf{P} = \left(\rho, P, v^i, B^i\right), \qquad (8.55)$$

from $\mathbf{U}$. The reason is that, in relativity, the velocity is present in all components of the conserved state vector, as is the geometry in the form specified previously. Deriving the primitive variables from the conserved variables becomes a complex nonlinear algebraic problem. Considerable effort must be devoted to this process, as described in Noble et al. (2006), with special attention devoted to various limiting cases.

As before, the conservative update scheme is

$$U_i^{n+1} = U_i^n - \frac{\Delta t}{\Delta x}\left(F_{i+1/2}^{n+1/2} - F_{i-1/2}^{n+1/2}\right). \qquad (8.56)$$

Fluxes can be solved using the same sort of approximate Riemann techniques as in nonrelativistic hydrodynamics, although, as one might expect, things can be considerably more complex depending on the level of approximation used. In the end, the overall timestep requires deriving the primitive variables from the conserved state, solving for the fluxes at the zone faces, and updating the conserved state using those fluxes. Schematically, this is

$$\mathbf{U}^n \to \mathbf{P} \to \mathbf{F}(\mathbf{P}) \to \mathbf{U}^{n+1}. \qquad (8.57)$$

There are several advantages to solving the equations in this conservative form. One is that the system easily adapts to arbitrary coordinate systems. For HARM, this means using the Kerr-Schild metric for the Kerr hole, which is not singular at the horizon. (The cost is that the radial and time coordinates are mixed, and the t component of the stress-energy is no longer simply identified with the energy flux.) The conservative approach is also appropriate for use in full GR codes where the spacetime is time dependent.

One of the reasons that it is important to have several codes that use different algorithms as well as several different code versions that use the same or a similar algorithm is that it is difficult to assess the reliability of the results of a given numerical simulation in isolation. When they are written, codes are tested against a set of standard test problems, particularly ones that have analytic solutions. In other contexts, code results can be compared against experimental data, but this is difficult in the case of black-hole accretion. *Convergence testing* is where the same problem is run at multiple resolutions and one observes how the results change with resolution. The difficulty for three-dimensional accretion is that such simulations are underresolved at the best resolution that can presently be employed. This makes true convergence studies impractical for the moment. Thus, the recent increase in the number of research groups employing GRMHD codes to investigate black-hole accretion is essential to gaining reliable insights from simulations.

We now turn to the use of these numerical techniques to study accretion. First we will look at local, nonrelativistic simulations that are carried out with nonrelativistic codes such as ZEUS or ATHENA. Finally, we will consider global simulations done in the full Kerr metric.

8.5 Local simulations: the shearing box

The MRI tells us why accretion disks are locally unstable, and why they should become turbulent, which, in turn, tells us why they should accrete. But that is only the beginning. What levels of stress are produced by various physical conditions? To what extent does this stress behave like the parameterized models? What factors determine the turbulent energies? Some observations suggest that disks experience significant changes in the stress level; can MRI turbulence account for that? Questions such as these must be answered through well-considered numerical experiments.

The first simulations of the MRI were carried out in a local limit, which reduces the complexity of the full accretion problem to a minimum. In the *shearing box* model (Hawley and Stone, 1995), one considers a small patch of a disk centered at some radius R_o and boosts into a local corotating frame corresponding to an orbital velocity of $R_o\Omega$. Define the variables $x = R - R_o$, with $x \ll R_o$, $y = R_o\phi$, and z as a local Cartesian frame. Assume that the background angular velocity is of a power-law form, $\Omega(R) \propto R^{-q}$; a Keplerian shear corresponds to $q = 1.5$. The y velocity then is $x\,(Rd\Omega/dR)_o = -q\Omega x$. We can write the momentum evolution equation in this frame as

$$\frac{\partial \rho \mathbf{v}}{\partial t} + \nabla \cdot (\rho \mathbf{v}\mathbf{v} - \mathbf{B}\mathbf{B}) + \nabla\left(P + \frac{1}{2}B^2\right) = 2q\rho\Omega^2\mathbf{x} - 2\mathbf{\Omega} \times \rho\mathbf{v} - \Omega^2\mathbf{z}. \qquad (8.58)$$

The term $2q\rho\Omega^2\mathbf{x}$ corresponds to the gravitational and centrifugal forces, $\Omega^2\mathbf{z}$ is the vertical gravity, and $2\mathbf{\Omega} \times \rho\mathbf{v}$ is the Coriolis force. The term *vertically stratified* shearing box is used when vertical gravity is included. Many simulations ignore the vertical gravity term for simplicity when studying the basic properties of MRI turbulence.

The shearing box gets its name from the radial boundary conditions. Although we want to restrict the simulation domain to a small, local patch of a disk, we don't want to limit the possible radial motions through the imposition of reflecting or outflow boundary conditions. Often local simulations that are intended to represent a larger domain employ periodic boundary conditions, but the presence of the background shear complicates things. We assume that the box is surrounded by boxes that are identical but that slide relative to one another according to the relative background shear. The radial boundary conditions are thus *shearing periodic*. In addition, the angular velocity of a fluid element must be adjusted by the relative shear value as it moves through the boundary. It should be noted that stresses across the radial boundary will do work on the fluid in the box, so if there is a positive stress, the energy within the box will increase with time. If the turbulence is dissipated as heat, the temperature in the box will increase unless one uses an isothermal equation of state or adds a cooling term to the energy equation.

8.5.1 *Two-dimensional simulations*

The simplest local simulation can be done in the axisymmetric limit. One can then use periodic boundary conditions in the radial direction, modified only by the need to adjust the angular momentum of fluid elements as they move through that boundary. The first such simulations were done by Hawley and Balbus (1992). These simulations demonstrated that the MRI could lead to turbulence within the disk as the nonlinear outcome. A surprising additional result was that the MRI operating on a uniform vertical field produced coherent motions dubbed a "channel flow" for its appearance as two vertically separated radial flows. Goodman and Xu (1994) subsequently demonstrated that in this case the MRI modes are exact (exponentially growing) solutions of the nonlinear fluid equations. They also found that these modes were themselves subject to instabilities that, in three dimensions, subsequently lead to a breakdown of the channel flows into turbulence (Hawley *et al.*, 1995). The possible role of the channel flows in the saturation of the full nonlinear MRI remains a subject of study today (Latter *et al.*, 2009; Sano, 2007).

What limits the usefulness of axisymmetric simulations is that, in the absence of net flux supported by currents *outside* of the simulation domain, the magnetic field must eventually decay due to Cowling's antidynamo theorem. The poloidal field can be pulled and stretched, but only within the (x, z) plane, and so must eventually reconnect or diffuse away. Mathematically, the toroidal (y) component of the vector potential that generates the poloidal field is conserved in axisymmetry, except for losses due to diffusion. The simplicity of the axisymmetric system, and the ability to use much higher resolution, nevertheless makes it useful for limited studies of MRI properties (e.g., Guan and Gammie, 2008).

8.5.2 *Three-dimensional simulations*

The first three-dimensional shearing box simulations (Hawley *et al.*, 1995, 1996) found that the presence of a net magnetic field and its orientation play a role in setting the amplitude of the MRI turbulence. Without a net field, the complete decay of the field is possible (and guaranteed in axisymmetry, as remarked earlier). Net vertical fields give the largest turbulent energies, with the energy level approximately proportional to the background field. Net toroidal fields behave similarly, but with a smaller energy for the same background field strength. In terms of the α parameter, typical values range from $\alpha \sim 0.01$ to 0.1, with the net vertical field cases at the high end and the net toroidal field toward the low end. Values of α that are larger or smaller by another power of 10 have occurred in various simulations.

It makes sense that the presence of a background net field would play an important role in establishing the average energy and stress in the fully developed turbulence. A net field guarantees the continued presence of driving MRI modes on scales set by that background field. Shearing box studies have also examined the role of other effects such as computational domain size (e.g., Pessah *et al.*, 2007) and resolution, and others have looked at physical parameters such as background field strength and gas pressure. An investigation of the influence of gas pressure carried out by Sano *et al.* (2004), for example, found an extremely weak pressure dependence. Even here, the influence of the gas pressure depends on the magnetic field geometry. Blackman et al. (2008) examined the results of an ensemble of shearing box simulations taken from the literature and found that $\alpha\beta$ is generally constant and equal to ~ 0.1, where β is the ratio of thermal to magnetic pressure. In other words, the stress is proportional to magnetic pressure. This follows from the high degree of correlation within the magnetic fields in the turbulence, that is, B_ϕ and B_r are almost always aligned so as to produce a Maxwell stress with the correct sign. B_ϕ tends to be the dominant field component (due to the background shear amplification) and some multiple of the average B_r. Then

$$\langle B_\phi \rangle \langle B_r \rangle \propto \langle B_\phi^2 \rangle, \tag{8.59}$$

where the brackets indicate a suitable average over time and space. Of course, knowing that the stress is proportional to the magnetic pressure still leaves one with the problem of determining what establishes the magnetic pressure in the disk (or, alternatively, β). Considerable uncertainty remains on this point.

In the MRI turbulence, the Maxwell stress exceeds the Reynolds stress by a factor of a few. This is to be expected because angular momentum transport by the MRI is driving the turbulence in the first place. The ratio of the Maxwell to the Reynolds stress is a function of the background shear parameter q (Abramowicz et al., 1996; Hawley et al., 1999; Pessah et al., 2006). As $q \to 2$, the hydrodynamic stabilizing effect of the Coriolis force becomes smaller and the Reynolds stress increases. Of course, for $q > 2$ the flow is Rayleigh unstable, and even a purely hydrodynamic system becomes turbulent.

The idea of *nonlinear* hydrodynamic instabilites has been proposed to drive turbulence despite the disk's stability to the Rayleigh criteria, and such instabilities are observed in fluid experiments. The linear shear flows where this is observed are, however, equivalent to the $q = 2$ case where the epicyclic frequency is zero (Balbus and Hawley, 1998). In the absence of restoring Coriolis forces, the linear terms sum to zero; the terms remaining are the nonlinear ones, and the flow can be destabilized. Indeed, purely hydrodynamic shearing box simulations show that when $q \approx 2$, hydrodynamic turbulence can develop from finite velocity perturbations. For $q < 2$, any imposed velocity fluctuations quickly die out and the flow becomes laminar (Balbus et al., 1996). In the absence of a magnetic field, a Keplerian flow, in particular, is stable.

Recently there has been considerable work done on the influence of physical viscosity ν and Ohmic resistivity η on MRI-driven turbulence. Previous work had investigated in some detail the influence of resistivity (Hawley et al., 1996; Sano et al., 1998; Fleming et al., 2000; Sano and Inutsuka, 2001; Ziegler and Rüdiger, 2001; Sano and Stone, 2002), finding that resistivity leads to a decrease in turbulence, independent of the magnetic field configuration. Further, the effect of resistivity on the turbulence is larger than one might expect from the linear MRI relation (Fleming et al., 2000). Fromang et al. (2007), Lesur and Longaretti (2007), and Simon and Hawley (2009) have shown that both resistivity and viscosity are important, particularly in the ratio of one to the other, that is, the magnetic Prandtl number, $P_r = \nu/\eta$. Larger Prandtl numbers produced larger α values. This effect is particularly intriguing since, as Balbus and Henri (2008) point out, the Prandtl number can vary strongly in the vicinity of a black hole. If the Prandtl number dependence seen in numerical simulations carried over to the behavior of real systems where the viscous and resistive scales are microscopic compared to the the disk, then one could have a mechanism to account for dramatic state changes in such systems.

Greater realism in the local simulations is achieved by including vertical gravity in the shearing box. The first such stratified shearing box simulations were carried out by Brandenburg et al. (1995) and Stone et al. (1996). As with the unstratified boxes, the MRI operates to generate significant stress. The simulations of Miller and Stone (2000) showed that a good deal of magnetic field can rise buoyantly to establish a magnetically dominated corona. It seems to be a fairly generic result that the magnetic pressure scale height exceeds that of the gas pressure. Near the equator, β is substantially greater than 1, but it drops off with height. Stratified simulations have also provided suggestions that some sort of coherent dynamo action might be taking place. In the simulations, a net toroidal flux is generated that subsequently rises through the disk. The sign of the flux changes across the equator and also changes on a given side of the equator as a function of time.

A particularly interesting set of stratified box simulations were carried out by Stone and Balbus (1996), who looked at the angular momentum transport that might occur in a purely hydrodynamic system that had vertical convection. They found that the vertical convective motions themselves did not produce a significant Reynolds stress. Indeed, what net angular momentum transport they did observe was *inward* rather than outward. In vertical convection, there is no correlation between the velocity fluctuation components u_r and u_ϕ. Specific angular momentum tends to be conserved, and any mixing (including numerical mixing) of fluid elements tends to reduce the angular momentum gradient, that is, transport angular momentum inward. This result demonstrates that the presence of turbulence alone is not sufficient to guarantee a positive stress.

Vertical gravity does add important elements to the evolution of the MRI turbulence in ways that are only now being more fully investigated. For example, in nonstratified simulations without a net field, the turbulence saturation level turns out to depend on the numerical resolution. Fromang and Papaloizou (2007) found that in an unstratified shearing box with no net field, increased grid resolution led to decreased levels of stress. The implication was that the converged, infinite-resolution stress value would be zero. This result seems to be a consequence of the simplicity of the unstratified shearing box model and its lack of any length scale beside that of a grid zone. For the unstratified box, the addition of physical viscous and resistive scales to the nonstratified shearing box problem restores convergence to nonzero stress levels. Fromang et al. (2007), Shi et al. (2010), and Davis et al. (2010) have carried out resolution studies in a stratified shearing box and have found convergence to a nonzero stress level even without the addition of viscosity or resistivity. What seems to be important is the presence of a physical length scale other than a grid zone. The addition of vertical gravity adds such a length scale in the form of the vertical scale height.

One of the most important developments in local MRI simulations has been the addition of radiative transfer to the equations (e.g., Turner, 2004; Hirose et al., 2006). With radiation transport included, the local stratified shearing box becomes a truly representative slice of an accretion disk, with both the dynamics and the thermodynamics captured in the simulation.

Such simulations offer the potential for rapid progress. The recent work of Hirose et al. (2009) provides an illustration of the power of these simulations by answering a decades-old puzzle. Shakura and Sunyaev (1973) showed that radiation pressure will exceed gas pressure in black-hole accretion disks when the accretion rate is even a small fraction of the Eddington rate. However, if one adopts the α formulation for the stress levels in the disk, radiation-dominated disks become thermally unstable (Lightman and Eardley, 1974; Shakura and Sunyaev, 1976). Essentially the stress, and hence heating, increase faster than the disk's ability to radiate that heat away. This thermal instability goes away if the stress formula is modified, and it has been long suspected that it is simply an artifact of the α prescription. The ability to carry out local simulations of radiation-dominated disks has allowed the question to be examined directly. What Hirose et al. (2009) found is that even when the radiation pressure exceeds the gas pressure by a factor of 10, there is no thermal runaway. At least two factors come into play. First, although stress is related to the total pressure, the causal relationship is stress to pressure, not the other way around. In other words, the stress is what it is and the pressure follows. An increase in pressure will not, by itself, necessarily lead to an increase in stress. The second factor is that when the heating rate is greater than can be radiated, the excess energy is transported vertically by radiative advection, cooling the disk as needed to maintain thermal equilibrium. Dissipation in the disk is concentrated near, but off, the midplane. They also found that the energy dissipation was not significant within the corona.

This brings us to the question of just how much MRI-driven turbulence behaves in a manner that is consistent with α disk theory. Balbus and Papaloizou (1999) showed that for a stress to behave in a manner consistent with α models, the same stress value that transports the angular momentum outward must also be the stress value that accounts for the thermalization of the turbulence. In other words the turbulence needs to dissipate into heat locally and on a prompt time scale. Simon *et al.* (2009) carried out a detailed analysis of the flow of energy within a shearing box simulation. They found that the time scale over which thermalization occurs is $\sim \Omega^{-1}$ with magnetic dissipation dominating over kinetic dissipation. In their radiation-dominated simulations, Hirose *et al.* (2009) found that the time and vertically averaged stress is indeed proportional to the total pressure, not the gas or radiation pressure separately. Despite this, there is no thermal instability because changes in pressure by themselves do not result in changes in stress. These results suggest that for average, steady-state disk models, the α prescription is appropriate. What is not appropriate, however, is using α for time-dependent models or stability analysis, or in any context where an explicit causal dependence of the stress on the pressure is implied.

8.6 Global simulations of black-hole accretion and jets

Global simulations compute the full three-dimensional structure of an accretion disk by allowing the self-consistent turbulent stresses that arise from the MRI to drive its evolution. The reason that global disk simulations have only recently been attempted is because of the computational resources required for even relatively simple models. A three-dimensional grid with N cells in each direction requires N^3 grid updates each timestep, and the number of timesteps required goes up as the resolution increases, owing to the Courant condition limit on Δt. Achieving adequate spatial resolution is difficult; the important length scales within the disk vary widely. A disk extends radially over distances that are large compared to the black hole's horizon. This further implies a wide range in time scales, as orbital periods are proportional to $R^{3/2}$ and the accretion time from large radius will be very large compared to Δt. Because the sound speed is generally much less than the orbital speed in a disk, the vertical thickness of the disk, H, will be much less than the radius R, that is, the disk is thin, $H/R \ll 1$. Finally, there is the problem of resolving the turbulence itself. Even in the local shearing box models, the available resolutions are too small to resolve the full turbulent cascade, and the global problem only makes things worse! The outer scale of the turbulent eddies is on order of or less than H, and to resolve the turbulence properly probably would require a substantial range of scales below H. The outer scales are the ones that are most important for the turbulent stresses, so the lack of resolution may not be fatal, at least so long as unstable MRI modes continue to be adequately resolved. The thermodynamics used in the present global simulations are also relatively simple: for example, limited simple adiabatic equations of state, optically thin radiative cooling, and dissipation controlled by numerical losses at the grid scale rather than some true dissipative physics.

Thus, although we are still a long way from a detailed first-principles simulation of a black-hole accretion disk, we are nevertheless making progress in answering specific and important questions about accretion dynamics. Here we review the status of some of these questions and what has been learned to date.

8.6.1 *Stress at the ISCO*

The total efficiency of accretion depends on how effectively the process is able to convert orbital energy into heat. In accretion disks, this is accomplished by turbulent stresses that remove angular momentum from fluid elements. The turbulence then dissipates into heat that, in the thin disk model, is radiated locally to achieve steady-state thermodynamic equilibrium. The classic assumption in the black-hole thin disk model (Novikov and Thorne, 1973) has been that this stress ends at the ISCO, which corresponds to the orbit inside of which a fluid element can fall into the black hole without further loss of angular momentum. Although this assumption is probably reasonable for hydrodynamic torques (Thorne, 1974), magnetic fields can exert force over extended distances. Gammie (1999) and Krolik (1999) showed that, in principle, magnetic fields could operate effectively within the ISCO, even for cold, thin disks. This question of stress at the ISCO is ideally suited for investigation through direct simulation.

The first magnetized global simulations found that, indeed, significant magnetic stress can continue right through and inside the ISCO (Hawley and Krolik, 2001; Reynolds and Armitage, 2001; Hawley and Krolik, 2002; Machida and Matsumoto, 2003; Gammie et al., 2004; Krolik et al., 2005). The additional stress can decrease the specific angular momentum of the material accreted into the hole by 6% to 8% for low-spin holes to as much as 42% in a maximally spinning case (Krolik et al., 2005). However, all of these first simulations were for relatively thick disks with $H/R \sim 0.1$ to 0.2. There was some question as to the behavior of the magnetic stress if the vertical scale height of the disk at the location of the ISCO were reduced. For example, Reynolds and Fabian (2008) carried out a pseudo-Newtonian simulation of a disk with $H/R \sim 0.05$ and found that the stress in the plunging region was smaller than that found in the pseudo-Newtonian simulations of Hawley and Krolik (2001) and Hawley and Krolik (2002) for thicker disks. Using a version of the relativistic HARM code, Shafee et al. (2008) found a significantly reduced stress inside the ISCO for an $H/R \sim 0.07$ disk accreting onto a Schwarzschild hole. In their simulation, the specific angular momentum accreted into the hole was only 2% below the ISCO value. Noble et al. (2009) used a simple cooling function added to their HARM3D code to compute a disk with $H/R \sim 0.1$ accreting into an $a/M = 0.9$ hole. They found enhanced stress inside the ISCO that led to a 6% increase in luminosity at infinity.

A more systematic study was carried out by Noble et al. (2010), who ran sets of HARM3D simulations of disks with different H/R values accreting into a Schwarzschild black hole. They found that the time-averaged radial dependence of the fluid-frame electromagnetic stress is almost independent of the disk thickness. The stress continues to rise through the ISCO, dropping to zero only right outside of the black hole horizon (Fig. 8.7). The Reynolds stress was also significant near the ISCO, but it did show some dependence on disk thickness. The enhanced stress reduced the net angular momentum per unit mass accreted into the hole by 7% to 15% less than the angular momentum of the ISCO. This study also provided evidence of the importance of adequate resolution. Models where the fastest growing MRI modes appeared to be underresolved showed a reduced stress at and inside the ISCO.

One factor that does seem to reduce or even eliminate any significant magnetic stress inside the ISCO is the absence of significant poloidal flux. Beckwith et al. (2008a) computed disks around an $a/M = 0.9$ black hole that began with different field topologies: a large-scale dipole loop, two loops forming a net quadrupole, and a purely toroidal initial field. Whereas the dipole and quadrupole cases had similar disks and similar enhanced electromagnetic stress at the ISCO, the toroidal field case showed almost no additional stress there.

To summarize, simulations have found that magnetic stresses can and do operate at and inside the ISCO. The amount of additional stress depends on the field strength and topology there, which, in turn, will depend on factors elsewhere within the disk. This additional stress has the potential to increase the efficiency of accretion, η, beyond the standard value associated with the binding energy of the ISCO itself. Increased stress will increase the effective emitting area of the disk and the characteristic temperature of the disk spectrum. This has implications for attempts to infer disk spin from observations of disk continuum spectra (e.g., Done and Davis, 2008).

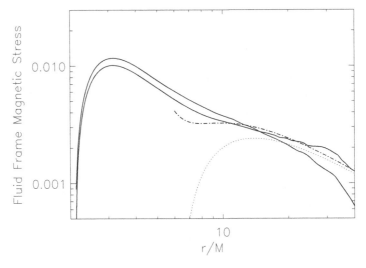

FIGURE 8.7. Fluid frame electromagnetic stress for a thin disk simulation from Noble et al. (2010). The dotted curve corresponds to the stress profile for the steady-state disk model of Novikov and Thorne (1973) with zero stress at the ISCO. The dot-dashed line is the prediction of the semianalytic model of Agol and Krolik (2000), which includes stress at, but not inside, the ISCO. Reproduced by permission of the AAS.

8.6.2 Jets and outflows

Jets are seen in protostellar systems, accreting binary systems and in AGN. As Livio (2000) has argued, jets must be a natural outcome of the combination of accretion, rotation, and magnetic fields. What one would hope for from simulations, then, is to learn how jets arise under fairly general circumstances, and to understand more thoroughly the factors that permit or prevent jet formation and determine jet power and collimation.

The most promising theories of jet formation depend on the presence of a large-scale, organized poloidal field. In the Blandford-Payne model (Blandford and Payne, 1982), a large-scale, mostly vertical magnetic field is anchored in and rotates with the disk. If the field lines are angled outward sufficiently with respect to the disk, there will be a net outward force that accelerates matter along the rotating field lines. Simulations have demonstrated that this mechanism can be effective (see the review by Pudritz et al., 2007), but they have to include a preexisting vertical field in the initial conditions.

Another jet model is the Blandford-Znajek mechanism (Blandford and Znajek, 1977), where the jet is powered by the rotating spacetime of a black hole. Radial magnetic field lines extend outward from the horizon through the *funnel* region along the hole's rotation axis, so called because centrifugal forces will tend to keep the region empty of accreting matter. Field-line rotation is created by frame dragging, and this drives an outgoing Poynting flux. Global GRMHD simulations have, under the right circumstances, produced reasonably powerful and stable Poynting flux jets from relatively simple initial conditions. The key to their formation is to establish a long-lived, large-scale field along the axis of the black hole. In the simulations, field loops originally contained entirely within the accretion flow expand dramatically within a cone around the rotation axis, successfully supporting jets carrying substantial Poynting flux to large distances (De Villiers et al., 2005; Hawley and Krolik, 2006; McKinney, 2006). In these models, a large-scale dipole field becomes attached to the black hole. The rotation of the black hole generates an EMF that drives a Poynting flux outward at relativistic velocities. The jet power matches well with the predictions of the Blandford-Znajek model (McKinney and Gammie, 2004; McKinney, 2005). The Poynting flux jet has very little mass within it, as centrifugal forces keep the axial funnel evacuated. The power of the Poynting flux jet is

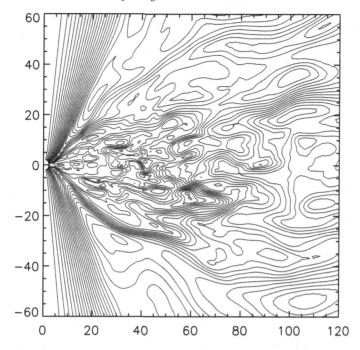

FIGURE 8.8. Poloidal field lines in a global simulation of black-hole accretion. A large-scale dipole field has been established within the low-density axial funnel above and below the hole. Such a field can support a Poynting flux jet if the black hole is rotating. Axes show distance from the black hole in units of M. The simulation shown here is described in Beckwith *et al.* (2009). Reproduced by permission of the AAS.

determined by the spin of the black hole and the strength of the magnetic field. What establishes the field strength? In the simulations, the magnetic pressure within the funnel is more or less comparable to the pressure in the inner disk.

Unbound outflows can also develop outside the axial funnel along the centrifugal boundary. These outflows have the only significant unbound mass flux and an outgoing velocity of about $\sim 0.5c$ (De Villiers *et al.*, 2005; Hawley and Krolik, 2006).

The formation of the Poynting flux jets depends on several factors. First, jet formation requires a consistent sense of vertical field to be brought down to the event horizon. Figure 8.8 shows an example of such a field in a black-hole accretion simulation. This readily occurs for simulations that begin with a large dipole loop. It is less likely for quadrupolar initial conditions, and not at all for toroidal field initial conditions (Beckwith *et al.*, 2008a). A consistently strong jet seems to require a net vertical field in the disk midplane whose sign remains consistent for at least an inner-disk inflow time. How is this accomplished in nature? Although there have been a number of dynamo models put forward suggesting how some form of inverse cascade might occur within the MHD turbulence to generate a large-scale field where none existed before (e.g., Tout and Pringle, 1996), these scenarios remain speculative, with little supporting evidence seen in the global simulations done so far. An alternative is that net vertical field is brought in from an external source at large distance by the accretion process itself. The geometric concentration of the field would ensure that even a very weak vertical field would be quite strong at the hole itself. This notion is described in the next subsection.

All of these jet simulations to date assume rather simple initial conditions and that the spin axis of the disk is aligned with that of the hole. This ensures that the centrifugally evacuated funnel coincides with the hole's spin axis. Also, because of numerical difficulties with the coordinate axis singularity along the axis, most of the simulations have placed a "cut-out" around the axis. One exception is the simulation of McKinney and Blandford

(2009), who computed a black hole jet in three dimensions while allowing flow across the grid axis. This is an important step forward in investigating the stability of Poynting flux jets to various instabilities that could destroy their collimation. Fragile *et al.* (2009) have made use of a modified "cubed-sphere" coordinate system to evolve tilted disks. The cubed sphere eliminates the axial coordinate singularity, but at the cost of a more complex grid structure.

8.6.3 *Radial advection of field*

The key ingredient in Poynting flux jet formation seems to be to get a reasonably strong dipole field attached to the black hole. In the simulations to date where such fields develop, the field has most often been supplied from the initial conditions, which consist of a large-scale dipole loop in the initial torus. We can hardly count on such a condition occurring in nature.

The simplest way to develop a net field at the hole is to carry one in through the process of accretion itself. The question as to whether or not this can happen, however, has a long history of controversy. Some (e.g., Thorne *et al.*, 1986) have argued that flux freezing would ensure that accretion brings net magnetic flux to the event horizon of the black hole, and random walk statistics would guarantee that at any given time the magnitude of this net flux would be substantial. The counterargument is that magnetic reconnection should be efficient enough to eliminate large-scale field. Van Ballegooijen (1989) described this as a balance between effective turbulent viscosity (driving accretion) and an effective turbulent resistivity (allowing field to diffuse outward through the disk). The ratio of the two terms, an effective magnetic Prandtl number, would determine whether or not field was concentrated at the hole. The expectation is that the viscous and diffusive terms would be comparable, limiting the amount of field carried in.

The problem with this formulation is that we do not know a priori what that ratio should be, or whether the real mechanics of angular momentum transport through magnetic stresses and magnetic diffusivity through reconnection can be correctly described in such terms. In an effort to approach the problem directly, Beckwith *et al.* (2009) followed the evolution of an initial torus threaded by a vertical field orbiting a Schwarzschild hole. The vertical field within the disk created vigorous MRI turbulence that drove strong accretion, but the net field itself did not accrete with the gas. The amount of reconnection within the disk, a natural consequence of the MRI turbulence, was too great. Nevertheless, vertical field found its way to the black hole. This happened due to fluid motions outside of the disk body. This *coronal mechanism* derives from magnetic stresses within the corona driven by orbital shear that allow for rapid accretion of low-density matter that stretches field radially inward. If these field hairpins reconnect across the equator, there is a discontinuous jump in the location of net magnetic flux. The coronal mechanism establishes a dipole magnetic field in the evacuated funnel around the orbital axis with a field intensity regulated by a combination of the magnetic and gas pressures in the inner disk. The local picture of flux advection, based on the ratio of competing diffusion coefficients, is replaced by a global picture of flux motion.

8.6.4 *Evolution of black-hole spin*

The spin of a black hole can have significant effects. The spin determines the location of the ISCO and with it the total luminosity and efficiency from the accretion disk. A spinning black hole is a source of energy and can, as described previously, power Poynting flux jets. As there are no real observational certainties about black-hole spin, we are left mostly with speculation in attempting to estimate what the likely distribution of spins would be for a population of black holes. Through the usual arguments regarding the conservation of angular momentum, one might expect that stellar-mass black holes formed in supernovas would be born with large spins. This does seem to be the case where the stellar remnant is a neutron star. With supermassive black holes, the situation is less clear because the *origin* of supermassive black holes is less clear. However it is born, once a black hole begins to interact with its environment, its spin can be affected by the accretion process. Under what circumstances would a disk spin up or spin down, and can we determine what limits a black hole's spin?

The spin evolution of a black hole was considered by Thorne (1974), who showed that although the accreting matter would tend to spin up the hole to its natural limit, the capture of photons from the disk, preferentially with retrograde orbits, would limit the spinup to $a/M = 0.998$. Simulations have found that the upper limit could be considerably smaller for holes that are spun up primarily through magnetized accretion (Gammie et al., 2004; McKinney and Gammie, 2004; Krolik et al., 2005; Beckwith et al., 2008a). The critical factor is the presence of magnetic fields threading the horizon that carry angular momentum off to infinity (as the hole captures Alfvén waves with retrograde angular momentum). McKinney and Gammie (2004) computed a series of simulations with similar initial conditions but with Kerr holes of different spin and showed how, in principle, the process might work. In that set of simulations, the equilibrium spin was $a/M \sim 0.93$. Although that particular number is, no doubt, specific to the conditions of this set of simulations, the fact that it is so much less than the traditional value of 0.998 emphasizes why this is an issue that warrants further study.

8.6.5 *Tilted disks*

One of the unique features of a rotating black hole is that it exerts a torque on surrounding matter through the dragging of inertial reference frames. This torque causes a precession of the orbital plane, also known as Lense-Thirring precession. Bardeen and Petterson (1975) pointed out that if at large distance the mean angular momentum of an accretion disk was misaligned with the spin axis of a black hole, this precession, coupled with dissipation inside the disk, could cause the disk's orbital plane to move into the black hole's equatorial plane as it approaches the black hole. This is significant because, although misalignment of the accreting matter's and the black hole's spin axes might be generic, the Bardeen-Petterson effect could provide the means to ensure that they are aligned in the near-hole region. Whether or not the disk and hole are aligned is a critical issue for understanding jet formation and launching and for determining the spin of a hole through observations of the disk's inner edge.

Because past work on this problem has mostly been based on the α model, the issue is unresolved. Simulations, however, can calculate the internal dynamics of an accretion disk and provide a much better picture of this effect. This is a difficult problem as it requires simulation of the full 2π domain extended over a large radius. Fragile and collaborators have begun initial efforts to study the precession effects using full GRMHD simulations (Fragile et al., 2007; Fragile and Blaes, 2008; Fragile, 2009). So far, these simulations have shown dramatic differences between tilted and untilted disks. Misalignment can strongly influence the inner radius of the accretion disk while creating internal shocks that create additional angular momentum transport and dissipation near the hole, possibly creating quasi-periodic signals. These results demonstrate the potential importance of this issue, and in the coming years it is hoped that simulations make rapid progress investigating this largely unexplored area of black-hole accretion physics.

8.6.6 *Some observational implications*

Black-hole systems are observed through the photons they emit (although in the future this may be supplemented by observations of gravitational radiation). Currently, global simulation codes are a long way from including either first-principles dissipation and photon production, or full radiation transport through both optically thick and thin regimes. However, interesting information can still be obtained by applying approximations for photon emission and then tracing the path of outgoing radiation through the relativistic spacetime of the black hole.

One of the first attempts at determining the observational implications of global simulations used the output from a simulation of a disk around a Schwarzschild black hole to study the effects of the MHD turbulent fluctuations on possible variability of the thermal radiation from the disk (Schnittman et al., 2006). The results showed that the power spectrum of the light curve is a steep power-law over a broad range of frequencies, but integration over independently radiating regions of the disk surface keeps the amplitude of the observed fluctuations relatively small. More recently, Noble and Krolik (2009) used data from a HARM3D simulation to study variability in the coronal luminosity. They found that the intrinsic variability of the underlying disk MHD turbulence drives variability with a power spectrum that is roughly consistent with a power law index of -2.

Beckwith et al. (2008b) made a simple connection between local internal stress and disk luminosity to estimate how emission from the inner part of a disk would be affected by stresses at and inside the ISCO. They found that for slowly spinning black holes ($a/M \leq 0.9$), radiation from well inside the ISCO could significantly alter the observed continuum, particularly when the system is viewed at high inclination. For rapidly spinning black holes, the contrast is smaller, as most of the additional photons created by the enhanced stress are captured by the hole. Additional stress has the potential to increase the characteristic temperature of the accretion flow by roughly a factor of 2 while enhancing the total luminosity by a value ranging from a few percent to order unity.

The Beckwith et al. (2008b) study is limited by the use of the stress as a proxy for disk emission. Noble et al. (2009) improved on this by using the energy-conserving HARM3D code to equate local heating to local dissipation of kinetic and magnetic energy. They use an optically thin cooling function and ray-trace the photon trajectories to compute the luminosity seen at infinity for a disk with spin $a/M = 0.9$. They find that although there is significant additional dissipation inside the ISCO releasing as much as 20% additional energy, photon capture and gravitational redshifting limit the net luminosity increase at infinity to about 6% above the prediction of the standard model.

Recently, Mościbrodzka et al. (2009) used results from a HARM simulation to model the emission from the galactic center, Sgr A*, using a Monte Carlo technique. The inclusion of more detailed emission processes, such as synchrotron emission and absorption and Compton scattering, brought a new level of realism to this sort of effort. Fragile and Meier (2009) carried out an energy-conserving axisymmetric GRMHD simulation with similar emission processes included to explore the transition from a hot, radiatively inefficient flow to a magnetically dominated inner accretion flow. Work of this nature shows great promise of rapidly increasing our understanding of the various states of black-hole accretion.

8.7 Summary

For over three decades, our theoretical understanding of accretion disk systems has been based on one-dimensional time-steady models. The standard paradigm for disk accretion has been an optically thick, vertically thin, Keplerian disk with an internal stress given by αP. Although these analytic models have served us well to understand many properties of accretion systems, their limitations are well known. Observations indicate that many accreting systems are dynamic, with time variability on many scales, including intervals too short to be due solely to secular changes in the accretion rate, the only process accessible to analytic disk models. Accretion systems exhibit multiple thermodynamic and radiative states, many more than can be accounted for by the simple thermal disk continuum spectrum. Observations have greatly outpaced theory. Recent progress, however, suggests that theory may finally be in a position to catch up. Simulations of accretion disks and the physical processes associated with accretion are now making great strides in extending our understanding of black hole phenomena.

We now know that *magnetized* Keplerian disk flow is linearly unstable, and the outcome of that instability is turbulence. Correlated magnetic and velocity fluctuations within the MHD turbulence act to transport angular momentum outward. The magnetic Maxwell stress level exceeds the hydrodynamic Reynolds stress, but both contribute. The total stress is generally proportional to the pressure, with α values ranging from 0.001 to 0.1 under different circumstances. There are, however, large variations in α in both time and space; average stress and pressure are better described as correlated rather than strictly proportional. In radiation-pressure–dominated disks, the stress is correlated with the total pressure rather than just the gas pressure or some other quantity. Despite this, radiation-pressure–supported disks are thermally stable because, while the stress is related to the pressure, the causal relationship is from stress to pressure, not the other way around. An increase in pressure (say, from a momentary loss of cooling efficiency) does not produce an increase in stress. An increase in stress will produce an increase in pressure when the resulting increased turbulence is dissipated as heat. So while much of the overall α disk model is supported by simulation results, there are very real limitations to extending such analytic models beyond their original time- and space-averaged context.

Although we now know the origin of the stress, and know that it will be related to the total pressure, we do not yet know what factors determine the magnetic field strength and hence the stress levels within a disk. The presence of a weak field and the MRI guarantee MHD turbulence, but what are the other controlling factors? Computational investigations of this point are made somewhat difficult by the need to distinguish between *physical* influences and *numerical* influences associated with the limitations of the numerical experiments. Examples of the latter include effects due to shearing box size and aspect ratio, resolution, and presence or absence of a net field. Of these, the presence or absence of a vertical field has the largest influence; the largest stress levels are seen when there is a net vertical field present. The overall energy in the turbulence is also greatly influenced by how it is dissipated. The presence of resistivity and viscosity in the simulations greatly influences the average turbulent energy levels. Resistivity can depress or even eliminate MHD turbulence, whereas the turbulent energy increases with $P_r = \nu/\eta$. It is unclear whether these simulation results would necessarily hold in nature, where the viscous and resistive length scales are much smaller than other characteristic lengths. That is, we can only simulate modest Reynolds numbers, but the Reynolds numbers in disks are huge.

Disks are magnetized, but at what level and with what overall topology? The background shear favors the development of toroidal field over poloidal, and simulations consistently find that toroidal fields dominate the magnetic energy. Beyond that, however, things are still rather uncertain. Does the MRI operating in a disk constitute a magnetic dynamo, and, if so, what are its properties? This question has received considerable attention, but without clear resolution. It seems that under some circumstances, the MRI can maintain magnetic field despite the presence of dissipation, and in that sense it constitutes a dynamo. But of greater interest is whether some sort of more organized field, such as an overall dipole, might be created. Results from stratified shearing box simulations suggest that toroidal fields of opposite sign can be generated above and below the midplane. In global simulations, large-scale fields have developed, such as the dipole field attached to the black hole. So far, however, the generation of such fields can always be ascribed to the particular initial conditions that were used. It is possible that the field strength and topology depend on that found at large distance from the black hole. But it remains to be seen whether any such large-scale field can be dragged in by the accretion process.

The total luminosity and efficiency of black-hole accretion depend on how much orbital energy is extracted, thermalized, and then radiated. The energy available near the black hole is large, but the time available to thermalize and radiate any heat becomes small as gas nears the horizon. It is also difficult for photons to escape from this region. Thus, the question of the level of the stress at the ISCO is critical. Although simulations have shown that additional magnetic stress *can* be important there, we don't know if it *must*, or under what circumstances it will be present. Increased stress can lead to higher characteristic disk temperatures and greater overall efficiency, but we require improved studies to relate this enhanced stress to emitted photons. A related issue is the degree to which enhanced magnetic torques may limit the value of a/M for holes whose spin comes from the accretion process.

Black-hole systems can produce more than heat and light – they often produce winds and jets. The power to drive these outflows could come from the rotational energy within the accretion disk, or it could come from the rotating spacetime of the black hole itself. Simulations have demonstrated that powerful outflows can be produced by large-scale poloidal fields threading either the disk or the hole. Simulations have not yet been able to answer the question as to when and how such fields will occur. It is nevertheless something of an achievement that relatively stable and powerful jets have developed within global simulations without being put in "by hand" in the initial conditions. This adds some level of confidence that jets are indeed a natural outcome of magnetized accretion.

In the realm of black-hole accretion, we are in the midst of a transformation from a discipline marked by formal analytic models with weak ties to observations, to one of detailed and specific calculations capable of predicting the fundamental physical processes that occur in disks and tying them directly to observational diagnostics. Along the way toward this goal, the results obtained from simulations will continue to provide new insights. Many challenges remain: we need novel algorithms and implementations to improve simulation realism, new techniques for extracting observational implications from simulations, and new theory inspired by simulation results that be enabled by the petascale (or larger) computational platforms that will become available in the next few years.

8.8 Acknowledgments

This work was supported by NSF grant AST-0908869 and NASA grant NNX09AD14G. The author thanks the organizers of the XXI Winter School in Astrophysics for their hospitality during the school, and also collaborators Julian Krolik, Kris Beckwith, and Jake Simon for their assistance in preparing these lectures.

REFERENCES

Abramowicz, M., Brandenburg, A., and Lasota, J.-P. 1996. The dependence of the viscosity in accretion discs on the shear/vorticity ratio. MNRAS, **281**, L21–L24.

Agol, E., and Krolik, J. H. 2000. Magnetic stress at the marginally stable orbit: altered disk structure, radiation, and black hole spin evolution. ApJ, **528**(Jan.), 161–170.

Anninos, P., Fragile, P. C., and Salmonson, J. D. 2005. Cosmos++: relativistic magnetohydrodynamics on unstructured grids with local adaptive refinement. ApJ, **635**(Dec.), 723–740.

Antón, L., Zanotti, O., Miralles, J. A., Martí, J. M., Ibáñez, J. M., Font, J. A., and Pons, J. A. 2006. Numerical 3+1 general relativistic magnetohydrodynamics: a local characteristic approach. ApJ, **637**(Jan.), 296–312.

Balbus, S. A. 1995. General local stability criteria for stratified, weakly magnetized rotating systems. ApJ, **453**, 380–383.

Balbus, S. A. 2003. Enhanced angular Momentum transport in accretion disks. Ann. Rev. Astron. Astrophys., **41**, 555–597.

Balbus, S. A., and Hawley, J. F. 1992a. A powerful local shear instability in weakly magnetized disks. IV. Nonaxisymmetric perturbations. ApJ, **400**, 610–621.

Balbus, S. A., and Hawley, J. F. 1992b. Is the Oort A-value a universal growth rate limit for accretion disk shear instabilities? ApJ, **392**, 662–666.

Balbus, S. A., and Hawley, J. F. 1998. Instability, turbulence, and enhanced transport in accretion disks. Reviews of Modern Physics, **70**, 1–53.

Balbus, S. A., and Henri, P. 2008. On the magnetic Prandtl number behavior of accretion disks. ApJ, **674**(Feb.), 408–414.

Balbus, S. A., and Papaloizou, J. C. B. 1999. On the dynamical foundations of alpha disks. ApJ, **521**, 650–658.

Balbus, S. A., Hawley, J. F., and Stone, J. M. 1996. Nonlinear stability, hydrodynamical turbulence, and transport in disks. ApJ, **467**, 76–86.

Bardeen, J. M., and Petterson, J. A. 1975. The Lense-Thirring effect and accretion disks around Kerr black holes. ApJ, **195**(Jan.), L65–L67.

Beckwith, K., Hawley, J. F., and Krolik, J. H. 2008a. The influence of magnetic field geometry on the evolution of black hole accretion flows: similar disks, drastically different jets. ApJ, **678**(May), 1180–1199.

Beckwith, K., Hawley, J. F., and Krolik, J. H. 2008b. Where is the radiation edge in magnetized black-hole accretion discs? MNRAS, **390**(Oct.), 21–38.

Beckwith, K., Hawley, J. F., and Krolik, J. H. 2009. Transport of large scale poloidal flux in black-hole accretion. ApJ, **707**, 428–445.

Blackman, E. G., Penna, R. F., and Varnière, P. 2008. Empirical relation between angular momentum transport and thermal-to-magnetic pressure ratio in shearing box simulations. New Astronomy, **13**(May), 244–251.

Blaes, O. M., and Balbus, S. A. 1994. Local shear instabilities in weakly magnetized disks. ApJ, **421**, 163–177.

Blandford, R. D., and Payne, D. G. 1982. Hydromagnetic flows from accretion discs and the production of radio jets. MNRAS, **199**(June), 883–903.

Blandford, R. D., and Znajek, R. L. 1977. Electromagnetic extraction of energy from Kerr black holes. MNRAS, **179**(May), 433–456.

Brackbill, J.U., and Barnes, D.C. 1980. The effect of nonzero product of magnetic gradient and B on the numerical solution of the magnetohydrodynamic equations. J. Comp. Phys., **35**, 426–430.

Brandenburg, A., Nordlund, A., Stein, R. F., and Torkelsson, U. 1995. Dynamo-generated turbulence and large-scale magnetic fields in a Keplerian shear flow. ApJ, **446**(June), 741–754.

Colella, P. 1990. Multidimensional upwind methods for hyperbolic conservation laws. J. Comp. Phys., **87**(Mar.), 171–200.

Colella, P., and Woodward, P. R. 1984. The piecewise parabolic method (PPM) for gas-dynamical simulations. J. Comp. Phys., **54**, 174–201.

Crawford, J. A., and Kraft, R. P. 1956. An intrepretation of AE Aquarii. ApJ, **123**(Jan.), 44–53.

Davis, S. W., Stone, J. M., and Pessah, M. E. 2010. Sustained magnetorotational turbulence in local simulations of stratified disks with zero net magnetic flux. ApJ, 713(Apr.), 52–65.

De Villiers, J., Hawley, J. F., Krolik, J. H., and Hirose, S. 2005. Magnetically driven accretion in the Kerr metric. III. Unbound outflows. ApJ, **620**(Feb.), 878–888.

De Villiers, J.-P., and Hawley, J. F. 2003. A numerical method for general relativistic magnetohydrodynamics. ApJ, **589**(May), 458–480.

Done, C., and Davis, S. W. 2008. Angular momentum transport in accretion disks and its implications for spin estimates in black hole binaries. ApJ, **683**(Aug.), 389–399.

Duez, M. D., Liu, Y. T., Shapiro, S. L., and Stephens, B. C. 2005. Relativistic magnetohydrodynamics in dynamical spacetimes: Numerical methods and tests. Phys. Rev. D, **72**(2), 024028+.

Evans, C. R., and Hawley, J. F. 1988. Simulation of magnetohydrodynamic flows: a constrained transport method. ApJ, **332**, 659–677.

Fleming, T. P., Stone, J. M., and Hawley, J. F. 2000. The effect of resistivity on the nonlinear stage of the magnetorotational instability in accretion disks. ApJ, **530**(Feb.), 464–477.

Fragile, P. C. 2009. Effective inner radius of tilted black-hole accretion disks. ApJ, **706**(Dec.), L246–L250.

Fragile, P. C., and Blaes, O. M. 2008. Epicyclic motions and standing shocks in numerically simulated tilted black-hole accretion disks. ApJ, **687**(Nov.), 757–766.

Fragile, P. C., and Meier, D. L. 2009. General relativistic magnetohydrodynamic simulations of the hard state as a magnetically dominated accretion flow. ApJ, **693**(Mar.), 771–783.

Fragile, P. C., Blaes, O. M., Anninos, P., and Salmonson, J. D. 2007. Global general relativistic magnetohydrodynamic simulation of a tilted black-hole accretion disk. ApJ, **668**(Oct.), 417–429.

Fragile, P. C., Lindner, C. C., Anninos, P., and Salmonson, J. D. 2009. Application of the cubed-sphere grid to tilted black-hole accretion disks. ApJ, **691**(Jan.), 482–494.

Fromang, S., and Papaloizou, J. 2007. MHD simulations of the magnetorotational instability in a shearing box with zero net flux. I. The issue of convergence. A&A, **476**(Dec.), 1113–1122.

Fromang, S., Papaloizou, J., Lesur, G., and Heinemann, T. 2007. MHD simulations of the magnetorotational instability in a shearing box with zero net flux. II. The effect of transport coefficients. A&A, **476**(Dec.), 1123–1132.

Gammie, C. F. 1999. Efficiency of magnetized thin accretion disks in the Kerr metric. ApJ, **522**(September), L57–L60.

Gammie, C. F., and Balbus, S. A. 1994. Quasi-global linear analysis of a magnetized disc. MNRAS, **270**(Sept.), 138–152.

Gammie, C. F., McKinney, J. C., and Tóth, G. 2003. HARM: a numerical scheme for general relativistic magnetohydrodynamics. ApJ, **589**(May), 444–457.

Gammie, C. F., Shapiro, S. L., and McKinney, J. C. 2004. Black hole spin evolution. ApJ, **602**(Feb.), 312–319.

Goodman, J., and Xu, G. 1994. Parasitic instabilities in magnetized, differentially rotating disks. ApJ, **432**(Sept.), 213–223.

Guan, X., and Gammie, C. F. 2008. Axisymmetric shearing box models of magnetized disks. ApJS, **174**(Jan.), 145–157.

Hawley, J. F., and Balbus, S. A. 1992. A powerful local shear instability in weakly magnetized disks. III – Long-term evolution in a shearing sheet. ApJ, **400**(Dec.), 595–609.

Hawley, J. F., and Krolik, J. H. 2001. Global MHD simulation of the inner accretion disk in a pseudo-Newtonian potential. ApJ, **548**(Feb.), 348–367.

Hawley, J. F., and Krolik, J. H. 2002. High-resolution simulations of the plunging region in a pseudo-Newtonian potential: dependence on numerical resolution and field topology. ApJ, **566**(Feb.), 164–180.

Hawley, J. F., and Krolik, J. H. 2006. Magnetically driven jets in the Kerr metric. ApJ, **641**(Apr.), 103–116.

Hawley, J. F., and Stone, J. M. 1995. MOCCT: a numerical technique for astrophysical MHD. Computer Physics Communications, **89**(Aug.), 127–148.

Hawley, J. F., Balbus, S. A., and Winters, W. F. 1999. Local hydrodynamic stability of accretion disks. ApJ, **518**(June), 394–404.

Hawley, J. F., Gammie, C. F., and Balbus, S. A. 1995. Local three-dimensional magnetohydrodynamic simulations of accretion disks. ApJ, **440**(Feb.), 742–763.

Hawley, J. F., Gammie, C. F., and Balbus, S. A. 1996. Local three-dimensional simulations of an accretion disk hydromagnetic dynamo. ApJ, **464**(June), 690–703.

Hirose, S., Krolik, J. H., and Blaes, O. 2009. Radiation-dominated disks are thermally stable. ApJ, **691**(Jan.), 16–31.

Hirose, S., Krolik, J. H., and Stone, J. M. 2006. Vertical structure of gas pressure–dominated accretion disks with local dissipation of turbulence and radiative transport. ApJ, **640**(Apr.), 901–917.

Jin, L. 1996. Damping of the shear instability in magnetized disks by Ohmic diffusion. ApJ, **457**, 798–804.

Koide, S., Shibata, K., Kudoh, T., and Meier, D. L. 2001. Numerical method for general relativistic magnetohydrodynamics in Kerr space-time. Journal of Korean Astronomical Society, **34**(Dec.), 215–224.

Komissarov, S. S. 2004. General relativistic magnetohydrodynamic simulations of monopole magnetospheres of black holes. MNRAS, **350**(June), 1431–1436.

Krolik, J. H. 1999. Magnetized accretion inside the marginally stable orbit around a black hole. ApJ, **515**(April), L73–L76.

Krolik, J. H., Hawley, J. F., and Hirose, S. 2005. Magnetically driven accretion flows in the Kerr metric. IV. Dynamical properties of the inner disk. ApJ, **622**(Apr.), 1008–1023.

Latter, H. N., Lesaffre, P., and Balbus, S. A. 2009. MRI channel flows and their parasites. MNRAS, **394**(Apr.), 715–729.

Lesur, G., and Longaretti, P.-Y. 2007. Impact of dimensionless numbers on the efficiency of magnetorotational instability induced turbulent transport. MNRAS, **378**(July), 1471–1480.

Lightman, A. P., and Eardley, D. M. 1974. Black holes in binary systems: instability of disk accretion. ApJ, **187**(Jan.), L1–L3.

Livio, M. 2000. Astrophysical jets. Pages 275–297 of: Holt, S. S., and Zhang, W. W. (eds.), *American Institute of Physics Conference Series*. American Institute of Physics Conference Series, vol. 522.

Lynden-Bell, D. 1969. Galactic nuclei as collapsed old quasars. Nature, **223**(Aug.), 690–694.

Machida, M., and Matsumoto, R. 2003. Global three-dimensional magnetohydrodynamic simulations of black hole accretion disks: X-ray flares in the plunging region. ApJ, **585**(Mar.), 429–442.

McKinney, J. C. 2005. Total and jet Blandford-Znajek power in the presence of an accretion disk. ApJ, **630**(Sept.), L5–L8.

McKinney, J. C. 2006. General relativistic magnetohydrodynamic simulations of the jet formation and large-scale propagation from black-hole accretion systems. MNRAS, **368**(June), 1561–1582.

McKinney, J. C., and Blandford, R. D. 2009. Stability of relativistic jets from rotating, accreting black holes via fully three-dimensional magnetohydrodynamic simulations. MNRAS, **394**(Mar.), L126–L130.

McKinney, J. C., and Gammie, C. F. 2004. A measurement of the electromagnetic luminosity of a Kerr black hole. ApJ, **611**(August), 977–995.

Meier, D. L. 2005. Magnetically dominated accretion flows (MDAFS) and jet production in the lowhard state. Astrophys. Space Sci., **300**(Nov.), 55–65.

Mignone, A., Bodo, G., Massaglia, S., Matsakos, T., Tesileanu, O., Zanni, C., and Ferrari, A. 2007. PLUTO: a numerical code for computational astrophysics. ApJS, **170**(May), 228–242.

Miller, K. A., and Stone, J. M. 2000. The formation and structure of a strongly magnetized corona above a weakly magnetized accretion disk. ApJ, **534**(May), 398–419.

Mościbrodzka, M., Gammie, C. F., Dolence, J. C., Shiokawa, H., and Leung, P. K. 2009. Radiative models of SGR A* from GRMHD simulations. ApJ, **706**(Nov.), 497–507.

Noble, S. C., and Krolik, J. H. 2009. GRMHD prediction of coronal variability in accreting black holes. ApJ, **703**(Sept.), 964–975.

Noble, S. C., Gammie, C. F., McKinney, J. C., and Del Zanna, L. 2006. Primitive variable solvers for conservative general relativistic magnetohydrodynamics. ApJ, **641**(Apr.), 626–637.

Noble, S. C., Krolik, J. H., and Hawley, J. F. 2009. Direct calculation of the radiative efficiency of an accretion disk around a black hole. ApJ, **692**(Feb.), 411–421.

Noble, S. C., Krolik, J. H., and Hawley, J. F. 2010. Dependence of inner accretion disk stress on parameters: the Schwarzschild case. ApJ, **711**(Feb.), 959–973.

Novikov, I. D., and Thorne, K. S. 1973. Astrophysics of black holes. Pages 343–450 of: De Witt, C. (ed.), *Black Holes (Les Astres Occlus)*. Gordon and Breach.

Pessah, M. E., Chan, C.-K., and Psaltis, D. 2006. The signature of the magnetorotational instability in the Reynolds and Maxwell stress tensors in accretion discs. MNRAS, **372**(Oct.), 183–190.

Pessah, M. E., Chan, C.-k., and Psaltis, D. 2007. Angular momentum transport in accretion disks: scaling laws in MRI-driven turbulence. ApJ, **668**(Oct.), L51–L54.

Pudritz, R. E., Ouyed, R., Fendt, C., and Brandenburg, A. 2007. Disk winds, jets, and outflows: theoretical and computational foundations. Pages 277–294 of: Reipurth, B., Jewitt, D., and Keil, K. (eds.), *Protostars and Planets V*. University of Arizona Press, Tucson.

Reynolds, C. S., and Armitage, P. J. 2001. A variable efficiency for thin-disk black-hole accretion. ApJ, **561**(Nov.), L81–L84.

Reynolds, C. S., and Fabian, A. C. 2008. Broad iron-Kα emission lines as a diagnostic of black hole spin. ApJ, **675**(Mar.), 1048–1056.

Sano, T. 2007. The evolution of channel flows in MHD turbulence driven by magnetorotational instability. Astrophys. Space Sci., **307**(Jan.), 191–195.

Sano, T., and Inutsuka, S.-I. 2001. Saturation and thermalization of the magnetorotational instability: recurrent channel flows and reconnections. ApJ, **561**(Nov.), L179–L182.

Sano, T., and Stone, J. M. 2002. The effect of the Hall term on the nonlinear evolution of the magnetorotational instability. II. Saturation level and critical magnetic Reynolds number. ApJ, **577**(Sept.), 534–553.

Sano, T., Inutsuka, S.-I., and Miyama, S. M. 1998. A saturation mechanism of magnetorotational instability due to ohmic dissipation. ApJ, **506**(Oct.), L57–L60.

Sano, T., Inutsuka, S.-I., Turner, N. J., and Stone, J. M. 2004. Angular momentum transport by magnetohydrodynamic turbulence in accretion disks: gas pressure dependence of the saturation level of the magnetorotational instability. ApJ, **605**(Apr.), 321–339.

Schnittman, J. D., Krolik, J. H., and Hawley, J. F. 2006. Light curves from an MHD simulation of a black-hole accretion disk. ApJ, **651**(Nov.), 1031–1048.

Shafee, R., McKinney, J. C., Narayan, R., Tchekhovskoy, A., Gammie, C. F., and McClintock, J. E. 2008. Three-dimensional simulations of magnetized thin accretion disks around black holes: stress in the plunging region. ApJ, **687**(Nov.), L25–L28.

Shakura, N. I., and Sunyaev, R. A. 1973. Black holes in binary systems. Observational appearance. A&A, **24**, 337–355.

Shakura, N. I., and Sunyaev, R. A. 1976. A theory of the instability of disk accretion on to black holes and the variability of binary X-ray sources, galactic nuclei and quasars. MNRAS, **175**(June), 613–632.

Shi, J.-M., Krolik, J. H., and Hirose, S. 2010. What is the numerically converged amplitude of MHD turbulence in stratified shearing boxes? ApJ, 708(Jan.), 1716–1727.

Simon, J. B., and Hawley, J. F. 2009. Viscous and resistive effects on the magnetorotational instability with a net toroidal field. ApJ, **707**(Dec.), 833–843.

Simon, J. B., Hawley, J. F., and Beckwith, K. 2009. Simulations of magnetorotational turbulence with a higher-order Godunov scheme. ApJ, **690**(Jan.), 974–997.

Stone, J. M., and Balbus, S. A. 1996. Angular momentum transport in accretion disks via convection. ApJ, **464**(June), 346–372.

Stone, J. M., and Norman, M. L. 1992a. ZEUS-2D: A radiation magnetohydrodynamics code for astrophysical flows in two space dimensions. I – The hydrodynamic algorithms and tests. ApJS, **80**(June), 753–790.

Stone, J. M., and Norman, M. L. 1992b. ZEUS-2D: a radiation magnetohydrodynamics code for astrophysical flows in two space dimensions. II. The magnetohydrodynamic algorithms and tests. ApJS, **80**(June), 791–818.

Stone, J. M., Hawley, J. F., Gammie, C. F., and Balbus, S. A. 1996. Three-dimensional magnetohydrodynamical simulations of vertically stratified accretion disks. ApJ, **463**(June), 656–673.

Stone, J. M., Gardiner, T. A., Teuben, P., Hawley, J. F., and Simon, J. B. 2008. Athena: a new code for astrophysical MHD. ApJS, **178**(Sept.), 137–177.

Tassoul, J.-L. 1978. *Theory of Rotating Stars*. Princeton, NJ, Princeton University Press.

Thorne, K. S. 1974. Disk-accretion onto a black hole. II. Evolution of the hole. ApJ, **191**(July), 507–520.

Thorne, K. S., Price, R. H., and MacDonald, D. A. 1986, Black holes: The membrane paradigm. New Haven, CT, Yale University Press, 380, p.

Tout, C. A., and Pringle, J. E. 1996. Can a disc dynamo generate large-scale magnetic fields? MNRAS, **281**(July), 219–225.

Turner, N. J. 2004. On the vertical structure of radiation-dominated accretion disks. ApJ, **605**(Apr.), L45–L48.

van Ballegooijen, A. A. 1989. Magnetic fields in the accretion disks of cataclysmic variables. Pages 99–106 of: Belvedere, G. (ed.), *Accretion Disks and Magnetic Fields in Astrophysics*. Astrophysics and Space Science Library, vol. 156.

Wardle, M. 1999. The Balbus-Hawley instability in weakly ionized discs. MNRAS, **307**, 849–856.

Wilson, J. R. 1975. Some magnetic effects in stellar collapse and accretion. New York Academy of Sciences Annals, **262**(Oct.), 123–132.

Ziegler, U., and Rüdiger, G. 2001. Shear rate dependence and the effect of resistivity in magneto-rotationally unstable, stratified disks. A&A, **378**(Nov.), 668–678.

Appendix A. Piazzi Smyth, the Cape of Good Hope, Tenerife, and the siting of large telescopes

BRIAN WARNER

Abstract

This chapter is an excerpt from an ad hoc lecture given at the Winter School dedicated to Charles Piazzi Smyth.

Introduction

FIGURE AA.1. Self portrait of Piazzi Smyth: from the visitors' book of Dr John Lee, in the Museum for the History of science, Oxford.

The first major observatory built on Tenerife, albeit temporary, was that constructed by Charles Piazzi Smyth, Astronomer Royal for Scotland, in 1856. The lead-up to this event, and its long-lasting impact, are of considerable significance in the history of modern astronomy.

Piazzi Smyth was born in 1819, the son of an extraordinary seaman, William Henry Smyth, who among other things was a founder of the Royal Astronomical Society, the Royal Geographical Society, and the Numismatic Society. He was a leading naval hydrographer of his time, known as "Mediterranean Smyth," and owned his own observatory. He was a great friend of Giuseppe Piazzi, the Italian astronomer who in 1801 discovered the first asteroid, and made him godfather of his second son. The young Charles Piazzi Smyth grew up in an unusual family, both artistic and scientific, and was so well versed in astronomy and mathematics that in 1835, when only 16 years old, he was appointed

FIGURE AA.2. Photograph (calotype) of the Cape Observatory, taken by Piazzi Smyth in 1842. From Royal Society of Edinburgh.

as first assistant to Thomas Maclear, the director of the observatory at the Cape of Good Hope.

The Cape Observatory's task in the 1830s was largely to repeat the survey made in 1751 by Nicolas Louis Lacaille, who had measured an arc of meridian and found the Earth's radius in the southern hemisphere apparently to differ significantly from that in the north. From 1835 until 1845, when he left to become Astronomer Royal for Scotland, Piazzi was thus principally engaged with Maclear in a survey of the southwestern part of the Cape. This involved much climbing of the highest mountains in order to make theodolite measurements used in the triangulation process. During this work, Piazzi became greatly impressed with the quality of star images viewed from tops of mountains; and he made use of the dark skies to study the shape and extent of the Zodiacal Light.

Piazzi was no doubt already familiar with the prophetic statement in Newton's Opticks (1704):

> Long Telescopes may cause Objects to appear brighter and larger than short ones can do, but they cannot be so formed as to take away that confusion of the Rays which arises from the Tremors of the Atmosphere. The only Remedy is a most serene and quiet Air, such as may perhaps be found on the tops of the highest Mountains above the grosser Clouds.

Combined with his personal experience, this stimulated him to propose an expedition to a suitable mountaintop in a good climate, which he carried out (on his honeymoon) by visiting Tenerife in 1856, funded by the British Admiralty and aided by the loan of the ship *Titania*, the personal property of the railway engineer Robert Stephenson.

Piazzi observed from 14 July to 20 August 1856 at Guajara, some 8 km from the main volcanic peak, and then moved to higher altitude at Alta Vista, where he observed until 19 September. The results were as he had expected: the seeing was outstanding, as judged by observations of double stars through the 7-inch Cook refractor and the smaller Sheepshanks telescope that he had taken with him. He reported on the expedition in a 10-volume scientific treatise and in his book *Teneriffe: an Astronomer's Experiment*, which incidentally was the first book illustrated with stereoscopic photographs (he had been the pioneering photographer in South Africa – starting in 1839, the year that the invention of photography was announced). This book has had a permanent effect on the choice of sites for large telescopes – Tenerife became the forerunner of all mountain observatories. E. S. Holden, in an overview of mountain observatories in 1896, expressed the opinion that "There is no doubt that this expedition served to attract general attention to the matter of choosing suitable sites for observatories; and also to spread the idea that all

FIGURE AA.3. Piazzi Smyth's sketch of himself observing the zodiacal light in 1845 in the Hottentot Hollands mountains, north of Cape Town, with brown hyaena night assistants. Courtesy of the South African Astronomical Observatory.

mountain-stations possessed striking advantages." The first permanently occupied mountaintop observatory was that funded by James Lick, at Mount Hamilton in California, to house the 36-inch refractor, at the time the largest in the world – and with Holden as the first director. Later, the first giant reflectors, at Mount Wilson and Mount Palomar, were similarly sited. Thus, when the observatories on La Palma and Tenerife were established, they were based on a principle that had originally been grown in the Canary Islands – from a seed sown at the Cape of Good Hope.

FIGURE AA.4. Stereoscopic photograph of the Sheepshanks telescope mounted at Mount Guajara. Teide peak in the distance. From Piazzi Smyth's book.

Further reading

Brück, H. A., and Brück, M. T. 1988. *The Peripatetic Astronomer: The Life of Charles Piazzi Smyth*. Inst. Physics Publ., Bristol.

Holden, E. S. 1986. *Mountain Observatories in America and Europe*. Smithsonian Institution, Washington, D.C.

Smyth, C. P. 1858. *Teneriffe, an Astronomer's Experiment, or Specialities of a Residence above the Clouds*. Reeve, L. (ed.), London.

Warner, B. 1983. *Charles Piazzi Smyth, Astronomer Artist: The Cape Years, 1835–1845*. Balkema, A. A. (ed.). University of Cape Town.

Printed in the United States
By Bookmasters